MONTE CARLO CALCULATIONS
IN NUCLEAR MEDICINE

Medical Science Series

MONTE CARLO CALCULATIONS IN NUCLEAR MEDICINE
Applications in Diagnostic Imaging

Edited by

Michael Ljungberg
Sven-Erik Strand

Department of Radiation Physics, Lund University, Lund, Sweden

Michael A King

Department of Nuclear Medicine, University of Massachusetts
Medical Center, Worcester, MA, USA

Institute of Physics Publishing
Bristol and Philadelphia

British Library Cataloguing-in-Publication Data

A catalogue record for this book is available from the British Library.

ISBN 0 7503 0479 0

Library of Congress Cataloging-in-Publication Data are available

Series Editors:

R F Mould, Croydon, UK
C G Orton, Karamanos Cancer Institute, Detroit, USA
J A E Spaan, University of Amsterdam, The Netherlands
J G Webster, University of Wisconsin-Madison, USA

Published by Institute of Physics Publishing, wholly owned by The Institute of Physics, London

Institute of Physics Publishing, Dirac House, Temple Back, Bristol BS1 6BE, UK

US Office: Institute of Physics Publishing, The Public Ledger Building, Suite 1035, 150 South Independence Mall West, Philadelphia, PA 19106, USA

Typeset and printed in the UK by J W Arrowsmith Ltd, Bristol BS3 2NT

The Medical Science Series is the official book series of the International Federation for Medical and Biological Engineering (IFMBE) and the International Organization for Medical Physics (IOMP).

IFMBE

The IFMBE was established in 1959 to provide medical and biological engineering with an international presence. The Federation has a long history of encouraging and promoting international cooperation and collaboration in the use of technology for improving the health and life quality of man.

The IFMBE is an organization that is mostly an affiliation of national societies. Transnational organizations can also obtain membership. At present there are 42 national members, and one transnational member with a total membership in excess of 15 000. An observer category is provided to give personal status to groups or organization considering formal affiliation.

Objectives

- To reflect the interests and initiatives of the affiliated organizations.
- To generate and disseminate information of interest to the medical and biological engineering community and international organizations.
- To provide an international forum for the exchange of ideas and concepts.
- To encourage and foster research and application of medical and biological engineering knowledge and techniques in support of life quality and cost-effective health care.
- To stimulate international cooperation and collaboration on medical and biological engineering matters.
- To encourage educational programmes which develop scientific and technical expertise in medical and biological engineering.

Activities

The IFMBE has published the journal *Medical and Biological Engineering and Computing* for over 34 years. A new journal *Cellular Engineering* was established in 1996 in order to stimulate this emerging field in biomedical engineering. In *IFMBE News* members are kept informed of the developments in the Federation. *Clinical Engineering Update* is a publication of our division of Clinical Engineering. The Federation also has a division for Technology Assessment in Health Care.

Every three years, the IFMBE holds a World Congress on Medical Physics and Biomedical Engineering, organized in cooperation with the IOMP, and the IUPESM. In addition, annual, milestone, regional conferences are organized in different regions of the world, such as the Asia Pacific, Baltic, Mediterranean, African and South American regions.

The administrative council of the IFMBE meets once or twice a year and is the steering body for the IFMBE. The council is subject to the rulings of the General Assembly which meets every three years.

For further information on the activities of the IFMBE, please contact: Jos A E Spaan, Professor of Medical Physics, Academic Medical Center, University of Amsterdam, PO Box 22660, Meibergdreef 9, 1105 AZ, Amsterdam, The Netherlands. Tel: 31 (0) 20 566 5200. Fax: 31 (0) 20 691 7233. E-mail: IFMBE@amc.uva.nl. WWW: http://vub.vub.ac.be/~ifmbe.

IOMP

The IOMP was founded in 1963. The membership includes 64 national societies, two international organizations and 12 000 individuals. Membership of IOMP consists of individual members of the Adhering National Organizations. Two other forms of membership are available, namely Affiliated Regional Organization and Corporate Members. The IOMP is administered by a Council, which consists of delegates from each of the Adhering National Organizations; regular meetings of Council are held every three years at the International Conference on Medical Physics (ICMP). The Officers of the Council are the President, the Vice-President and the Secretary-General. IOMP committees include: developing countries, education and training; nominating; and publications.

Objectives

- To organize international cooperation in medical physics in all its aspects, especially in developing countries.
- To encourage and advise on the formation of national organizations of medical physics in those countries which lack such organizations.

Activities

Official publications of the IOMP are *Physiological Measurement*, *Physics in Medicine and Biology* and the *Medical Science Series*, all published by Institute of Physics Publishing. The IOMP publishes a bulletin *Medical Physics World* twice a year.

Two Council meetings and one General Assembly are held every three years at the ICMP. The most recent ICMPs were held in Kyoto, Japan (1991),

Rio de Janeiro, Brazil (1994) and Nice, France (1997). The next conference is scheduled for Chicago, USA (2000). These conferences are normally held in collaboration with the IFMBE to form the World Congress on Medical Physics and Biomedical Engineering. The IOMP also sponsors occasional international conferences, workshops and courses.

For further information contact: Gary D Fullerton, Professor, University of Texas HSC – San Antonio, Department of Radiology, 7703 Floyd Curl Drive, San Antonio, TX 78284-7800, USA. Tel: (210) 567-5550. Fax: (210) 567 5549. E-mail: fullerton@uthscsa.edu.

CONTENTS

LIST OF CONTRIBUTORS

Pedro Andreo (Ch 4)†
Head Dosimetry and Medical
  Radiation Physics Section
Division of Human Health
  (RIHU)
International Atomic Energy
  Agency
PO Box 200
A-1400 Vienna
Austria
E-mail: p.andreo@iaea.org

Marie-Jose Belanger (Ch 9)
Harvard-MIT Division of Health
  Sciences and Technology
MIT
Cambridge, MA 02139
USA
E-mail: belanger@wcbicl.mit.edu

Magnus Dahlbom (Ch 19)
Division of Nuclear Medicine
Department of Molecular and
  Medical Pharmacology
UCLA School of Medicine
Los Angeles, CA 90095
USA
E-mail:
mdahlbom@mail.nuc.ucla.edu

Daniel J de Vries (Ch 10, 13)
Department of Nuclear Medicine
University of Massachusetts
  Medical School
Worcester, MA 01655
USA
E-mail:
devries@wachusett.ummed.edu

Andrew B Dobrzeniecki (Ch 9)
Whitaker College of Health
  Sciences and Technology
MIT
Cambridge, MA 02139
USA
E-mail: andy@rad.mit.edu

Lars Eriksson (Ch 19)
Department of Clinical
  Neurophysiology
Karolinska Hospital and
  Karolinska Institute
S-104 01 Stockholm
Sweden
E-mail: lars.eriksson@neuro.ks.se

†Work carried out at: Department of Radiation Physics, Lund University, Lund, Sweden.

Kjell Erlandsson (Ch 16)
Department of Radiation Physics
Lund University Hospital
S-221 85 Lund
Sweden
E-mail:
kjell.erlandsson@radfys.lu.se

Peter D Esser (Ch 5)
Department of Radiology
College of Physicians and
    Surgeons
Columbia University
New York
USA
E-mail:
esserpe@cucis.cis.columbia.edu

Robert L Harrison (Ch 7, 18)
Imaging Research Laboratory
Department of Radiology
NW-040 University Hospital
Seattle, WA 98195
USA
E-mail:
roberth@u.washington.edu

David R Haynor (Ch 2)
Department of Radiology, RC-05
University of Washington
Seattle, WA 98195
USA
E-mail:
haynor@u.washington.edu

Marie Foley Kijewski (Ch 13)
Harvard Medical School
Boston, MA 02155
USA
and
Department of Radiology
Brigham & Women's Hospital
Boston, MA 02115
USA
E-mail:
kijewski@bwh.harvard.edu

Michael A King (Ch 14)
Department of Nuclear Medicine
University of Massachusetts
    Medical Center
55 Lake Avenue North
Worcester, MA 01655
USA
E-mail:
michael.king@ummed.edu

Kenneth F Koral (Ch 12)
Department of Internal Medicine
(Division of Nuclear Medicine)
University of Michigan Medical
    Center
3480 Kresge III
Ann Arbor, MI 48109-0552
USA
E-mail: kenkoral@umich.edu

Tom Lewellen (Ch 7)
Imaging Research Laboratory
Department of Radiology
NW-040 University Hospital
Seattle, WA 98195
USA
E-mail: tkldog@u.washington.edu

Michael Ljungberg (Ch 1, 4, 11,
    15, 20)
Department of Radiation Physics
Lund University Hospital
S-221 85 Lund
Sweden
E-mail:
michael.ljungberg@radfys.lu.se

Robert S Miyaoka (Ch 18)
Division of Nuclear Medicine
Department of Radiology
University of Washington
Medical Center
Seattle, WA 98195
USA
E-mail:
rmiyaoka@u.washington.edu

Stephen C Moore (Ch 10, 13)
Nuclear Medicine Service
V A Medical Center
W. Roxbury, MA 02132
USA
and
Harvard Medical School
Boston, MA 02215
USA
E-mail:
scmoore@bwh.harvard.edu

Stefan P Mueller (Ch 13)
Harvard Medical School
Boston, MA 02155
USA
and
Abteilung Nuklearmedizin
Universitaetsklinik Essen
Germany
E-mail:
stefan.mueller@uni-essen.de

Tomas Ohlsson (Ch 16)
Department of Radiation Physics
Lund University Hospital
S-221 85 Lund
Sweden
E-mail:
tomas.ohlsson@radfys.lu.se

Bill C Penney (Ch 13)
Department of Nuclear Medicine
University of Massachusetts
Medical School
Worcester, MA 01655
USA
E-mail:
bobrienp@midway.uchicago.edu

Yani Picard (Ch 17)
Department of Nuclear Medicine
    and Radiobiology
Université de Sherbrooke
3001 12th Avenue North
Sherbrooke (Quebec)
Canada J1H 5N4
E-mail:
yani@tep.mednuc.usherb.ca

Mark F Smith (Ch 8)
National Institutes of Health
Building 10, Room 1C401
10 Center Drive, MSC 1180
Bethesda, MD 20892-1180
USA
E-mail: smith@nmdhst.cc.nih.gov

Sven-Erik Strand (Ch 6, 20)
Department of Radiation Physics
Lund University Hospital
S-221 85 Lund
Sweden
E-mail:
sven-erik.strand@radfys.lu.se

Chris J Thompson (Ch 17)
Montreal Neurological Institute
3801 University Street
McGill University
Montreal, QC H3A 2B4
Canada
E-mail:
chris@rclvax.medcor.mcgill.ca

Tin-Su Pan (Ch 14)
General Electric Company
Applied Science Laboratory
PO Box 414, NB-922
Milwaukee, WI 53201
USA
E-mail: pant@med.ge.com

Steven Vannoy (Ch 7)
Imaging Research Laboratory
Department of Radiology
NW-040 University Hospital
Seattle, WA 98195
USA
E-mail:
svannoy@u.washington.edu

Jacquelyn C Yanch (Ch 9)
Department of Nuclear
  Engineering

and

Whitaker College of Health
  Sciences and Technology
MIT
Cambridge, MA 02139
USA
E-mail: jcy@wcbicl.mit.edu

George Zubal (Ch 3)
Imaging Science Research
  Laboratories
Department of Diagnostic
  Radiology
Yale University School of
  Medicine
New Haven, CT 06510
USA
E-mail: george.zubal@yale.edu

PREFACE

Nuclear medicine has historically been the field in which most of the early Monte Carlo calculations in medical radiation physics were done. One can remember, for example, the early works of Hal O Anger using the Monte Carlo method in the early 1960s to calculate the performance parameters of his newly developed scintillation camera. Today we see an increasing number of scientific papers referring to the Monte Carlo method as the method of choice in evaluating different topics in nuclear medicine. While Monte Carlo simulation does not replace experimental measurements, it does, however, offer a unique possibility for gaining understanding of the underlying physics that forms images. It also provides substantial help to researchers developing methods for improvement in these images.

This book has been put together for three purposes: first, to explain the Monte Carlo method; second, to introduce the reader to some Monte Carlo software packages developed and used by different research groups; and third, to give the reader a detailed idea of some possible applications of Monte Carlo in current research in SPECT and PET. A significant part of this book describes the physics and technology behind simulated imaging detectors and systems. The reason for this is to familiarize the reader with the systems and to give an understanding of the complexity of the systems that Monte Carlo programs are intended to simulate. The text is intended for educational use at both graduate and undergraduate levels and as a reference book on the Monte Carlo method in diagnostic nuclear medicine.

The editors are happy that so many high-quality contributions could be included in the book. Editing the material showed that this method is widespread in different areas for nuclear medicine imaging. The present book has focused on the diagnostic aspect of the Monte Carlo technique, but its use in radionuclide therapy is growing as well.

We would like to thank all those colleagues who have given us invaluable help over the years in developing our knowledge in this area. Professor Bertil Persson's introduction to the Monte Carlo technique several years ago marked the start of our earlier work in Lund. Also, financial support from different sources is highly appreciated, especially from national funds such as the Swedish Cancer Foundation and local funds such as the Gunnar Nilsson Foundation, the Mrs Berta Kamprad Foundation and the John and Augusta Persson Foundation.

CHAPTER 1

INTRODUCTION TO THE MONTE CARLO METHOD

Michael Ljungberg

1.1 INTRODUCTION

In the literature, we see today an increasing number of scientific papers which use Monte Carlo as the method of choice for the evaluation of a range of topics in nuclear medicine such as the determination of scatter distributions, collimator design and the effects of various parameters upon image quality. So what is the Monte Carlo method and why is it so commonly used as a tool for research and development?

Monte Carlo numerical simulation methods can be described as statistical methods that use random numbers as a base to perform simulation of any specified situation. The name was chosen during the World War II Manhattan Project because of the close connection to games based on chance and because of the location of a very famous casino in Monte Carlo.

In most Monte Carlo applications, the physical process can be simulated directly. It only requires that the system and the physical processes can be modelled from known probability density functions (pdfs). If these pdfs can be defined accurately, the simulation can be made by random sampling from the pdfs. A large number of simulations of histories (e.g. photon or electron tracks) are necessary to obtain an accurate estimate of the parameters to be calculated.

Generally, simulation studies have several advantages over experimental studies. For any given model, it is very easy to change different parameters and investigate the effect of these changes on the performance of the system under investigation. Thus the optimization of an imaging system can be

1

aided greatly by the use of simulations. Also, one can study the effects of parameters that cannot be measured experimentally. For example it is impossible to measure the scatter component of radiation emitted from a distributed source independently of the unscattered component. By using a Monte Carlo technique incorporating the known physics of the scattering process it is possible to simulate scatter events from the object and determine their effect on the final image. Hence, a simulation program can help the understanding of the underlying processes since all details of the simulation are accessible.

Overview papers on the Monte Carlo method and its applications in different fields of radiation physics have been given elsewhere by, for example, Raeside [1], Turner [2] and, recently, Andreo [3]. Here we will outline only the basic methodology and how this may be applied to nuclear medicine problems.

1.2 THE RANDOM NUMBER GENERATOR

A fundamental part of any Monte Carlo calculation is the random number generator. Basing the numbers on detection of true random events, such as radioactive decay, can be done but this is generally very cumbersome and time consuming. On the other hand, true random numbers cannot be calculated since they, by definition, are randomly distributed and, as a consequence of this, are unpredictable. However, for practical considerations, a computer algorithm can be used to generate uniformly distributed random numbers from calculated seed numbers. An example of such an algorithm is the linear congruental algorithm where a series of random numbers I_n is calculated from a first seed value I_0, according to the relationship

$$I_{n+1} = (aI_n + b) \bmod(2^k) \tag{1.1}$$

where a and b are constants and k is the integer word size of the computer. If b is equal to zero then the random number generator is called a multiplicative congruental random number generator. The following FORTRAN statements describe the random number generator in (1.1). SEED is the initial value and RAN is the real random number in the range [0,1].

```
REAL FUNCTION RAN(SEED)
PARAMETER (IA=7141, IC=54773, IM=259200)
SEED = MOD(INT(SEED)*IA+IC,IM)
RAN = SEED/IM
END
```

It is important to realize that using the same value of SEED will give the same sequence of random numbers. Thus, when comparing different simulations, one needs to randomly change the initial value of SEED. This can be done, for example, by triggering a SEED value from a call to the system clock or by storing the value of the previous SEED immediately before exit from the previous simulation and then using this value as an initial value in the next simulation. This approach avoids obtaining the same results if a previous simulation is repeated. Repetition can in some cases be advantageous, for example, in a debugging procedure where small systematic errors can be difficult to spot if errors occur between simulations for statistical reasons.

An effect of this form of digital data representation in a computer is that there is a risk that the initial seed number can appear later in the random number sequence. If this occurs, then it is said that the random number generator has 'looped'. Although the subsequent numbers are still randomly distributed, they are copies of the values generated earlier in the sequence. The severity of this effect depends on the application. The length of the sequence for the linear congruental generator is 2^k if b is odd. For the multiplicative congruental generator the sequence length is $2^k - 2$.

1.3 SAMPLING TECHNIQUES

In all Monte Carlo calculations, some *a priori* information about the process to be simulated is needed. This information is usually expressed as probability distribution functions, pdfs, for the different processes. For example, when simulating photon interactions, the total and partial cross-section data represent such information used to calculate the path length and interaction type. From this information, a random choice can be made on which type of interaction will occur or how far a photon will go before the next interaction.

A pdf is defined over the range of $[a, b]$. The function is ideally integrable so that the function can be normalized by integration over its entire range. To obtain a stochastic variable that follows a particular pdf two different methods can be used.

1.3.1 The distribution function method

A cumulated distribution function cpdf (x) is constructed from the integral of pdf(x) over the interval $[a, x]$ according to

$$\text{cpdf}(x) = \int_a^x \text{pdf}(x') \, \mathrm{d}x'. \tag{1.2}$$

A random sample x is then sampled by replacing cpdf(x) in (1.2) with a uniformly distributed random number in the range of $[0,1]$ and solving for

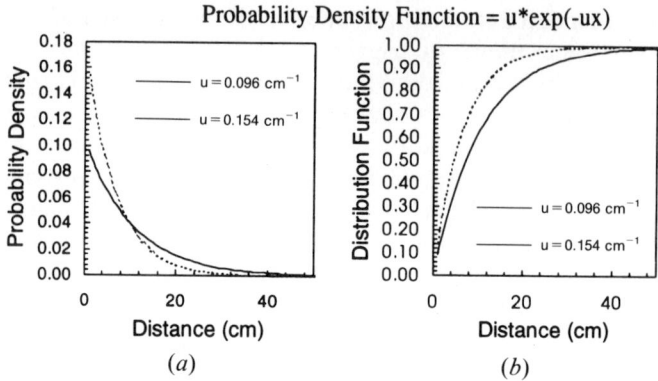

Figure 1.1   *Two exponential pdfs and their related calculated cpdfs.*

x. Two examples of a pdf(x) and the corresponding cpdf(x) are shown in figure 1.1.

1.3.2 The 'rejection' method

Occasionally, the distribution function method is cumbersome to use due to mathematical difficulties in the calculation of the inverse of the cpdf. In these cases, one can use the rejection method which can basically be described by three steps.

Step 1.   Let the probability distribution function, pdf(x), be bounded in the range $[a, b]$. Calculate a normalized function pdf$^*(x) =$ pdf$(x)/$max$[$pdf$(x)]$ so the maximum value of pdf* is equal to unity.

Step 2.   Sample a uniform distributed value of x within the range $[a, b]$ from the relation $x = a + R_1 (b - a)$ where R_1 is a random number.

Step 3.   Let a second random number R_2 decide whether the sampled x should be accepted. This choice is made by calculating the function value of pdf$^*(x)$ from the sampled x value and then checking whether $R_2 <$ pdf$^*(x)$. If this relation is fulfilled, then x is accepted as a properly distributed stochastic value. Otherwise, a new x value needs to be sampled, according to the procedure in step 2.

1.3.3 Mixed methods

A combination between the two methods described above can be used to overcome potential problems in developing algorithms based on either of the two methods alone. Here, the pdf(x) is the product of two probability distribution functions $\text{pdf}_A(x) \cdot \text{pdf}_B(x)$. The different steps in using this method are the following.

Step 1. Let $\text{pdf}_A(x)$ be normalized so that the integral of $\text{pdf}_A(x)$ over the range $[a, b]$ is unity.

Step 2. Let $\text{pdf}_B(x)$ be normalized so that the maximum value of $\text{pdf}_B(x)$ is equal to unity.

Step 3. Choose an x value from $\text{pdf}_A(x)$ using the distribution function method.

Step 4. Apply the rejection method to $\text{pdf}_B(x)$ using the sampled value x from step 3 and check whether or not a random number R is less than $\text{pdf}_B(x)$. If not, then return to step 3.

1.4 EXAMPLES OF SAMPLING IN PHOTON INTERACTION SIMULATIONS

Since this book will focus mainly on Monte Carlo applications for photon transport, describing the basic steps in simulating a photon path can be educative.

1.4.1 Cross-section data

Data on the scattering and absorption of photons are fundamental for all Monte Carlo calculations since the accuracy of the simulation depends on the accuracy in the probability functions, i.e. the cross-section tables [4–6]. Photon cross sections for compounds can be obtained quite accurately (except at energies close to absorption edges) as a weighted sum of the cross sections for the different atomic constituents.

A convenient computer program developed to generate cross sections and attenuation coefficients for single elements as well as compounds and mixtures as needed is XCOM [7]. This program calculates data for any element, compound or mixture, at energies between 1 keV and 100 GeV. The program includes a database of cross sections for the elements. The total cross sections, attenuation coefficients, partial cross sections for incoherent scattering, coherent scattering, photoelectric absorption and pair production in the field of the atomic nucleus and in the field of the atomic electrons are calculated. For compounds, the quantities tabulated are partial and total mass interaction coefficients, which are equal to the product of the corresponding cross sections and the number of target molecules per unit mass of the material. The sum of the interaction coefficients for the individual processes is equal to the total attenuation coefficient. In XCOM, a comprehensive database for all elements over a wide range of energies has been constructed by combining incoherent and coherent scattering cross sections from [8] and

cm²/g

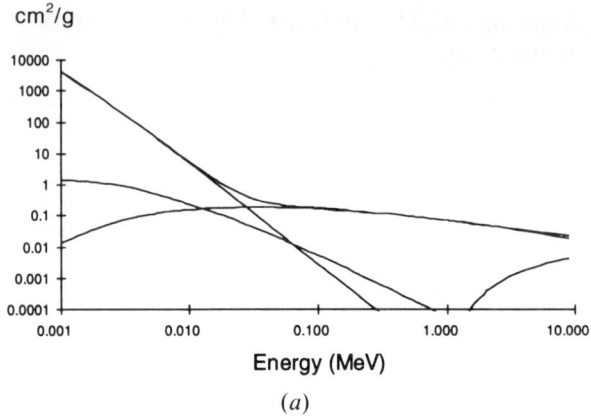

Energy (MeV)

(*a*)

cm²/g

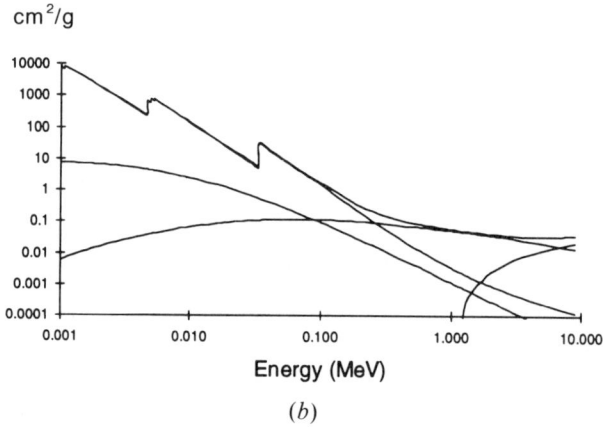

Energy (MeV)

(*b*)

Figure 1.2 *Total and partial attenuation coefficients, obtained from the XCOM program, for (a) water and (b) NaI.*

[9], photoelectric absorption from [10], and pair production cross sections from [11]. Figure 1.2 shows differential and total attenuation coefficients for water and NaI. Note the discontinuities around 8 keV and 30 keV due to the L and K absorption edges in iodine.

An aspect which deserves further attention is the fact that there exists a variation in the physical cross-section tables included in available Monte Carlo codes. This is of special importance when comparing results from different codes. The use of different cross-section data and approximations will usually yield different results and the accuracy of the results is not always obvious.

1.4.2 Photon path length

The path length of a photon in the material must be calculated to determine the next point of interaction in the volume. Generally, this distance depends upon the photon energy and the material density and composition. The distribution function method can be used to sample the distributed photon path length x. If the probability function is given by

$$p(x) = \mu \exp(-\mu x) \tag{1.3}$$

then the probability that a photon will travel distance d or less is given by

$$P(d) = \int_0^d \mu \exp(-\mu x) \, dx = [-\exp(-\mu x)]_0^d = 1 - \exp(-\mu d). \tag{1.4}$$

To sample the path length a uniform random number R is substituted for $P(d)$ and the problem is solved for d.

$$R = P(d) = [1 - \exp(-\mu d)]$$
$$d = -\frac{1}{\mu} \ln(1 - R) = -\frac{1}{\mu} \ln(R). \tag{1.5}$$

Since $(1 - R)$ is also a random number and has the same distribution as R, one can simplify the calculation, according to (1.5).

1.4.3 Selection of the type of photon interaction

The probability for a certain interaction type to occur is given by the partial attenuation coefficients. These are tabulated for different energies and materials. The sum of the partial attenuation coefficients for photoelectric effect (τ), Compton interaction (σ_{incoh}), coherent interaction (σ_{coh}) and pair production (κ) is called the linear attenuation coefficient $\mu = \tau + \sigma_{incoh} + \sigma_{coh} + \kappa$, or mass attenuation coefficient if normalized by the density. To select a particular interaction type during the simulation, a uniform random number R is sampled and if the condition $R < \tau/\mu$ is true, then a photoelectric interaction has occurred. If this condition is false, then the same value of R is used to test whether $R < (\tau + \sigma_{incoh})/\mu$. If this is true then one continues with a Compton interaction. If not, then the test $R < (\tau + \sigma_{incoh} + \sigma_{coh})/\mu$ will determine whether a coherent interaction has taken place. If all conditions are false, then pair production will be simulated. Obviously this will only occur if the photon energy is greater than 1.022 MeV.

1.4.4 Incoherent photon scattering

Incoherent scattering, commonly denoted Compton scattering, means an interaction between an incoming photon and an atomic electron where the photon loses energy and changes direction. The energy of the scattered photon, hv', depends upon the initial photon energy, hv, and the scattering angle θ (relative to the incident path), according to

$$hv' = \frac{hv}{1 + (hv/m_0 c^2)[1 - \cos \theta]}.$$ (1.6)

One very commonly used method to sample the energy and direction of a Compton-scattered photon is the algorithm developed by Kahn [12]. This algorithm is based on the Klein–Nishina cross-section equation assuming the electrons to be free and at rest in the scatterer.

$$d_e \sigma(\theta) = \left(\frac{r_0^2}{2}\right)\left(\frac{\lambda}{\lambda'}\right)^2\left(\frac{\lambda}{\lambda'} + \frac{\lambda'}{\lambda} - 1 + \cos^2 \theta\right) d\Omega.$$ (1.7)

Kahn's sampling method is based on a mixed method and is shown in a Fortran statement below. The mathematical proof for the algorithm has been described by, for example, Raeside [1].

```
      ALPHA  = HV / 511.
      TEST   = (2.*ALPHA+1.) / (2.*ALPHA + 9.)

1     RANDOM = 2. * RAN(SEED)

      IF(RAN(SEED).LT.TEST) THEN
         UU = 1. + ALPHA * RANDOM
         IF(RAN(SEED) .GT. 4.*(UU-1.)/(UU*UU) ) GOTO 1
         COSTET = 1 - RANDOM
      ELSE
         UU = (2. * ALPHA + 1.) / (ALPHA * RANDOM + 1.)
         COSTET = 1. - (UU-1.)/ALPHA
         IF(RAN(SEED) .GT. 0.5*(COSTET*COSTET + (1. / UU)) ) GOTO 1
      ENDIF
```

For situations where the incoming photon energy is of the same order as the binding energy of the electron, the assumption of a free electron at rest becomes less justified. The cross section for this occurrence is given by

$$d_a \sigma_{incoh}(\theta) = d_e \sigma_{KN}(\theta)S(x, Z)$$ (1.8)

where $S(x, Z)$ is the incoherent scattering function [8], Z is the atomic number and $x = (\sin \theta/2)/\lambda$ is the momentum transfer parameter that varies with the photon energy and scatter angle. It can be shown [13] that

$$\frac{d_a \sigma_{incoh}(\theta)/d\Omega}{{}_a\sigma_{incoh}} = \frac{d_e \sigma_{KN}(\theta)/d\Omega}{{}_e\sigma_{KN}} \frac{S(x, Z)}{S_{max}(x, Z)} K(hv, Z)$$ (1.9)

where $K(hv, Z)$ is constant for a fixed Z and energy. A scattering angle is sampled from 'a normalized distribution of (1.7)' using, for example, the Kahn method. A momentum transfer parameter, x, is then calculated and θ (obtained from the sampled x) is accepted only if a random number $R < [S(x, Z)/S_{max}(x, Z)]$. Otherwise, a new scattering angle is sampled.

1.4.5 Coherent photon scattering

Coherent scattering is an interaction between an incoming photon and the whole atom and where the direction of the photon is changed but without energy loss. This type of interaction leads to the photon being scattered mostly in the forward direction. The sampling technique for a coherent scattering is based on the Thompson cross section multiplied by the atomic form factor [9] $F(x, Z)$,

$$d_a \sigma(\theta) = (d_e \sigma_{Th}(\theta))F^2(x, Z) = (r_0^2/2)(1 + \cos^2 \theta)F^2(x, Z)2\pi \sin \theta \, d\theta.$$
$$(1.10)$$

It can be shown [13] that the probability of a photon being scattered into the interval $d\theta$ around θ is given by

$$P(\theta) \, d\theta = K(hv, Z)G(\theta)f(x^2, Z) \tag{1.11}$$

where $K(hv, Z)$ is constant for a fixed energy and atomic number, $G(\theta)$ has a fixed range and

$$f(x^2, Z) = \frac{F^2(x, Z)}{\displaystyle\int_0^{x_{max}^2} F^2(x, Z) \, dx^2}. \tag{1.12}$$

A value of x^2 is sampled from a pre-calculated distribution function of $f(x^2, Z)$. From this value, a scattering angle, θ, can be calculated and accepted provided that the relation $R < G(\theta)$ is fulfilled.

1.4.6 Coordinate calculations

After sampling a new photon path length and direction, the Cartesian coordinates to the next interaction point needs to be calculated. This can be achieved by a geometrical consideration where the new coordinates (x', y', z') in the Cartesian coordinate system are calculated from the photon path length and direction cosines, according to

$$x' = x + du' \qquad y' = y + dv' \qquad z' = z + dw' \tag{1.13}$$

where d is the distance between the previous point (x, y, z) and the new point of interest (x', y', z'). Assuming θ and ϕ are the polar and azimuthal angles in the Cartesian coordinate system and Θ and Φ are the polar and azimuthal angles defining the direction change, relative to the initial path of

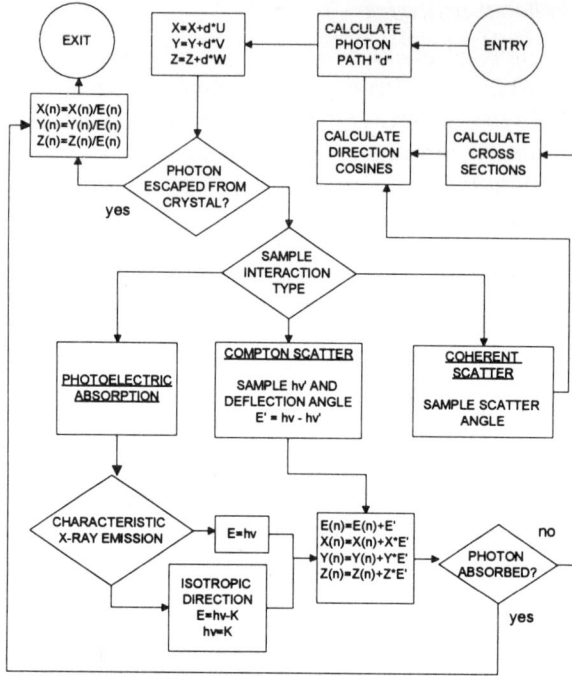

Figure 1.3   *A flowchart describing the basic steps for simulation of photon interactions in a defined volume.*

the photon, then the new direction cosines (u', v', w'), necessary to calculate (x', y', z'), are calculated from

$$u' = \sin\theta'\cos\phi' = u\cos\Theta + \sin\Theta\,(w\cos\Phi\cos\phi - \sin\Phi\sin\phi)$$
$$v' = \sin\theta'\sin\phi' = v\cos\Theta + \sin\Theta\,(w\cos\Phi\sin\phi + \sin\Phi\cos\phi)$$
$$w' = \cos\theta' = w\cos\Theta - \sin\Theta\,(\sin\theta\cos\Phi)$$
$$v = \sin\phi\sin\theta \tag{1.14}$$
$$u = \sin\theta\cos\phi$$
$$w = \cos\theta$$

1.4.7 Example of a calculation scheme

Figure 1.3 shows a flowchart of a photon simulation in a volume including photoelectric absorption, incoherent and coherent scattering and simulation of characteristic x-ray emission at a site of photoelectric absorption. The probability for such an occurrence is given by the fluorescence yield.

1.5 VARIANCE REDUCTION METHODS

In some cases Monte Carlo simulations can be very time consuming. This is particularly true where most of the photon histories generated by direct Monte Carlo simulation are likely to be rejected, for example when simulating a point source at a large distance from a small detector so the probability that a photon will hit the detector is small.

Variance reduction techniques can then be applied to improve the efficiency of the simulation and hence the statistical properties of the images produced. These techniques are based on calculating a photon history weight, W, that is applied to a photon. This weight represents the probability that the photon passes through a particular history of events. These techniques are based on assuming probability functions for the physical processes, either to reduce the variance per history or to speed up the simulation to allow more histories to be simulated during the same CPU time. To correct for this, the photon is associated with a weight W which is the quotient between the true probability density and the fictitious probability density that the photon goes through a particular series of events. A detailed description of these various methods is given in Chapter 2.

REFERENCES

[1]   Raeside D E 1976 Monte Carlo principles and applications *Phys. Med. Biol.* **21** 181–97

[2]   Turner J E, Wright H A and Hamm R N 1985 A Monte Carlo primer for health physicists *Health Phys.* **48** 717–33

[3]   Andreo P 1991 Monte Carlo techniques in medical radiation physics *Phys. Med. Biol.* **36** 861–920

[4]   Hubbell J H 1969 *Photon Cross Sections, Attenuation Coefficients and Energy Absorption Coefficients from 10 keV to 100 GeV (Natl Stand. Ref. Data Ser. 29)*

[5]   McMaster W H, Del Grande N K, Mallett J H and Hubbell J H 1969 Compilation of x-ray cross sections, *Report* UCRL-50174 (abstract)

[6]   Storm E and Israel H I 1970 Photon cross sections from 1 keV to 100 MeV for elements $Z = 1$ to $Z = 100$ *Nucl. Data Tables* A **7** 565–681

[7]   Berger M J and Hubbell J R 1987 *XCOM: Photon Cross-sections on a Personal Computer* NBSIR 87-3597 (Washington, DC: National Bureau of Standards)

[8]   Hubbell J H, Veigle J W, Briggs E A, Brown R T, Cramer D T and Howerton R J 1975 Atomic form factors, incoherent scattering functions and photon scattering cross sections *J. Phys. Chem. Ref. Data* **4** 471–616

[9]   Hubbell J H and Overbo I 1979 Relativistic atomic form factors and photon coherent scattering cross section *J. Phys. Chem. Ref. Data* **8** 69–105

[10] Scofield J H 1973 Photoionization cross sections from 1 to 1500 keV, *Report* UCRL-51326 (abstract)

[11] Hubbell J H, Gimm H A and Overbo I 1980 Pair, triplet and total atomic cross sections (and mass attenuation coefficients) for 1 MeV–100 GeV photons in elements $Z = 1$ to 100 *J. Phys. Chem. Ref. Data.* **9** 1023–147

[12] Kahn H 1956 Application of Monte Carlo, *Report* RM-1237-AEC (abstract)
[13] Persliden J 1983 A Monte Carlo program for photon transport using analogue
       sampling of scattering angle in coherent and incoherent scattering processes
       Comput. Program. Biomed. **17** 115–28

CHAPTER 2

VARIANCE REDUCTION TECHNIQUES

David R Haynor

2.1 WHY ARE VARIANCE REDUCTION TECHNIQUES NECESSARY?

The popularity of the Monte Carlo method in nuclear medicine research arises from the possibility of accurately simulating the physical processes undergone by particles originating in and passing through heterogeneous media and then interacting with collimator and detector systems. In a typical nuclear medicine acquisition, nearly all the photons which arise within a patient are absorbed within the patient, miss the camera entirely, are absorbed in the collimator or pass through the detector without interacting. Consider, for example, an acquisition in which 25 mCi of activity is administered to a patient and 500 000 counts are accumulated in a planar image over a period of 1 min. In that period of time, $0.025 \times 2.2 \times 10^{12} = 5.5 \times 10^{10}$ disintegrations take place within the whole body; only 5×10^5, or approximately one disintegration in 110 000, gives rise to a detected event. A so-called *analogue* Monte Carlo simulation, in which particle histories are generated and followed according to the underlying physics, would follow the same statistics: only one complete photon history out of every 110 000 generated by the Monte Carlo code would give rise to a detected event. Assuming our interest lies primarily in the detected events, this brute force computation would be extremely inefficient. Put another way, the fluxes of detected particles would be extremely small, and the variance of the flux estimates would be high. The Monte Carlo methods known collectively as *variance reduction techniques* have been developed to deal with this problem.

For variance reduction techniques to be effective, a *weight* must be attached to each photon history that is generated by the Monte Carlo code.

13

This is discussed in more detail below, but roughly speaking the weight of a history may be thought of as the number of 'real-world' particles represented by that history. In the case of an analogue simulation, all history weights are the same. When variance reduction techniques are used, this is no longer the case. Variance reduction techniques are effective because they increase the ratio of detected events to histories over that obtained with analogue Monte Carlo simulations. They achieve this by simulating a different random process, one which is obtained from the analogue process by modifying it so that the histories which arise are enriched in those that give rise to detectable events. In order that the particle fluxes estimated from the new process be accurate estimates of the true (analogue) fluxes, the weight of each history must be adjusted. Generally, this is done each time a simulated physical event (particle emission, scattering, interaction with the collimator, etc) occurs during a history. As a consequence, each history ends up with a unique weight. Because the statistical quality of the simulated detected events is adversely affected by widely varying history weights, variance reduction methods must also include techniques that attempt to equalize the weights attached to individual histories as much as possible.

In summary, variance reduction techniques must do three things: (i) enrich the generated histories with those that give rise to detected events; (ii) adjust the history weights correctly and (iii) ensure that the final history weights do not vary too widely. The remainder of this chapter will be devoted to making these three notions more precise and to illustrating the principles of variance reduction with typical examples that are used in contemporary nuclear medicine simulations. It should be emphasized that the basic principles of variance reduction techniques are not new; they have been well understood since the first appearance of the Monte Carlo technique. The reader is referred to [1–4] for a fuller discussion of the basic principles of variance reduction methods.

2.2 WEIGHTS AND THEIR MANAGEMENT

2.2.1 Weights and the Monte Carlo evaluation of definite integrals

It is easier to understand the basis of variance reduction techniques if we start with the problem of estimating the expected value of a random variable by the Monte Carlo method. The extension to estimating the solutions of transport equations, such as those that describe the birth, scattering and absorption or detection of photons, is then straightforward (see the next section). For the purposes of exposition, let D (for 'detector') be a random variable whose expectation we wish to estimate. D is a function on a probability space X, which we take to be a set equipped with an integration method and a probability density $p(x)$ with the properties that $p(x) \geqslant 0$ for all x in X and $\int_X p(x)\,\mathrm{d}x = 1$. The expectation of D is then defined by the equation $E(D) = \int_X D(x)p(x)\,\mathrm{d}x$. The analogue Monte Carlo approach to the calculation of $E(D)$ is to draw $N \geq 1$ values of x from the space X, according to the distribution $p(x)$, and then to estimate $E(D)$ from the average of the sampled values of D

$$E(D) \approx D_1 = \frac{1}{N}\sum_{i=1}^{N} D(x_i). \tag{2.1}$$

D_1 is an unbiased estimator of $E(D)$, i.e., $E(D_1) = E(D)$. The variance of D_1 may be calculated from (2.1)

$$\mathrm{var}(D_1) = \frac{\mathrm{var}(D)}{N} = \frac{\left(E(D^2) - (E(D))^2\right)}{N}$$

$$= \frac{\int_X D^2(x)p(x)\,\mathrm{d}x - \left(\int_X D(x)p(x)\,\mathrm{d}x\right)^2}{N}. \tag{2.2}$$

Now suppose that we sample from X using a different probability density, but that we still wish to estimate $E(D)$ using the same number of particles in the second Monte Carlo simulation as in the first. If our new probability density function is given by $q(x)$, with $q(x) \geq 0$ for all x in X and $\int_X q(x)\,\mathrm{d}x = 1$, then we have

$$E(D) = \int_X D(x)p(x)\,\mathrm{d}x = \int_X D(x)[p(x)/q(x)]q(x)\,\mathrm{d}x.$$

This gives us a new way of estimating $E(D)$

$$E(D) \approx D_2 = \frac{1}{N}\sum_{i=1}^{N} D(x_i')\frac{p(x_i')}{q(x_i')}$$

$$= \frac{1}{N}\sum_{i=1}^{N} D(x_i')w(x_1') \tag{2.3}$$

where the primes serve to remind us that the x_i' are sampled using the new probability density $q(x)$. The factor $w(x_i') = p(x_i')/q(x_i')$ is just the *weight* that must be attached to the ith particle history as a result of our sampling from $q(x)$ instead of $p(x)$. Note that the weight factor depends only on the history; it does not depend on the random variable D, except for the requirement that, if $D(x)$ is nonzero, $w(x)$ must be finite; that is, $q(x)$ must be nonzero. In other words, if we have a whole family of 'detectors' D, as would be the case in a nuclear medicine simulation, our sampling density $q(x)$ must be positive for all 'detectable' histories. However, $q(x)$ can vanish whenever x is not detectable. This concentration of the sampling density on the particle histories of potential interest is the key to the success of variance reduction techniques.

A possible advantage of using the weighted estimator D_2 instead of D_1 (assuming that q is chosen appropriately) may be seen when we compare the variance of the two estimators. In most nuclear medicine simulations, the random variable D we are interested in represents the flux of a small specified subset of all the particles generated, such as the flux into a particular crystal in a block detector system or into a particular xy position in an Anger camera. In this case, $D(x)$ is either zero or unity for all histories x. The second term in the square brackets of (2.2) will then be the square of the first term. Since the first term will be small, the second term will be substantially smaller, and we may approximate var(D_1) and var(d_2) by using the first term in the square brackets of (2.2) only. We obtain

$$\text{var}(D_1) \approx \frac{1}{N} \frac{\int_x D^2(x) p(x)\, dx}{N} \qquad \text{var}(D_2) \approx \frac{1}{N} \frac{\int_x D^2(x) w^2(x) q(x)\, dx}{N}.$$

$$(2.4)$$

We may approximate the integrals in (2.4) with their Monte Carlo estimates

$$\text{var}(D_1) \approx \frac{1}{N^2} \sum_{i=1}^{N} D^2(x_i) \qquad \text{var}(D_2) \approx \frac{1}{N^2} \sum_{i=1}^{N} D^2(x_i') w^2(x_i'). \qquad (2.5)$$

To better understand (2.5), consider again a random variable D which takes on only the values zero and unity. Without loss of generality, assume that it is precisely the first M terms of the summation for var(D_1) and the first M' terms of the summation for var(D_2) that are nonzero. Then we have, after some manipulation,

$$\text{var}(D_1) \approx \frac{M}{N^2} \qquad (2.6)$$

$$\text{var}(D_2) \approx \frac{M}{N^2} \left[\left(M' \sum_{i=1}^{M'} w^2(x_i') \right) \Bigg/ \left(\sum_{i=1}^{M'} w(x_i') \right)^2 \right] \left[\frac{1}{M'M} \left(\sum_{i=1}^{M'} w(x_i') \right)^2 \right]. \qquad (2.7)$$

The terms in the square brackets on the right-hand side of (2.7) can be simplified. The Cauchy–Schwartz inequality states that, for any real numbers $x_1, x_2, \ldots, x_n$, we have $(\sum_{i=1}^{n} x_i)^2 \leqslant n \sum_{i=1}^{n} x_i^2$; it follows that the first right-hand term in square brackets is at least unity, and we write it as $(\text{QF})^{-1}$, where QF is the *quality factor* for the simulation. The quality factor is always at most unity, and is equal to unity if and only if all the histories have equal weights (i.e., we are performing an analogue simulation). The simulation is assumed to be unbiased, and $(\sum_{i=1}^{M'} w(x_i'))/N$ is precisely the D_2 estimate for $E(D)$ (see (2.3)); therefore, it must be approximately equal to M/N, the D_1 estimate for $E(D)$ (see (2.2)). Making the simplifications, we obtain

$$\frac{\text{var}(D_2)}{\text{var}(D_1)} \approx \frac{M}{\text{QF}\, M'} \qquad \text{with} \quad \text{QF} = \frac{(\sum_{i=1}^{M'} w(x_i'))^2}{M' \sum_{i=1}^{M'} w^2(x_i')} \qquad (2.8)$$

our sought-after result. While several approximations were made in deriving (2.8), it is nonetheless quite useful. It demonstrates that it is possible to reduce the variance of our estimate for $E(D)$ by substituting D_2 for D_1, if two conditions can be met: (i) the number of histories that end in detections that are derived by sampling from $q(x)$, M', must be substantially larger than M, the number derived from sampling from $p(x)$; and (ii) the resulting dispersion of weights, as measured by the quality factor, must not become too great. The reduction in variance is not measured by the ratio M/M', as would be expected with pure Poisson statistics (analogue Monte Carlo). Instead, the effective number of particles in the D_2 simulation is reduced by the QF, which never exceeds unity. In principle, the QF for a simulation is different for each random variable D that is to be estimated; in our experience with nuclear medicine problems, a single QF, obtained by evaluating (2.8) for *all* histories, works well and is a useful yardstick for assessing different variance reduction strategies.

2.2.2 Weights in the simulation of random walk processes

In the section above, the concept of weights was introduced for sampling from a single probability distribution. In a Monte Carlo simulation in nuclear medicine, there typically is a series of random events which, collectively, constitute a single simulated photon history. The particle must be *born*, i.e., assigned an initial energy and direction. It then travels a random distance in the assigned direction and reaches the next point of interaction. Assuming the particle has not escaped the object entirely, it then undergoes either photoelectric absorption, coherent scattering or Compton scattering. In the first case, it is absorbed at the point of interaction. In the second or third case, a new direction and energy are generated and the process repeated. Finally, if the photon escapes the object entirely, it may miss the collimator/detector entirely or it may collide with the front face of the detector. Interaction with the collimator and detector may be treated either deterministically or stochastically.

In applying variance reduction techniques to the simulation of random walk processes, various techniques will be used in the simulation of different events (particle birth, absorption, collimator interaction etc). An initial weight is attached to each particle history as it is generated. The history weight is then updated after each event by multiplying it by the correct weight factor for the variance reduction technique, if any, used at that step of the simulation. At each step, the principles outlined above are used to calculate the correct weights. The final value of the weight is assigned to the history as a whole, and is used for calculation of the overall QF of the simulation (2.8). Thus, good control of weight variation (QF close to unity) ideally requires good control of the weight variations produced at each step of the simulation. The use of Russian roulette, particle splitting and weight windows makes it possible to correct to some extent for the effects of excessive weight variations introduced by sampling techniques.

2.3 APPLICATIONS OF VARIANCE REDUCTION TECHNIQUES IN NUCLEAR MEDICINE SIMULATIONS

In the preceding section, we discussed the principles of the application of variance reduction techniques to processes modelled by a random walk, such as photon passage through matter. We will now illustrate the application of these techniques to nuclear medicine simulations through a series of examples. These examples are arranged roughly according to complexity, rather than in temporal sequence. While these examples are described in the context of single-photon imaging, they are easily adapted to positron imaging as well.

2.3.1 Photoelectric absorption

Our first, and simplest example, relates to the handling of photoelectric absorption. For imaging simulations, photoelectric absorption is an undesired outcome, since it cannot lead to a detected event. To calculate the weight factor, suppose that, given the photon energy E and its current location in space (which determines the material property at that point and, therefore, the probability of absorption), the particle will be absorbed with probability α and scattered with probability $1 - \alpha$. Consider an event space X consisting of two events: $X = \{$absorption, scatter$\}$. The physics assigns probabilities $p(x) = \{\alpha, 1 - \alpha\}$ to the points of X. Because none of our detectors will detect histories which end in the 'absorption' state, we take for our sampling probability $q(x) = \{0,1\}$. According to the principles described above, then, each history must be multiplied by the weight $p(\text{scatter})/q(\text{scatter}) = 1 - \alpha$, and only histories which do not end in absorption are generated. This step is performed each time an interaction within the object being simulated (the patient or phantom) occurs. Typically, α is small, and the variation in weights that occurs as a result of this modification is also small. By the same token, the effective enrichment in particles (M'/M) is also small, although it becomes more significant for particles undergoing larger numbers of interactions.

2.3.2 Sampling from complex angular distributions: coherent scattering and the use of form factors in Compton scattering

For higher photon energies $(E \geq 100 \text{ keV})$, coherent (Rayleigh) scattering can be neglected, and the classical Klein–Nishina formula for the outgoing angular distribution of a photon scattering of a free electron may be used. At lower photon energies, these approximations are less valid (see [2] for a more detailed discussion). The coherent cross section amounts to about 10% of the total cross section at 50 keV. Moreover, the distribution of coherently scattered photons is very different from those that scatter via a Compton process: it is strongly peaked in the forward direction, although there is no perfectly forward coherent scattering. The exact calculation of the coherent distribution is difficult (it is a many-electron effect) and so semi-empirical form factors $C(Z, E, \mu)$ (μ is the cosine of the scattering angle), which are multiplied by the classical Thomson cross section $T(E, \mu)$ to yield the final cross section $p(x)$, are used. Similarly, for low photon energies, the Klein–Nishina cross section for scattering, $K(E, \mu)$, is modified by a form factor that depends on the Z of the scattering material, E and μ: $p(E, \mu) = I(Z, E, \mu)K(E, \mu)$.

Sampling may be performed from these empirical angular distributions in either of two related ways. The first method is simpler, but produces more variation in particle weights than the second method. In the first method, we sample first from $q(x) = K(E, \mu)$ or $T(E, \mu)$ to obtain μ, using classical algorithms such as the Kahn method for sampling from $K(E, \mu)$ [5–7]. The weight of the particle history is then simply multiplied by $I(Z, E, \mu)$ for incoherent scattering or by $C(Z, E, \mu)$ for coherent scattering. Note that the method of weight calculation described above requires that $q(x)$ be a true density; thus, $I(x)$ must be normalized so that $\int I(Z, E, \mu)K(E, \mu) \, d\mu = 1$ for all E and Z. The second method sets $q(x)$ equal to a discretized approximation to the true semi-empirical scattering distribution $p(x)$. For a given Z and E, $q(\mu)$ is defined to be proportional to $p(E, Z, \mu_i)$ for $\mu_i \leq \mu \leq \mu_{i+1}$, where $0 = \pi_0 < \mu_1 < \cdots < \mu_n = 1$ form a set of grid points, and is normalized so that, $\int q(E, Z, \mu) \, d\mu = \sum p(E, Z, \mu_i)(\mu_{i+1} - \mu_i) = 1$. The discrete density $q(x)$ is then easily sampled from to obtain μ, and the particle history is multiplied by $p(E, Z, \mu)/q(E, Z, \mu)$ as before. By choosing a sufficiently fine discretization, the weight variation produced by this method can be made arbitrarily small. The second method will prove useful in designing a variance reduction technique for forced detection, below.

2.3.3 Detector interactions

In most nuclear medicine simulations, the interaction of an escaped photon with the detector crystal is treated only approximately. The deposition of energy within the crystal is assumed to occur at a single point along the photon path, the exact location being chosen according to the usual techniques for exponential sampling along a ray [1]. Variance reduction in this case is similar to the handling of photoelectric absorption described above: the weight of the history is simply multiplied by the probability of interaction within the crystal, and all photons whose paths intersect the crystal and which pass through the collimator will be detected. The variability in weights introduced at this step is related to the dispersion of the energy and inclination of the escaped photons, but is typically small.

2.3.4 Preventing particle escape

In an analogue simulation, particles which escape the object, but whose path does not intersect the detection system, are similar to particles that undergo photoelectric absorption: computational effort on them is wasted. If forced detection (below) is not in use, these particles may be handled in a similar way to those undergoing absorption. Specifically, after each interaction, the probability π of photon escape is calculated by integrating the total attenuation along the outgoing direction

$$\pi = \exp\left(-\int_0^{\text{escape}} A(E, x + t\boldsymbol{n})\, \mathrm{d}t\right)$$

where $A(E, \boldsymbol{r})$ is the total cross section for a photon with energy E at position $\boldsymbol{r}$, x is the current position and $\boldsymbol{n}$ the unit vector in the outgoing direction. Two copies of the history are now created. The first, with weight $W\pi$ (W is the weight prior to the scattering event), is allowed to escape and possibly interact with the detector. The second is assigned weight $W(1-\pi)$ and is forced to interact at some point in the object along the direction $\boldsymbol{n}$ before escaping. Particle histories are terminated in this scheme only when the particle energy drops below a set threshold. Preventing particle escape has two disadvantages. The fluctuation in weights resulting from the repeated multiplications may be substantial. Multiple detected particles may, in this scheme, arise from a single particle birth and consequently subsequent detected events may be highly correlated.

2.3.5 Collimator penetration/scatter

Most of the inefficiency of conventional gamma cameras arises from the effects of collimation. Accordingly, considerable care must be taken with the collimator simulation, or large numbers of photons will be lost. In single-photon imaging, collimators typically collimate both in the axial plane (φ direction) and in the azimuthal direction (θ direction). In positron imaging, collimators (if present at all) collimate primarily in the θ direction. In either case, however, it may be of interest to produce a stream of escaping photons all pointing primarily in the same φ direction. This corresponds to obtaining a single projection of a single-photon emission computed tomography (SPECT) or positron emission tomography (PET) image. This may be done by using techniques similar to those discussed below under 'forced detection'. Otherwise, it is clearly both more efficient and computationally much simpler to collect a full set of projections, or to study rotationally symmetric activity/attenuation distributions.

For low-energy photons, standard collimators are effectively 'black', i.e., there is little penetration of the collimator material itself. In this case, an excellent approximation to the geometric response of parallel hole collimators was given by Metz and Doi [8] and extended to cone and fan beam collimators in [9]. An effective aperture function for these collimators is given in these articles which may be used to multiply the weight of the photon as it intersects the front face of the collimator. Accurate, efficient modelling of collimator scatter in the general case in which the collimator is not treated as perfectly absorbing remains an unsolved problem. Although it is presumed that a tabular approach, based on detailed Monte Carlo simulations of photon transport and scatter in the collimator, would be effective, little work has been published on the parametric representation of scatter (for example, as a function of incidence angle).

Whether scatter is modelled or not, potential for serious weight dispersion arises at the collimation stage. This is because, even in the geometric model in which collimator scatter is neglected, particle weights may be multiplied by zero or extremely small positive numbers if the incident particle is sufficiently far from perpendicular to the collimator face (in the case of the parallel hole collimator) or from the local hole direction in the case of fan beam and cone beam collimators. The best solution to this problem, in the case of 'black' collimators, is to produce escaped photons that are heavily enriched in photons whose direction of travel is favourable for passing through the collimator. This may be done using the technique of forced detection described below.

2.3.6 Forced detection

Forced detection is a technique for increasing the number of photons that escape the object with a direction of travel that makes it likely that they will pass through the collimator [4, 10]. At each interaction point in a photon history, an attempt is made to extend the history with an additional scatter so that an escaped photon is generated with an appropriate direction of travel. Weighting of the photon 'descendant' is performed using a combination of the techniques described above, including corrections for the additional travel within the attenuator and appropriate sampling from the Klein–Nishina distribution. Modifications for coherent scattering and bound-electron form factors may be performed as described above. Properly done, forced detection allows an average of approximately one detected particle per computational history (in the case of 2π collection geometry as in SPECT or PET). Preliminary work in our laboratory shows that forced detection may also be applied to fan beam and cone beam simulation, with benefits similar to those seen with parallel hole collimators. The weights of unscattered particles and scattered particles (all of which undergo a terminal forced scatter into the collimator) will be quite different unless appropriate adjustments are made with stratification (see the next section).

2.3.7 Stratification

The techniques described above all increase M' in (2.8), at the expense of increasing weight dispersion. This latter effect decreases the QF, undercutting the apparent increase in M' and making the reduction in the variance of flux estimates less than expected (see (2.8)). This section and the next describe techniques for improving the QF of a simulation. The technique of stratification, described here, is applied at the time of particle birth, while the weight window technique (next section) may be applied at any point during a particle history [10].

The number of scatters undergone inside a patient by a photon emitted by a nuclear medicine isotope is small (typically zero to four). Consequently, the ultimate fate of a photon (detection versus no detection) is strongly influenced by the photon's position and direction at its birth (or at any later step in its history). Suppose that we crudely divide the object into a set of spatially compact regions $R_1, \ldots, R_n$ and divide the set of photon directions into $T_1, \ldots, T_m$ (in the case of 2π detection geometry, it is usually sufficient to consider only the photon's inclination, or θ value). We define each of the combinations (R_i, T_j) as a *stratification cell*. If we perform a preliminary run with a small number of histories, we can calculate a *productivity* π_{ij} for each stratification cell

$$\pi_{ij} = \frac{\text{RMS detected weight for particles born in } (R_i, T_j)}{\text{RMS original weight for particles born in } (R_i, T_j)}. \tag{2.9}$$

π_{ij} may be thought of as an estimate of the probability that a photon starting from a point in R_i in a direction lying in the set T_j will be detected. If the photon is born into the (ij)th stratification cell with weight W, the predicted total detected weight of the photon's descendants is approximately $W\pi_{ij}$. This suggests that, in order to achieve approximate equality of the weights of the detected particles, the *initial* weight of the particles in the (ij)th cell should be inversely proportional to π_{ij}. The weight may be adjusted by adjusting the initial sampling density. Specifically, if the activity in the (ij)th cell is A_{ij}, then the number of particles born in the (ij)th cell should be proportional to $A_{ij}\pi_{ij}$; each such particle will have an initial weight of C/π_{ij}, but an average detected weight of approximately C, where C is a constant whose value is adjusted according to the desired value of M' in (2.8). The more productive stratification cells will be sampled more frequently than less productive cells, but the final weight of each history at detection will be approximately equal (independent of starting cell).

In practice, the stratification can be rather coarse, and the estimates of the π_{ij} need only be rough in order to achieve a significant improvement in quality factor. Therefore, the initial run required to estimate their values can be small. The dependence of the productivity on the energy may also be neglected. As mentioned above, the productivities may be defined by

(2.9) separately for unscattered and scattered photons, and are generally rather different. Stratification is, therefore, performed separately for scattered and unscattered photons, using the appropriate productivities.

2.3.8 *Use of weight windows, splitting and Russian roulette*

The use of *weight windows* [4, 10] is similar in spirit to stratification. Using the productivities calculated via (2.9) for scattered photons, the expected value of the detected weight may be approximately calculated for a photon at each point in its history (again, neglecting the dependence of the productivity on photon energy). If the predicted detected weight is too large, the particle may be *split* into several particles, each carrying a fraction of the weight of the original particle. The daughter particles are then tracked independently. If the expected detected weight is too small, the weight of the particle may be increased by a factor κ (with probability $1/\kappa$) or the particle history may be terminated (with probability $1 - 1/\kappa$). This *Russian roulette* technique preserves expected values while preventing the code from spending too much time tracking particles that have small weights or that have wandered into unproductive stratification cells.

2.3.9 *Modifications for positron imaging*

The modifications of the techniques described above for positron imaging are straightforward. To simplify the description, let us call the two photons that arise from a positron annihilation the 'blue' and 'pink' photons. The productivity of a stratification cell must be defined by modifying (2.9) to include the product of the detected weights of blue and pink photons. In addition, in order to calculate expected values correctly, the possibility that several blue and several pink detection events might arise from a single annihilation must be considered. In this case, the coincidence events that arise from all possible blue–pink pairings are all valid (simulated) coincidences, although only one at most will be an unscattered coincidence.

2.4 CONCLUSIONS

A framework for understanding the goals of variance reduction techniques— to increase the number of detected events, while maintaining rough equality of weights—has been presented, and the major techniques that have been used in practice have been described. These techniques are used in currently available public domain Monte Carlo codes (see Chapter 7 by Lewellen *et al.*, in this volume) and typically result in effective speedups by factors of three to 100 over analogue methods after allowing for the reduction in quality factors produced by variance reduction methods and the increased computation time per history required. The exact speedup depends on the geometry of the detection system, and is produced by a large increase in the average number of detected particles produced per computational history. Since this average figure is on the order of one or more per particle birth, further improvement in computational efficiency will arise primarily through increases in QFs and improvements in collimator simulation.

REFERENCES

[1] Spanier J and Gelbard E M 1969 *Monte Carlo Principles and Neutron Transport Problems* (Reading, MA: Addison-Wesley) ch 2, 3
[2] Carter L L and Cashwell E D 1975 Particle-transport simulation with the Monte Carlo method *Oak Ridge National Laboratory Technical Report TID-26607*
[3] Bielajew A F and Rogers D W O 1988 Variance reduction techniques *Monte Carlo Transport of Electrons and Photons* ed T M Jenkins, W R Nelson and A Rindi (New York: Plenum) pp 407–20
[4] Los Alamos Scientific Laboratory 1979 *MCNP—a general Monte Carlo code for neutron and photon transport* Los Alamos Scientific Laboratory Technical Report LA-7396-M 44–9
[5] Zerby C D 1963 A Monte Carlo calculation of the response of gamma-ray scintillation counters *Methods in Computational Physics* vol 1, ed B Alder, S Fernbach and M Rotenberg (New York: Academic) 89–134
[6] Williamson J F and Morin R L 1983 Concerning an efficient method of sampling the coherent angular scatter distribution *Phys. Med. Biol.* **28** 991–2
[7] Chen C, Doi K, Vyborny C, Chan H and Holje G 1980 Monte Carlo simulation studies of the detectors used in the measurement of diagnostic x-ray spectra *Med. Phys.* **7** 627–35
[8] Metz C E and Doi K 1979 Transfer function analysis of radiographic imaging systems *Phys. Med. Biol.* **24** 372–84
[9] Tsui B M W and Gullberg G T 1990 The geometric transfer function for cone and fan beam collimators *Phys. Med. Biol.* **35** 81–93
[10] Haynor D R, Harrison R L and Lewellen T K 1991 The use of importance sampling techniques to improve the efficiency of photon tracking in emission tomography simulations *Med. Phys.* **18** 990–1001

CHAPTER 3

ANTHROPOMORPHIC PHANTOMS

I George Zubal

3.1 INTRODUCTION

Monte Carlo modelling of the transmission and attenuation of internal radiation sources has led to a better understanding of the image formation process in diagnostic radiology. Historically, the computer models of the physical distributions of radioisotopes and attenuating materials have been highly simplified in order to achieve reasonable turnaround times for Monte Carlo simulations. Due to the current availability of inexpensive memory and disk storage as well as ever increasing execution speeds, these software models may be made more complex in order to model the human anatomy more realistically and still permit statistical Monte Carlo computations to be completed within acceptable time limits. In this chapter, some of the more recent anthropomorphic phantoms which model the human anatomy are described.

The computer models we refer to, for example, have also been applied to better understand the image formation process in diagnostic radiology [1–4], particularly for analysing scatter and attenuation problems in nuclear medicine [5–11]. Since much higher statistics are necessary to model imaging simulations (compared to dosimetry simulations), the speed of computing individual gamma ray histories becomes of paramount importance for imaging physics calculations. The software phantoms modelled in these imaging simulations have historically been limited to simple point, rod and slab shapes of sources and attenuating media. Such simple geometries are useful in studying more fundamental issues of scatter and attenuation, but clinically realistic distributions cannot be adequately evaluated by such simple geometries. The intricate protuberances and convolutions of human internal structures can be important in evaluating imaging techniques. As the resolution of imaging equipment improves from year to year (current gamma camera

SPECT reconstructed resolutions approaching 5 mm), it is essential to enhance our computer models in order to insure that our simulations are representative of the reality we try to emulate.

3.2 EARLY ANTHROPOMORPHIC PHANTOMS

The essence of a phantom used for radiological calculations is the ability to mathematically capture two physical distributions: the geometrical characteristics of a radiation source (internal or external) and the spatial distribution of attenuating material.

In order to make three-dimensional (3D) anatomical data suitable for use in any such patient-oriented radiological calculations, we must be able to delineate the surfaces and internal volumes which define the various structures of the body. For most applications what we mean by various structures is those internal organs or physiological substructures which can be imaged and diagnosed through radiological procedures. More specifically in nuclear medicine, we refer to structures which are targeted by any of the numerous radiopharmaceuticals available to the clinician. Clear and obvious examples of such structures are cardiac ventricular volume or myocardium, liver, lungs, bones and kidneys. As new radiopharmaceuticals are developed, new structures need to be delineated and modelled in our simulations. For example, the advent of new neuroreceptor agents has created a recent interest in understanding the imaging characteristics of deep-seated structures in the brain (e.g., the striatum and caudate nucleus). A realistic anthropomorphic phantom will contain these structures through some method of segmenting their volumes. These segmented volumes can then be indexed to activity distributions or other physical characteristics (density or elemental composition).

In order to yield faster computation times, early phantoms were constrained to model a single slice extracted from a 3D volume. The advantage of single-slice models is that the mathematical computation may be limited to two dimensions. Radiation leaving the 2D slice may either be ignored; or more efficiently, radiation may be forced to remain within the slice of interest within the simulation. This has obvious advantages in speed of computation, but falls short in modelling the radiation (in particular scattered radiation) from slices above and below the slice of interest. A full 3D modelling of radiation transport is essential for a complete understanding of the detection process. Although innovative ideas can quickly be tested by modelling within two dimensions, it is essential to make a more accurate, 3D, evaluation of any proposed method. Concomitant with the increased resolution and realistic representation of internal structures, anthropomorphic phantoms have evolved into full 3D descriptions. Such anthropomorphic phantoms containing internal segmented structures can either be

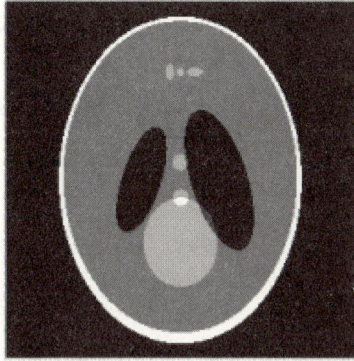

Figure 3.1   *The Shepp–Logan head phantom illustrating an early single-slice anthropomorphic representation of the head of a human by using simple ellipses.*

defined by mathematical (analytical) functions, or digital (voxel-based) volume arrays.

One of the earliest phantoms used for developing improved SPECT reconstruction techniques is referred to as the Shepp–Logan phantom. We see a representation of this phantom in figure 3.1. It is meant to represent a 'headlike' object with some internal structures. These are captured through the use of simple ellipses, which, when carefully placed on a 2D plane, work well to represent the essence of simple structures in the human head. The varying thickness of the skull as well as simple internal structures (striatum), including a small tumour, are easily recognized in the figure.

One of the earliest 3D computerized anthropomorphic phantoms was developed for estimating doses to various human organs from internal or external sources of radioactivity and served to calculate the S factors for internal dose calculations in nuclear medicine [12]. This mathematical phantom models internal structures as either ellipsoids, cylinders or rectangular volumes. Figure 3.2 shows an example of internal organs described through use of these mathematical methods. As we can see, the representation of internal organs is quite crude since the simple equations can only coarsely capture the most general description of the organ's position and geometry.

In order to represent the structures with more realistic detail, the equations would need to incorporate complicated higher-order terms—thereby rendering them cumbersome and computer intensive. For internal dosimetry purposes, such simple human model approximations serve quite sufficiently and have the advantage of allowing very fast calculation of the intersection of ray lines with the analytical surfaces which delineate the organs. A version of this mathematical phantom has been updated to include female organs [13]. There are additional versions of the phantom which are used for dedicated cardiac studies, where the structures adjacent to the heart and the

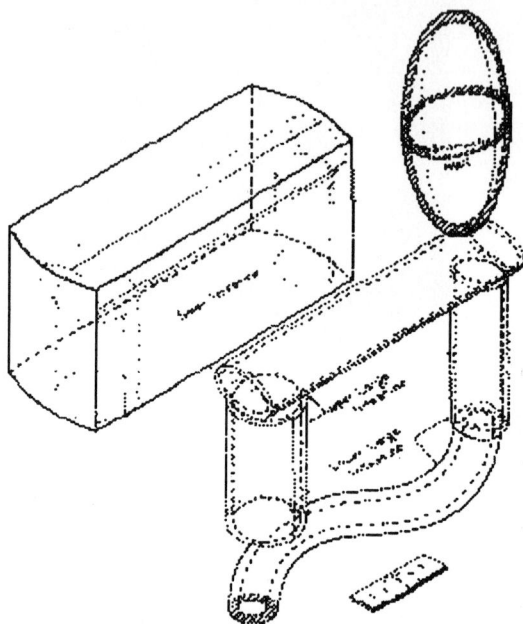

Figure 3.2   *Internal organs used in the MIRD phantom showing a 3D drawing of the internal surfaces. The objects described model the stomach, small intestine and upper and lower large intestine.*

heart itself have been more realistically modelled [14] (see figure 3.3). This mathematical phantom was specifically detailed in order to quickly generate projection data onto a simulated nuclear medicine camera for researching cardiac SPECT. The new reconstruction schemes can be tested and compared to a known (source distribution and attenuator) geometry.

3.3 VOXEL-BASED PHANTOMS

Transmission computerized x-ray tomography (CT) or magnetic resonance images (MRI) can supply us with the required high-resolution 3D human anatomy necessary to construct a volume-segmented phantom. By selecting a 'typical' adult male, a typical voxel-based phantom can be created. Such a typical male phantom has been developed for general use in radiological simulations [15]. At the time of the development of this phantom, MRI still suffered from phase artifacts caused by patient motion when imaging in the thorax. Additionally, CT images give us a good measure of the distribution of attenuating material within the patient. Hence, CT was considered the

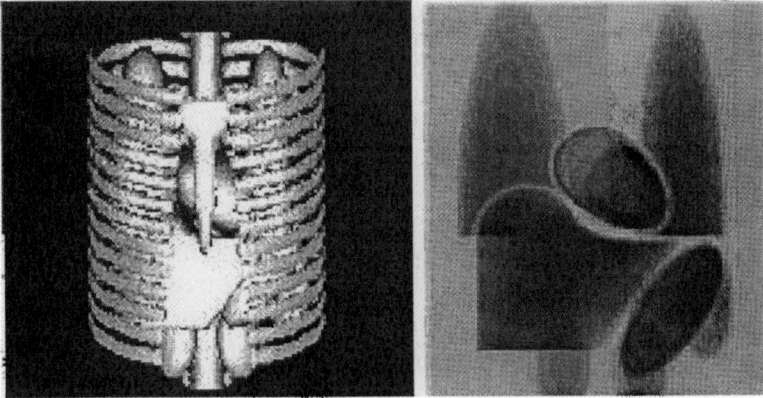

Figure 3.3 *An improved mathematical cardiac phantom, developed at Chapel Hill.*

better modality for imaging the torso for these applications. Although the mean energy used in CT imaging is below 100 keV, techniques exist which allow us to extrapolate the CT image values to higher energies. Thus, obtaining a realistic distribution of linear attenuation coefficients for various energies of radiation is a reasonably straightforward process. In nuclear medicine simulations, however, where a gamma ray emitting isotope distributes with high specificity within a particular organ or structure, we need to find a method of determining which image elements belong to which organ (segment the structures).

One of the most difficult and interesting image processing problems deals with this task of delineating organ outlines within a diagnostic image. Even high-contrast, high-resolution images (such as MRI or CT) are difficult to automatically segment into independent structures or organs. At the time of writing such computer vision algorithms, which intelligently locate surfaces of internal structures, are being investigated for very specific applications and have not been developed to the level where they can generally locate internal structures within 3D volumetric human diagnostic images with high accuracy. All too often, still, we must rely on human operators to manually draw and connect edges in order to designate internal human structures. Although this would be intolerable if applied diagnostically to individual patients, it is well worth the effort to create a very realistic typical representation of the human anatomy.

Intrinsic to this method of delineated internal structures of the human anatomy is the possibility of capturing the structural surfaces using several different techniques. Firstly, since we manually segment the structures by drawing or selecting pixel positions at the edge of each organ as seen on each slice, the array of x, y pixel positions can be stored and used to define the surfaces (when slices are stacked on top of each other). Given these x, y (and z) pixel positions and a substantial level of curve fitting, 2- or 3D analytical equations can be solved, in order to represent each organ in a closed mathematical formula. This returns to the earlier method used for dosimetric studies, but affords a higher level of resolution and the possibility of representing the structures more realistically through higher-order terms. Finally, we can construct a 3D array whose pixel values are attributed to the internal volume of each internal anatomical structure. This is achieved by a logical calculation, which determines the 'inside' and 'outside' of each closed manually drawn region. All of the 'inside' pixels are designated as the internal volume of a structure and are given a unique index number, which is stored as the value in the array. Such a 3D volume proves to be one of the handiest methods for storing and manipulating such anatomical data. However, it is not without its disadvantages: (i) changes in the geometry are fairly difficult to accomplish, since the whole process of manual drawing must be repeated; (ii) there is no quick and simple method to mathematically re-capture the original contours or surface of each structure; (iii) when calculating 3D trajectories of individual rays (or lines) within this 3D volume, we must tediously search voxel per voxel along the ray in order to find the point of intersection on the structure surfaces.

The segmented image information is therefore stored in three independent files. A variable-size file is created for each transverse slice and contains the x, y coordinates of each of the contours drawn on that slice for each organ. The slice number is retained in the name of the file and determines the z position. These contours serve as the input to the filling routine, which creates a fixed-size organ index image. The organ index image is a 512×512 byte matrix filled with integer values (organ index numbers) which delineate the internal structures of the body. The organ index image is therefore, in effect, the original CT transverse slice in which the Hounsfield numbers are replaced by integers corresponding to the organ index value. The list of organs is shown in table 3.1.

Table 3.1 *Organs for the torso phantom.*

Void outside phantom	Liver	Gas volume (bowel)
Skin/body fat	Gallbladder	Bone marrow
Brain	Left and right kidney	Lymph nodes
Spinal chord	Bladder	Thyroid
Skull	Oesophagus	Trachea
Spine	Stomach	Diaphragm
Rib cage and sternum	Small intestine	Spleen
Pelvis	Colon/large intestine	Urine
Long bones	Pancreas	Faeces
Skeletal muscle	Adrenals	Testes
Lungs	Fat	Prostate
Heart	Blood pool (all vessels)	Liver lesion

The 3D organ index slice images can be read into a $512 \times 512 \times 119$ voxel dimensional array, in which the x, y resolution is 1 mm/pixel and the slice thickness is 10 mm in the body and 5 mm in the head. The reduction from 129 to 119 slices is due to the overlap of information in the neck region.

Since the voxels in the original CT slices are not isotropic (the thickness of the slice is much larger than the width of a pixel), stacking these slices does not give us an anatomically correct geometry. For this reason the slices are repeated and interpolated in order to create a stack of slices which gives the anatomically correct relationship between the x, y and z dimensions.

3.4 ACCESS AND DISPLAY

We routinely transform the original segmented data into a $128 \times 128 \times 246$ matrix where the isotropic cubic voxel resolution is 4 mm on each side. In order to remove some of the blocky appearance created in the torso by duplicating voxels along the z direction, we applied 3D modal filtering. This makes the data more manageable and results in consistent voxel dimensions along all three axes. In our application of modal filtering a subvolume of $5 \times 5 \times 5$ voxels was selected out of the original phantom volume. The central voxel's filtered value was calculated as the mode (most often occurring) value from the selected 125 voxel subvolume.

The total storage capacities of the files are the following: original CT images, 29 Megabytes; x, y contours, 1 Megabyte; organ index matrices, 20 Megabytes. In order to appreciate the detail of this anthropomorphic phantom, we projected anterior and lateral views of selected structures from the $128 \times 128 \times 246$ volume. This was done by replacing selected index numbers with a positive integer value and setting other (unselected) structures to zero. The 3D volume was then collapsed parallel to the major axes onto two 2D matrices (each collapsed matrix = 128×246) by adding all integer values along rows of voxels. The final matrices were normalized and displayed using a grey-scale colour table and are shown in figures 3.4 and 3.5. These figures are rendered with the skin/fat voxels selected in order to show the outline of the patient's body; within this outline, various structures are selected for display.

3.5 DEDICATED BRAIN PHANTOM

The general concepts and data handling programs used for the previous torso phantom were applied to create a dedicated head phantom. Here MRI data proved to be superior to CT data since no motion artifacts are commonly associated with MRI scans and MRI provides slice data with isotropic voxel dimensions. A total of 124 slices were acquired from the head of a healthy volunteer. The same manual outlining and designation of anatomical

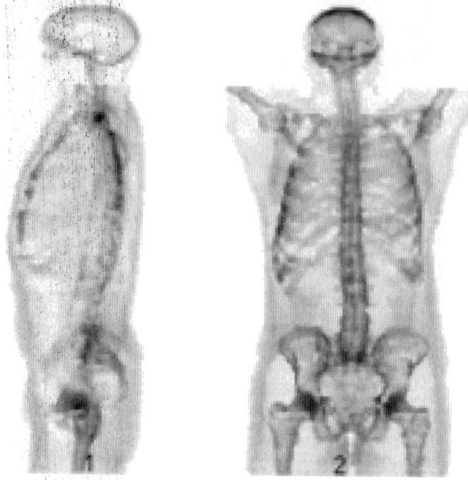

Figure 3.4   *Highlighted organs from the anthropomorphic voxel phantom showing the long bones, skull, ribs, spine and pelvis.*

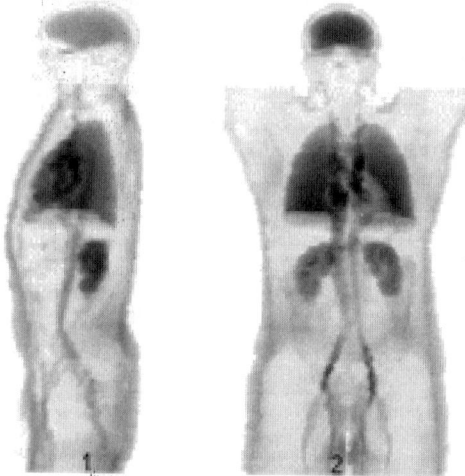

Figure 3.5   *Highlighted organs from the anthropomorphic voxel phantom showing the brain, lungs, kidneys and large arteries.*

structures was carried out as described for the torso phantom above. The organs or structures listed in table 3.2 were drawn on the 256×256 MRI images: an example of several views of the segmented brain phantom is shown in figure 3.6.

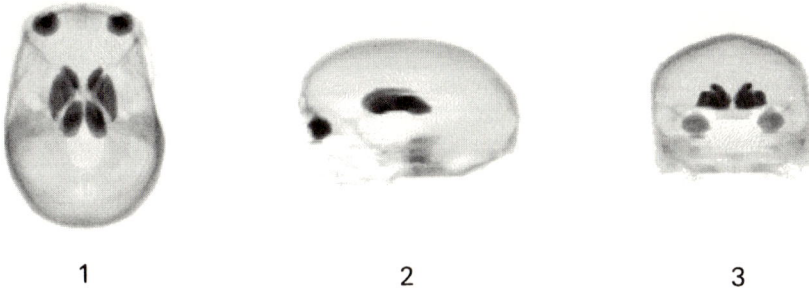

1   2   3

Figure 3.6 *Highlighted internal structures of the dedicated brain phantom showing the caudate nucleus, putamen, globus pallidus, thalamus and globe of the eye.*

Table 3.2 *Organs for the brain phantom.*

Skin	Medulla oblongata	Septum pellucidium
Brain	Fat	Optic nerve
Spinal cord	Blood pool	Internal capsule
Skull	Artificial lesion	Thalamus
Spine	Frontal lobes	Eyeball
Dens of axis	Bone marrow	Corpus collosum
Jaw bone	Pons	Special region frontal lobes
Parotid gland	Third ventricle	Cerebral falx
Skeletal muscle	Trachea	Temporal lobes
Lacrimal glands	Cartilage	Fourth ventricle
Spinal canal	Occipital lobes	Frontal portion eyes
Hard palate	Hippocampus	Parietal lobes
Cerebellum	Pituitary gland	Amygdala
Tongue	Cerebral fluid	Eye
Pharynx	Uncus (ear bones)	Globus pallidus
Oesophagus	Turbinates	Lens
Horn of mandible	Caudate nucleus	Cerebral aquaduct
Nasal septum	Zygoma	Lateral ventricles
White matter	Insula cortex	Prefrontal lobes
Superior sagittal	Sinuses/mouth cavity	Teeth
sinus	Putamen	Lesion

3.6 DISCUSSION

Voxel-based phantoms are becoming more prominent for Monte Carlo simulations to yield diagnostically realistic images of internal distributions of radiopharmaceuticals [6, 17]. Since we are able to model a known source distribution and known attenuator distribution, the Monte Carlo simulations give us projection data which not only closely resemble clinical data, but also include additional information not determinable in patient studies.

Such data sets can help to better understand the image formation process for clinically realistic models, and can prove especially interesting in testing and improving tomographic reconstruction algorithms [18].

New imaging devices can be investigated using *in vivo* simulations. The tumour detection capabilities of a novel coincidence counting probe system have been investigating using the anthropomorphic phantom described here [19, 20]. Early design changes can be realized before studies are conducted in living models. One of the advantages of developing this very realistic human model is that such simulations can decrease the necessity of conducting experimental studies using animal models—particularly primates. Dose calculations for internal and external radiation sources using this phantom can give new insights in the field of health physics and therapy.

The original impetus for developing the voxel-based anthropomorphic phantom described here was to simulate SPECT imaging geometries. Reconstructing SPECT source distributions from projection data is a difficult task since not only is the source distribution unknown, but the distribution of attenuating material surrounding the source is also unknown. Simulating projection data from software phantoms allows us to compare our reconstructed image to a known 'gold standard' where the source and attenuating material are completely known.

ACKNOWLEDGMENT

This work was performed under contract No DE FG02#030#88ER60724 with the US Department of Energy.

REFERENCES

[1] Barnea G and Dick C E 1986 Monte Carlo studies of x-ray scatterings in transmission diagnostic radiology *Med. Phys.* **16** 490–5

[2] Chan H P and Doi K 1985 Physical characteristics of scattered radiation in diagnostic radiology: Monte Carlo simulation studies *Med. Phys.* **12** 152–65

[3] Kanamori H, Nakamori N and Inone K 1985 Effects of scattered x-rays on CT images *Phys. Med. Biol.* **30** 239–49

[4] Dance D R and Day G J 1984 The computation of scatter in mammography by Monte Carlo methods *Phys. Med. Biol.* **29** 237–47

[5] Logan J and Bernstein H J 1983 A Monte Carlo simulation of Compton scattering in positron emission tomography *J. Comput. Assist. Tomogr.* **7** 316–20

[6] Floyd C E, Jaszczak R J, Harris C C and Coleman R E 1984 Energy and spatial distribution of multiple order Compton scatter in SPECT. A Monte Carlo investigation *Phys. Med. Biol.* **29** 1217–30

[7] Floyd C E, Jaszczak R J, Greer K L and Coleman R E 1985 Deconvolution of Compton scatter in SPECT *J. Nucl. Med.* **26** 403–8

[8]   Floyd C E, Jaszczak R J, Greer K L and Coleman R E 1986 Inverse Monte Carlo as a unified reconstruction algorithm for ECT *J. Nucl. Med.* **27** 1577–85

[9]   Floyd C E, Jaszczak R J, Greer K L and Coleman R E 1987 Brain phantom high resolution imaging with SPECT and I-123 *Radiology* **164** 279–81

[10]   Beck J W, Jaszczak R J, Coleman R E, Starmer C F and Nolte L W 1982 Analysis of SPECT including scatter and attenuation using sophisticated Monte Carlo modeling methods *IEEE Trans. Nucl. Sci.* **NS-29** 506–11

[11]   Acchiappati D, Cerullo N and Guzzardi R 1992 Assessment of the scatter fraction evaluation methodology using Monte-Carlo simulation techniques *Eur. J. Nucl. Med.* **15** 683–6

[12]   Snyder W, Ford M R and Warner G 1978 Estimates of specific absorbed fractions for photon sources uniformly distributed in various organs of a heterogeneous phantom *NM/MIRD Pamphlet* 5 (New York: Society of Nuclear Medicine)

[13]   Williams G, Zankl M, Abmayr W, Veit R and Drexler G 1986 The calculation of dose from external photon exposures using reference and realistic human phantoms and Monte Carlo methods *Phys. Med. Biol.* **31** 449–52

[14]   Tsui B Private communication

[15]   Zubal I G, Harrell C R, Smith E O, Rattner Z, Gindi G and Hoffer P B 1994 Computing three-dimensional segmented human anatomy *Med. Phys.* **212** 299–302

[16]   Zubal I G and Harell C H 1992 Voxel based Monte Carlo calculations of nuclear medicine images and applied variance reduction techniques *Image Vision Comput.* **10** 342–8

[17]   Zubal I G, Harrell C R and Esser P D 1990 Monte Carlo determination of emerging energy spectra for diagnostically realistic radiopharmaceutical distributions *Nucl. Instrum. Methods Phys. Res.* A **299** 544–7

[18]   Hademenos G S, King M A, Ljungberg M, Zubal I G and Harrell C R 1992 A scatter correction method for T1201 images: a Monte Carlo investigation *IEEE Nuclear Science Symp. and Medical Imaging Conf. (Orlando, FL, 1992)* vol 2, pp 1213–6

[19]   Saffer J R, Barrett H H, Barber H B and Woolfenden J M 1992 Surgical probe design for a coincidence imaging system without a collimator *Imae Vision Comput.* **10** 333–41

[20]   Hartsough N E, Barrett H H, Barber H B and Woolfenden J 1993 Intraoperative tumor detection relative performance of single-element dual-element and imaging probes with various collimators *IEEE Nucl. Science Symp. and Medical Imaging Conf. (San Francisco, CA, 1993)* vol 2 (Piscataway, NJ: IEEE) pp 1236–40

CHAPTER 4

GENERAL MONTE CARLO CODES FOR USE IN MEDICAL RADIATION PHYSICS

Pedro Andreo and Michael Ljungberg

4.1 INTRODUCTION

Historically nuclear medicine is the field where most of the early Monte Carlo calculations in medical radiation physics were performed and today their use continues to increase [1]. The advent of computer power in hospital departments, mainly in the form of workstations and fast personal computers, has made the Monte Carlo method a common tool in the different areas of research and clinical development where interaction of ionizing radiation with matter is of importance, namely diagnostic x-rays, nuclear medicine and radiation therapy, as well as radiation protection aspects in these fields.

An important aspect of the applications of the Monte Carlo technique to nuclear medicine is the simplicity of the simulation of the physics involved in most cases. The energy range of the photons used in this field allows minimum emphasis on the production of secondary charged particles and high-energy processes that yield an electromagnetic cascade; at the other extreme, very low-energy photon interactions can practically be ignored. This provides a sharp contrast with applications in other areas such as radiation therapy, where secondary electrons play a major role, or diagnostic x-rays, where photon elastic scattering might be of importance. Calculations of absorbed dose distributions for beta sources obviously fall outside this general characteristic, but even in these cases low-Z materials are involved. It is then possible to avoid the extra complication of intricate cross sections for elastic scattering or the treatment of bremsstrahlung processes in detail. On the other hand, detailed simulations of complicated geometry configurations is a field shared today by most other applications, and nuclear medicine is not an exception in adopting the enormous potential provided by some of the major 'public' Monte Carlo codes with their advanced geometry capabilities.

In the last years the two Monte Carlo codes EGS [2] and ITS [3] have become standard tools for the simulation of the transport of electron and photon beams in medical radiation physics, mainly in the area of dosimetry at therapeutic energies, and to a lesser extent at lower incident energies. The code MCNP [4], which is well known in other fields of radiation transport, has less strength in the medical community, but it is emerging with dynamism. Both ITS and MCNP have their electron physics based on the code ETRAN [5]. All the codes differ in a variety of aspects, among them the way in which electron interactions are grouped and the cross-section data generated by their accompanying data preparation programs. ETRAN, ITS and EGS today include ICRU 37 stopping powers [6] but, although updates in the treatment of the physics of bremsstrahlung interactions have also been implemented, the two codes EGS and ITS (or ETRAN) still differ in their treatment of electron energy loss and multiple scattering. More importantly, for low-energy applications such as those in nuclear medicine, data for some photon interactions are also different.

A general overview has been published elsewhere together with a review of the applications of the Monte Carlo technique in the different fields of medical radiation physics [1].

4.2 MONTE CARLO CODES IN THE PUBLIC DOMAIN

Table 4.1 lists alphabetically the Monte Carlo code systems cited by name throughout this chapter, together with a simplified description of the system. All codes are distributed free of charge and support to users is widely provided by the developers of each system, very often through electronic mail and in some cases (EGS and MCNP) with electronic lists of users. Most of them can run in a variety of computer platforms, including PCs equipped with specific FORTRAN compilers.

4.2.1 The EGS system

The EGS (electron gamma shower) Monte Carlo system was first developed at the Stanford Linear Accelerator Center for simulations of high-energy electromagnetic cascades. The earlier version 3 of EGS [7] has been in use for high-energy accelerator projects for many years but applications in medical physics were scarce [8]. Its high-energy background explains some of the characteristics of the physics in the code, compared to the ETRAN system which is the base for electron transport in the other codes. In EGS3, the treatment of multiple electron scattering is based on the theory developed

Table 4.1 *Alphabetical list of Monte Carlo systems in the public domain.*

MC system	General description
EGS4	Transport of electrons and photons in any material through user-supplied geometry. Package of subroutines plus block data with a flexible user interface. *Requires development of a user code in MORTRAN.* Class II electron transport. Different sources of data.
ETRAN	Transport of electrons and photons in any material through plane-parallel slabs and cylinders. User-selectable options to control output of different quantities or switch on/off certain interactions. Class I electron transport. NIST database for photon and electron cross sections and data.
ITS3 including *TIGER* *CYLTRAN* *ACCEPT*	Same physics as in ETRAN. Flexible input of data. Family of 8 codes with varying geometry (slabs cylindrical or combinatorial). Accurate electron cross sections down to 1 keV (P codes) and electromagnetic fields (M codes). Class I electron transport. NIST database.
MCNP4A	Continuous-energy coupled neutron–photon–electron transport. Time dependent. Arbitrary configuration of materials using generalized geometry. Class I electron transport, based on ITS 2.

by Molière, incorporating rather simple methods for simulating low-energy electrons and photons and including bremsstrahlung cross sections differential in energy only. According to a criterion generally adopted for electron transport [1, 9] the EGS system is a 'class II' code that treats knock-on electrons and bremsstrahlung photons individually. Such events require pre-defined energy thresholds and pre-calculated data for each threshold, determined with the cross-section generator PEGS. This includes the commonly used photon cross-section data derived by Storm and Israel [10]. In EGS, fluorescent photons created by photon interactions and corrections to the cross sections due to bounded electrons are not considered. The user has to define the threshold for electron production and the cut-off energy of electron transport to high values in order to avoid the time spent simulating electrons in those situations where the influence of electron transport is low.

EGS is written in the computer language MORTRAN, a FORTRAN pre-processor with powerful macro capabilities. The latest version, EGS4 [2], includes electron transport down to 1 keV and coherent scattering. The code is continuously being improved [14] including, for instance, the implementation of ICRU stopping powers [13], angular distribution of photoelectrons [11] and the algorithm PRESTA [12], that corrects electron path-lengths in order to take into account (i) the detour from multiple scattering of straight paths, (ii) lateral correlation according to Berger [9] and (iii) a boundary-crossing algorithm which allows very large step sizes except close to medium boundaries. PRESTA has been developed from the multiple-scattering theory of Molière but the algorithm employs Molière's theory down to '*e*' collisions, the mathematical rather than the physical limit of the theory [15–17].

Of interest for the simulation of low-energy photons are the binding corrections in Compton scattering available as a separate package developed at KEK in Japan [18]. Also, a replacement for the default EGS PHOTO routine with codes including K_{a1} and K_{b1} data is available through the package.

The EGS system is very flexible and powerful, but requires a good proportion of programming efforts and computational skills. It is not a stand-alone Monte Carlo code. Instead the user must link EGS subroutines to his/her own developed main code and routines that describe the geometry and the parameters needing to be scored to extract the desired quantities. The connections between the EGS 'core' and the user's own routines are maintained through FORTRAN COMMON blocks, MORTRAN macros and two user-callable subroutines, named 'HATCH' and 'SHOWER'.

An example of this interface is provided in appendix 4.1, adopted from the EGS manual series of tutorials [2]. The file must be 'mortraned', compiled ('fortraned') and linked prior to execution.

4.2.2 The ETRAN and ITS systems

The ETRAN Monte Carlo code [19] was originally developed at the National Bureau of Standards (now the National Institute of Standards and Technology) to simulate the transport of electrons and photons at low energies. The algorithms included in ETRAN are now used in other Monte Carlo systems such as ITS and MCNP. ETRAN has been used mainly in topics related to therapeutic beam dosimetry [20–22], and, for example, most of the data in *ICRU Report* 35 on electron dosimetry [23] were calculated using this code. Other examples refer to the beta-ray dosimetry [24, 25]. ETRAN is a 'class I' code that includes accurate simulation of electron multiple scattering using the Goudsmit–Saunderson theory and bremsstrahlung interactions that include cross sections differential in both energy and scattering angle. In addition, ETRAN includes characteristic x-rays from the K-shell and Auger electrons emitted during a photon interaction. Recently, coherent scattering and binding corrections [26] have been included [5].

Although ETRAN provided very accurate electron transport, it did not include geometries other than simple infinite media or plane-parallel slabs of different materials and cylindrical geometries. The ITS family of Monte Carlo codes [3] from Sandia National Laboratory have adopted most of the physics from ETRAN but provide good geometrical packages of increasing complexity. ITS consists of three codes, named TIGER, CYLTRAN and ACCEPT, which simulate electron and photon transport down to 1 keV in plane-parallel slabs, cylindrical geometry or any combination of the geometrical bodies included in its default combinatorial package, respectively. The code also includes characteristic x-rays from the K-shell and Auger electrons. The user interface is straightforward and programming is not required. As for EGS, the set of ITS codes can be installed on various computer platforms. The cross-section data, generated with an independent code XGEN, account for recent changes in physical data including the ICRU 37 electron stopping powers, numerical bremsstrahlung cross sections differential in photon energy, coherent scattering with binding corrections and binding corrections to incoherent scattering. Of interest for low-energy photon transport is the capability to eliminate electron transport with a single switch, making the code very efficient in this energy range. An example of input file to the code CYLTRAN is given in appendix 4.2.

4.2.3 The MCNP system

The MCNP Monte Carlo system is the code that has been available to the radiation physics community for the longest by far. Its name stands today for Monte Carlo *N*-particle transport (at first it stood for Monte Carlo neutron photon) and its long history has witnessed a variety of names (MCS, MCN, MCNG, MCP and MCNP) since its early origin at Los Alamos National Laboratory (LANL), back in the years of Ulam, von Neumann and Metropolis. Any reader curious about the history of the Monte Carlo method should read the introductory pages in the manual [4]. The association of the MCNP name with the simulation of neutrons, where it has no competitor, has probably provoked a lack of interest in the field of medical applications, most of its users presumably having been connected to military projects and the nuclear power industry until recent years.

Boron neutron capture therapy has probably been the catalyst for its use in medical physics, together with a decided policy of the Radiation Transport Group (X-6) at LANL for opening the domain of applications of their code. Although it is not the purpose of this chapter to provide details on the neutron physics included in MCNP, it is worth describing some of the important characteristics of the rest of the components of this powerful and versatile Monte Carlo system. Readers are however warned of the reduced knowledge of the present authors on MCNP, and certainly any comparison with the description of the other codes provided here will be unbalanced.

MCNP uses a built-in random number generator instead of that supplied with the computer operative system where the code is running. This provides a portable generator yielding the same random numbers in any computer platform although it should run more slowly than the standard generator in the computer. The system includes the most comprehensive set of techniques for variance reduction, 'from the trivial to the esoteric', and helpful plotting capabilities to analyse input geometry and scored results.

Photon transport in MCNP can be performed using two modalities, simple and detailed, and basically includes coherent (Thompson) and incoherent scattering with form factors to account for electron binding effects, photoelectric absorption followed by a very detailed treatment of fluorescence (as in ETRAN) and pair production. Some of the corrections can be switched off, as can the electron transport resulting from these interactions. Physical data for photons originate from a variety of sources, some of them dating back to the 1970s. They include a reduced set of atomic numbers and energies taken from an initial compilation of Storm and Israel prior to the common set in *Nuclear Data Tables* [10], expanded with the evaluated data from ENDF [27] and further completed with the Livermore EPDL [28] for very high energies.

In recent years MCNP has been extended to include the simulation of the transport of electrons [4]. As in ITS, this part of the code is based on the electron physics of the ETRAN system, and it is therefore a so-called 'class I' code where all the details described in preceding paragraphs apply. MCNP is, however, based on one of the early versions of ITS. This means that, among other data and algorithms superseded today, energy loss straggling is based on an incorrect algorithm. This questions the validity of the results for all the applications of MCNP based on the transport of high-energy photons or electrons whenever electron energy loss is of importance. In contrast, for low-energy photon simulations such as those in diagnostic x-rays or nuclear medicine MCNP appears as a strong candidate compared to EGS and ITS, especially when its powerful geometry capabilities are taken into account (see below).

At this point it should be mentioned that the interest in medical physics applications has recently motivated an unpublished compromise for implementing in MCNP the electron physics of the most updated version of the ETRAN and ITS codes. Possibilities for expanding the set of cross-section data to include (γ, n) reactions, of potential importance in high-energy photon therapy, are also under consideration. When these plans become a reality MCNP might become a Monte Carlo code in wide use (probably a 'bestseller') among medical physicists working in any field.

4.2.4 Other Monte Carlo codes

The so called 'all-particle method' approach [29, 30] which has been under development at the Lawrence Livermore National Laboratory (LLNL) for some years, might become of special interest in the near future. The system attempts to include the transport of neutrons, photons, electrons and other charged particles as well as the coupling between all species of particles. An important development in connection with this project has been the 'response history' method for electron transport [31] presented as an alternative to the conventional 'condensed history' or 'macroscopic' techniques [1, 9]. In parallel with coding the APM project has included the updating and extension of the comprehensive nuclear and atomic database developed at LLNL, which include for example the Evaluated Photon (or Electron) Data Libraries (EPDL and EEDL) [28, 32]. EPDL has become the new standard for the ENDF/B. The actual status of the development of the APM system or its availability is not known to the present authors, although intense efforts have been put into a less ambitious code called PEREGRINE, also developed at LLNL.

The LLNL database can be obtained through the common distributors and provides a standardized set which can be implemented in any Monte Carlo code. There exist plans, for example, to implement the database in the EGS system [39]. A graphical package called EPICSHOW has been developed to visualize the data that run on PCs and Unix workstations [33].

4.3 GEOMETRY IN THE MONTE CARLO CODES

It is well known that the coding for simulating a given geometry is the most challenging programming aspect in any Monte Carlo code. In general determination of the intersection of the track of a particle with a given surface is required. Mathematical tools for different types of surface are given as subroutines in the EGS package [34], which the user must connect with his own routines. Most of the EGS user-codes in medical physics calculations have, however, been based on simple cylindrical geometries. This has also been the case for calculations with other Monte Carlo codes, as discussed above for ETRAN and for the members of the ITS family of codes TIGER and CYLTRAN.

More sophisticated approaches are available, the combinatorial geometry package (CG) being perhaps the most commonly used due to its simplicity for the user. CG describes three-dimensional (3D) complex bodies using Boolean algebra with a few elementary bodies such as parallelepipeds (RPP), spheres (SPH), cylinders (RCC), wedges (WED) etc, which can be oriented arbitrarily in space. Although the use of CG is possible in the EGS system, its implementation has not received great attention until recently [35]. The ITS code ACCEPT is, on the other hand, fully based on this mathematical package, which does not require more than a few lines to construct complicated 3D geometries [34, 36].

A different and powerful geometry technique has been implemented in MCNP instead of pre-defined geometrical bodies. The user defines 3D configurations using geometrical cells bounded by first- and second-degree surfaces (and some special fourth-degree surfaces, such as elliptical tori), which are combined by using Boolean operators. Among the numerous geometry possibilities of the MCNP system it is of interest to mention the 'repeated structure geometry'. The user defines a 'unit' consisting of the cells and surfaces of any structure that appears more than once in a geometry. Such a 'unit' can be replicated to other spatial locations with a simple command 'LIKE m BUT'. In the words of the MCNP manual [4] 'The user specifies that a cell is filled with something called a universe. The U card identifies the universe, if any, to which a cell belongs. The FILL card specifies with which universe a cell is to be filled. A universe is either a lattice or an arbitrary collection of cells...' (readers should think of the complexity of a nuclear reactor core, for example, described with just a few commands). The enormous potential of this option in medical physics is still unexplored, but the present authors believe that the design of x-ray anti-scatter grids or nuclear medicine camera collimators can benefit considerably from reduction of the programming effort. An example of these powerful capabilities is given in appendix 4.3 for a hexahedral lattice and the reader is referred to Chapter 4 in the MCNP manual [4] for an impressive collection of simple descriptions of complicated geometries.

Complicated geometries are however difficult to deal with in practice. It is hard to position correctly in space many different components of a piece of equipment. A very helpful (and beautiful) geometry visualization package called Sabrina [37] has been developed at LANL in connection with MCNP, although the package can also be used with CG. Sabrina includes the option for providing the geometry coding directly for use in MCNP, can easily be adapted to yield input to ITS and can visualize particle tracks output by MCNP, zoom, rotate, etc. The package runs on virtually any Unix work-station and, as are all the codes described in this chapter, it is available free of charge from RSIC or NEA.

4.4 COMPARISON BETWEEN EGS AND ITS FOR LOW-ENERGY PHOTONS

Most correction and conversion factors used today in ionizing dosimetry have been determined using Monte Carlo methods. This gives great import-ance to the role of the technique at the time of comparing international procedures and standards. The two systems discussed in this section are actually involved in a comparison among the Monte Carlo codes used at primary-standard dosimetry laboratories of different countries under the auspices of the BIPM [38]. Different configurations for photon beams have been suggested by a working group of the CCEMRI(I), which starting from simple simulations will compare quantities scored under different geometries. No results are formally available yet from this comparison.

Results presented here correspond to calculations performed by the present authors using EGS and ITS for two of the CCEMRI(I) configura-tions (see figure 4.1). The aim is to compare the codes regarding algorithms for photon scattering (geometry *A*) and fluorescent x-ray effects and cross

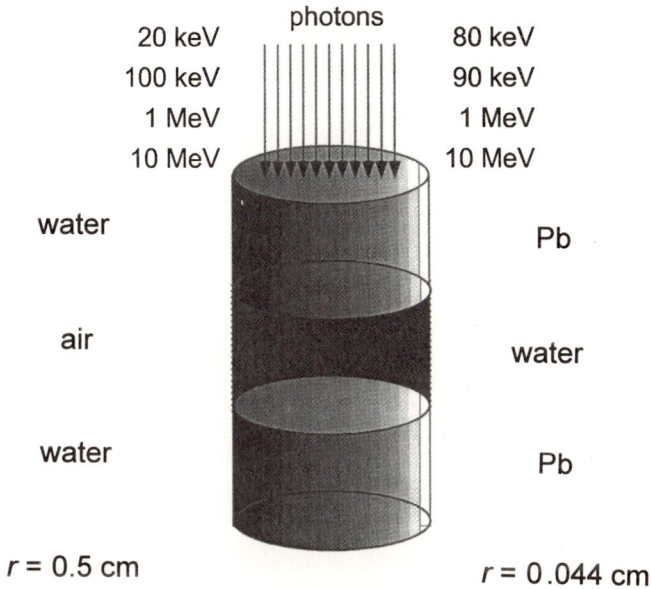

Figure 4.1 *Configurations A (water/air/water) and B (lead/water/lead) proposed by a working group of the CCEMRI(I) to compare Monte Carlo codes used at primary-standard dosimetry laboratories [8]. In these calculations the height of all subcylinders is 1 cm except for lead where it is equal to 0.0882 cm.*

sections near absorption edges (geometry *B*). No electron transport is simulated, in contrast to most other published comparisons, and therefore all discrepancies are exclusively due to the photon part of the codes. The results presented in table 4.2 can be used to benchmark those produced with other codes commonly used for the simulation of the transport of low-energy photons.

It has to be pointed out that the two systems have large differences in the physics of low-energy photons which play an important role in the behaviour of scattered radiation, especially for high atomic numbers. ITS includes cross sections for bound electrons, the effect of which is ignored in the default EGS package. The two codes include Rayleigh scattering and fluorescent effects, but it is clear that an updated database at low energies is needed in the EGS system. A comparison has not yet been made with the newly available bounded cross-section data for EGS.

Table 4.2 Comparison of results with the EGS4 and ITS3 Monte Carlo codes for the geometries A and B (figure 4.1) proposed by CCEMRI(1) [38]. Photon transport only. Results shown correspond to absorbed dose per incident fluence (10^{13} Gy cm^2) ± SEOM (%).

Photons		A			B			Photons
		1 (water)	2 (air)	3 (water)	1 (lead)	2 (water)	3 (lead)	
20 keV	ITS	13.39±0.0	8.234±0.2	6.081±0.0	110.7±0.0	0.287±0.2	8.260±0.0	**ITS 80 keV**
	EGS	13.05±0.0	7.460±2.0	6.107±0.1	108.8±0.2	0.293±0.5	8.371±0.2	**EGS**
	ratio	**1.026**	**1.104**	**0.996**	**1.017**	**0.980**	**0.987**	**ratio**
100 keV	ITS	3.996±0.0	3.254±0.2	3.389±0.1	96.92±0.0	0.0046±1.5	0.063±0.4	**ITS 90 keV**
	EGS	3.990±0.0	3.165±3.1	3.381±0.0	95.61±0.0	0.0351±1.4	0.741±0.4	**EGS**
	ratio	**1.002**	**1.028**	**1.002**	**1.014**	**0.131**	**0.085**	**ratio**
1 MeV	ITS	48.96±0.1	42.33±0.5	45.98±0.2	63.46±0.6	45.22±1.0	55.45±0.6	**ITS 1 MeV**
	EGS	48.88±0.0	41.55±1.6	45.75±0.0	63.58±0.1	45.01±0.1	55.39±0.0	**EGS**
	ratio	**1.002**	**1.019**	**1.005**	**0.998**	**1.005**	**1.001**	**ratio**
10 MeV	ITS	258.5±0.1	236.1±0.5	253.8±0.1	660.4±0.1	248.9±0.5	613.4±0.1	**ITS 10 MeV**
	EGS	257.5±0.0	234.4±1.7	252.5±0.0	643.8±0.0	244.8±0.1	600.1±0.1	**EGS**
	ratio	**1.004**	**1.007**	**1.005**	**1.026**	**1.017**	**1.022**	**ratio**

Options: 10^7 histories. Rayleigh scattering.
Photon cut-offs: 1 keV for $E_0 \leqslant 100$ keV; 10 keV for $E_0 > 100$ keV.
EGS4 v2.0 (NRCC code DOSRZ v17 including fluorescence, no bound cross sections)
ITS 3.0 (code CYLTRAN; P code for $E_0 \leqslant 100$ keV)

Electron transport off
$T_{c,\text{electrons}} = T_{0,\text{photons}}$
PHOTRAN switch

4.5 CONCLUSIONS

The availability of general Monte Carlo codes in 'the public domain', developed and tested world-wide in a variety of applications by multiple users, should provide confidence for their use as research tools in the different fields of medical physics. In general, results with the two most widely distributed Monte Carlo codes for electron and photon transport, EGS and ITS, show good agreement for medium energies and atomic numbers only, and whenever the geometrical configurations do not require a sophisticated algorithm. In these cases the use of one code or another should depend on personal preferences. However, the discrepancies reported here for a difficult case, as also for many other conditions different from the above comparison [1], demonstrate the need for being aware of the limitations of the two codes and the possible dissimilarity of results. No code is perfect, and depending on the problem at hand it is important to know if a weak aspect of a particular code might be relevant to the simulation being performed.

APPENDIX 4.1

```
"VERSION FROM SLAC"
%L
%E
!INDENT M 4;   "INDENT MORTRAN LISTING BY 4 PER NESTING LEVEL"
!INDENT F 2;   "INDENT FORTRAN OUTPUT BY 2 PER NESTING LEVEL"
"-------------------------------------------------------------------"
"STEP 1:   USER-OVERRIDE-OF-EGS4-MACROS                              "
"-------------------------------------------------------------------"
REPLACE {$MXMED} WITH {1}   "only 1 medium in the problem(default 10)"
REPLACE {$MXREG} WITH {3}   "only 3 geometric regions (default 2000)"
REPLACE {$MXSTACK} WITH {15}"less than 15 particles on stack at once  "
REPLACE {$EBIN} WITH {25}   "user parameter -# bins in energy spectrum"
"DEFINE A COMMON TO PASS INFORMATION TO THE GEOMETRY ROUTINE HOWFAR"
REPLACE {;COMIN/GEOM/;} WITH {;COMMON/GEOM/ZBOUND;}
"DEFINE A COMMON FOR SCORING IN AUSGAB"
REPLACE {;COMIN/SCORE/;} WITH {;COMMON/SCORE/EHIST,EBIN($EBIN);}
;COMIN/BOUNDS,GEOM,MEDIA,MISC,SCORE,THRESH/;
"        THE ABOVE EXPANDS INTO A SERIES OF COMMON STATEMENTS"
"-------------------------------------------------------------------"
"STEP 2 PRE-HATCH-CALL-INITIALIZATION"                               "
"-------------------------------------------------------------------"
$TYPE MEDARR(24) /$S'NAI',21*' '/; "PLACE MEDIUM NAME IN AN ARRAY"
"                          $S IS A MORTRAN MACRO TO EXPAND STRINGS"
"                          $TYPE IS INTEGER (F4) OR CHARACTER*4(F77)"
DO I=1,24[MEDIA(I,1)=MEDARR(I);]"THIS IS TO AVOID A DATA STATEMENT FOR"
"                          A VARIABLE IN COMMON"
"NMED AND DUNIT DEFAULT TO 1, I.E. ONE MEDIUM AND WE WORK IN CM"
/MED(1),MED(3)/=0;MED(2)=1;"REGIONS 1,3 ARE VACUUM, REGION 2, NAI"
ECUT(2)=0.7;"   TERMINATE ELECTRON HISTORIES AT 0.7 MEV IN THE PLATE"
PCUT(2)=0.1;"   TERMINATE   PHOTON HISTORIES AT 0.1 MEV IN THE PLATE"
%E
"-------------------------------------------------------------------"
"STEP 3   HATCH-CALL                                                 "
"-------------------------------------------------------------------"
;OUTPUT;('1START TUTOR3'//' CALL HATCH TO GET CROSS-SECTION DATA'/);
CALL HATCH;"   PICK UP CROSS SECTION DATA FOR NAI"
"          DATA FILE MUST BE ASSIGNED TO UNIT 12"
;OUTPUT AE(1)-0.511, AP(1);
(' KNOCK-ON ELECTRONS CAN BE CREATED AND ANY ELECTRON FOLLOWED DOWN TO'
/T40,F8.3,' MeV KINETIC ENERGY'/
 ' BREM PHOTONS CAN BE CREATED AND ANY PHOTON FOLLOWED DOWN TO',
/T40,F8.3,' MeV ');"NOTE, AE VALUES CAN OVER-RIDE ECUT VALUES"
"-------------------------------------------------------------------"
"STEP 4   INITIALIZATION-FOR-HOWFAR                                  "
"-------------------------------------------------------------------"
ZBOUND= 2.54;"    PLATE IS 2.54 CM THICK"
"-------------------------------------------------------------------"
"STEP 5   INITIALIZATION-FOR-AUSGAB                                  "
"-------------------------------------------------------------------"
DO I=1,$EBIN [ EBIN(I) = 0.0;]"ZERO SCORING ARRAY BEFORE STARTING"
BWIDTH = 0.2;  "ENERGY SPECTRUM WILL HAVE 100 KeV WIDTH"
"-------------------------------------------------------------------"
"STEP 6   DETERMINATION-OF-INCIDENT-PARTICLE-PARAMETERS              "
"-------------------------------------------------------------------"
"DEFINE INITIAL VARIABLES FOR 5 MEV BEAM OF PHOTONS NORMALLY INCIDENT"
"ON THE SLAB"
IQIN=0;"                 INCIDENT CHARGE - PHOTONS"
EIN=5.0;"                5 MEV KINETIC ENERGY"
/XIN,YIN,ZIN/=0.0;"      INCIDENT AT ORIGIN"
/UIN,VIN/=0.0;WIN=1.0;"  MOVING ALONG Z AXIS"
IRIN=2;"                 STARTS IN REGION 2, COULD BE 1"
WTIN=1.0;"               WEIGHT = 1 SINCE NO VARIANCE REDUCTION USED"
```

```
"-------------------------------------------------------------------"
"STEP 7    SHOWER-CALL                                              "
"-------------------------------------------------------------------"
NCASE=5000;   "INITIATE THE SHOWER NCASE TIMES                      "
DO I=1,NCASE [EHIST = 0.0; "ZERO ENERGY DEPOSITED IN THIS HISTORY   "
CALL SHOWER(IQIN,EIN,XIN,YIN,ZIN,UIN,VIN,WIN,IRIN,WTIN);
"INCREMENT BIN CORRESPONDING TO  ENERGY DEPOSITED IN THIS HISTORY "
IBIN= MIN0 (IFIX(EHIST/BWIDTH + 0.999), $EBIN);
IF(IBIN.NE.0) [EBIN(IBIN)=EBIN(IBIN)+1;]
]
"-------------------------------------------------------------------"
"STEP 8    OUTPUT-OF-RESULTS                                        "
"-------------------------------------------------------------------"
"PICK UP MAXIMUM BIN FOR NORMALIZATION                              "
BINMAX=0.0; DO J=1,$EBIN [BINMAX=MAX(BINMAX,EBIN(J));]
```

This is a portion of TUTOR3.MOR, a tutorial EGS4 user code which scores the spectrum of energy deposited in a 2.54 cm thick slab of NaI when a 5 MeV beam of photons is incident on it, i.e. it computes the fraction of histories which deposit a certain amount of energy in the slab. (From [2].)

APPENDIX 4.2

```
ECHO 1
TITLE
5.0 MEV NAI TEST PROBLEM
************************ GEOMETRY ****************************
*   MATERIAL   SUBZONES   THICKNESS
GEOMETRY 1
1 1 2.54
************************ SOURCE ******************************
PHOTONS
ENERGY 5.0
CUTOFFS 0.189 0.010
DIRECTION 0.0
*********************** OUTPUT OPTIONS *********************
PULSE-HEIGHT
NBINE 27
*********************** OTHER OPTIONS **********************
NO-COHERENT
NO-INCOH-BINDING
HISTORIES 5000
BATCHES 50
```

This is an example of input to the ITS code TIGER for the sample with the EGS code given in appendix 4.1. In addition to standard scores of various quantities, the use of the switch PULSE-HEIGHT outputs the spectrum of energy deposited in a 2.54 cm thick slab of NaI when a 5 MeV beam of photons is incident on it. Note the use of the switches NO-COHERENT and NO-INCOH-BINDING to avoid coherent photon scattering and binding effects to incoherent scattering which are otherwise included (default). The number of bins (NBINE) specifies the desired number (25) plus two, to account for both total absorption and escape. The switch BATCHES specifies that the total number of histories (5000) is divided into

50 batches, of 100 photon histories each, in order to obtain estimates of statistical uncertainties. (Adapted from [3].)

APPENDIX 4.3

```
simple lattice
1  0  -1 fill=1 imp:n=1
2  0  -301 302 -303 304 lat=1 u=1 imp:n=1 fill=-2:2 -2:2 0:0
   1 1 1 1 1 2 2 2 1 1 2 2 2 1 1 2 2 2 1 1 1 1 1 1 1
3  0  -10 u=2 imp:n=1
4  0  #3 imp:n=1 u=2
5  0  1 imp:n=0

  1 cz 45
 10 cz 8
301 px 10
302 px -10
303 py 10
304 py -10
```

This is an example of geometry description in MCNP to illustrate a hexahedral lattice. The set of instructions above describes a cylinder of radius 45 cm that contains a square lattice, with the inner 3×3 array of cells containing a small cylinder in each cell. (See figure 4.2.) (From MCNP manual, LA-12625-M [4].)

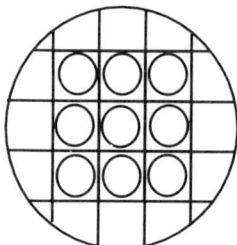

Figure 4.2   *Geometry generated by MCNP with the set of instructions given in this appendix.*

REFERENCES

[1]   Andreo P 1991 Monte-Carlo techniques in medical radiation physics *Phys. Med. Biol.* **36** 861–920
[2]   Nelson W R, Hirayama H and Rogers D W O 1985 The EGS4 code system *Stanford Linear Accelerator Center Report* SLAC–265
[3]   Halbleib J A, Kensek R P, Mehlhorn T A, Valdez G D, Seltzer S M and Berger M J 1992 *ITS Version 3.0: the Integrated TIGER Series of coupled Electron/Photon Monte Carlo Transport Codes* Report SAND91-1634 (Albuquerque, NM: Sandia National Laboratories)
[4]   Briesmeister J F (ed) 1993 *MCNP—a General Monte Carlo N-Particle Transport Code* (Los Alamos, NM: Los Alamos National Laboratory)

[5] Seltzer S M 1991 Electron–photon Monte Carlo calculations: the ETRAN code *Appl. Radiat. Isot.* **42** 917–41

[6] International Commission on Radiation Units and Measurements (ICRU) 1984 *Stopping Powers for Electrons and Positrons* ICRU Report 37

[7] Ford R L and Nelson W R 1978 *The EGS Code System: computer programs for the Monte Carlo simulation of electromagnetic cascade showers (version 3)* Report SLAC-210 (Standford, CA: Stanford Linear Accelerator Center)

[8] Nelson W R and Jenkins T M 1980 *Computer Techniques in Radiation Transport and Dosimetry* (New York: Plenum)

[9] Berger M J 1963 Monte Carlo calculation of the penetration and diffusion of fast charged particles *Methods in Computational Physics* eds B Alder, S Fernbach and M Rotenberg vol 1 (New York: Academic) pp 135–215

[10] Storm E and Israel H I 1970 Photon cross sections from 1 keV to 100 MeV for elements $Z = 1$ to $Z = 100$ *Nucl. Data. Tables* A **7** 565–681

[11] Bielajew A F and Rogers D W O 1986 *Photoelectron Angular Distribution in the EGS4 Code System* Report PIRS-0058 National Research Council of Canada (NRCC)

[12] Bielajew A F and Rogers D W O 1987 PRESTA—the parameter reduced electron-step algorithm for electron Monte Carlo transport *Nucl. Instrum. Methods* B **18** 165–81

[13] Duane S, Bielajew A F and Rogers D W O 1989 *Use of ICRU-37/NBS Collision Stopping Powers in the EGS4 System* Report PIRS-0173 National Research Council of Canada (NRCC)

[14] Bielajew A F, Hirayama H, Nelson W R and Rogers D W O 1994 *History Overview and Recent Improvements of EGS4* Report PIRS-0436 National Research Council of Canada (NRCC)

[15] Andreo P, Medin J and Bielajew A F 1993 Constraints of the multiple scattering theory of Molière in Monte Carlo simulations of the transport of charged particles *Med. Phys.* **20** 1315–25

[16] Bielajew A F, Wang R and Duane S 1993 Incorporation of single elastic scattering in the EGS4 Monte Carlo code system: tests of Molière theory *Nucl. Instrum. Methods* B **82** 503–12

[17] Bielajew A F 1994 Plural and multiple small-angle scattering from a screened Rutherford cross section *Nucl. Instrum. Methods* B **86** 257–69

[18] Namito Y, Ban S and Hirayama H 1994 Implementation of the Doppler broadening of a Compton scattered photon in the EGS4 code *Nucl. Instrum. Methods* A **349** 489–94

[19] Berger M J and Seltzer S M 1968 *Electron and Photon transport programs: I. Introduction and Notes on Program DATAPAC 4; II. Notes on program ETRAN. Reports NBS 9836* and *9837* (Gaithersburg, MD: National Bureau of Standards)
See also RSIC 1968 *ETRAN: Monte Carlo Code System for Electron and Photon transport through extended media. Report ORNL-RSIC CCC-107*

[20] Berger M J and Seltzer S M 1969 Calculation of energy and charge deposition of the electron flux in a water medium bombarded with 20 MeV electrons *Ann. NY Acad. Sci.* **161** 8–23

[21] Berger M J and Seltzer S M 1982 *Tables of energy-deposition distributions in water phantoms irradiated by point-monodirectional electron beams with energies from 1 to 60 MeV and applications to broad beams. Report NBSIR 82-2451 NBS*

[22] Berger M J, Seltzer S M, Domen S R and Lamperti P J 1975 Stopping-power ratios for electron dosimetry with ionization chambers (IAEA-SM-193/39) *Biomedical Dosimetry* (Vienna: IAEA) pp 589–609

[23]   International Commission on Radiation Units and Measurements (ICRU) 1984 *Radiation Dosimetry: Electron Beams with Energies between 1 and 50 MeV* ICRU 35 (Bethesda, MD: ICRU)

[24]   Berger M J 1971 Distribution of absorbed dose around point sources of electrons and beta particles in water and other media *J. Nucl. Med.* **12** (supplement 5) 5–23

[25]   Berger M J 1973 *Improved Point Kernels for Electrons and Beta-Ray Dosimetry* Report NBSIR 73-107 NBS

[26]   Seltzer S M 1988 An overview of ETRAN Monte Carlo methods *Monte Carlo Transport of Electrons and Photons* ed T M Jenkins, W R Nelson and A Rindi (New York: Plenum) pp 153–81

[27]   Hubbell J H, Veigele W J, Briggs E A, Brown R T, Cromer D T and Howerton R J 1975 Atomic form factors incoherent scattering functions and photon scattering cross sections *J. Phys. Chem. Ref. Data* **4** 471–538

[28]   Cullen D E, Chen M H, Hubbell J H, Perkins S T, Plechaty E F, Rathkopf J A and Scofield J H 1989 *Tables and graphs of photon-interaction cross sections from 10 eV to 100 GeV derived from the LLNL Evaluated Photon Data Library (EPDL) Part A: Z = 1 to 50 Part B: Z = 51 to 100* Report UCRL-50400 vol 6 part A and B rev 4 Lawrence Livermore National Laboratory

[29]   Cullen D E, Sterrett T P, Plechaty E F and Rathkopf J A 1988 *The All Particle Method: coupled neutron photon electron charged particle Monte Carlo calculations* Report UCRL-98975 Lawrence Livermore National Laboratory

[30]   Cullen D E, Ballinger C T and Perkins S T 1991 *The All Particle Method: 1991 Status Report* Report UCRL-JC-108061 Lawrence Livermore National Laboratory

[31]   Ballinger C T 1991 *The Response History Monte Carlo Method for Electron Transport* Report UCRL-ID-108040 Lawrence Livermore National Laboratory

[32]   Perkins S T, Cullen D E M and Seltzer S M 1991 *Tables and graphs of electron-interaction cross sections from 10 eV to 100 GeV derived from the LLNL Evaluated Electron Data Library (EEDL) Z = 1–100* Report UCRL-50400 vol 31 Lawrence Livermore National Laboratory

[33]   Cullen D E 1994 *Program EPICSHOW: a computer to allow interactive viewing of the EPIC data libraries* Report UCRL-ID-116819 Lawrence Livermore National Laboratory

[34]   Nelson W R and Jenkins T M 1988 Geometry methods and packages *Monte Carlo Transport of Electrons and Photons* ed T M Jenkins, W R Nelson and A Rindi (New York: Plenum) pp 385–406

[35]   Sato O, Iwai S, Nakamura M, Uehara T, Takagi S and Hirayama H 1994 *UCMARS—a user code with a multiple array system using combinatorial geometry for EGS4* Report KEK 94-12 National Laboratory for High-Energy Physics Japan

[36]   Halbleib J 1988 Structure and operation of the ITS code system *Monte Carlo Transport of Electrons and Photons* ed T M Jenkins, W R Nelson and A Rindi (New York: Plenum) pp 249–62

[37]   Van Riper K A 1993 *SABRINA user's guide* Report LA-UR-93-3696 Los Alamos National Laboratory (available from RSIC or NEA as PSR-0242/04 *SABRINA: Three Dimensional Geometry Visualization Code System* version 3 54

[38]   Comite Consultatif pour les Etalons de Mesure des Rayonnements Ionisants (Section I) (CCEMRI(I)) 1991 *Report to the Comite International des Poids et Mesures (N J Hargrave Rapporteur) 10th Meeting CCEMRI(I)* (Pavillon de Breteuil: BIPM)

[39]   Bielajew A F 1996 private communication

CHAPTER 5

AN INTRODUCTION TO SCINTILLATION DETECTOR PHYSICS

Peter D Esser

5.1 INTRODUCTION

Nuclear medicine applications of Monte Carlo (MC) simulation provide a bridge linking the physics of ionizing radiation in body tissues and scintillation detectors with clinical nuclear medicine images generated by the radiation. Before describing the different MC codes that simulate nuclear medicine imaging for planar, simple-photon emission computed tomography (SPECT) and positron emission tomography (PET) systems, it is essential to understand the underlying physical processes that form these images and, in particular, that the scintillation process is important for any simulations of clinical images. The reason is twofold: both to understand the complexity in obtaining an image of an activity distribution and also to realize that computer MC codes can give results very close to a real measurement.

Nuclear medicine practitioners utilize gamma-ray detector systems containing crystals with the ability to convert gamma rays into scintillations of lower-energy light photons. These luminescent emissions are in turn converted into electronic signals by arrays of photomultiplier tubes (PMTs) in gamma-ray imaging cameras. The systems generate a signal current from the PMTs whose magnitude is proportional to the gamma ray energy and whose spatial location within the detector arrays corresponds to the originating location of the event in the patient. However, to provide this information as a final image, sophisticated digital processing hardware and software are required for signal processing, attenuation correction and three-dimensional reconstruction of tomographic data. While the energy and spatial resolution

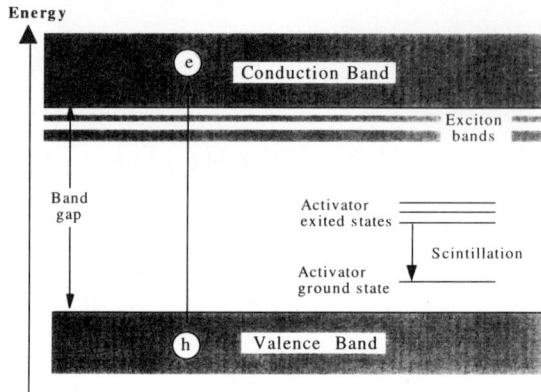

Figure 5.1   *A schematic representation of the energy band structure of NaI(Tl).*

of nuclear imaging systems has improved over many years, the core of the technology and our ability to image tracer distributions still continues to be scintillation-crystal-based detector systems using NaI(Tl) for planar and SPECT imaging and primarily BGO for PET. Accordingly, this article is devoted to a review of the scintillation processes in NaI(Tl) and BGO.

5.2 SCINTILLATORS: NaI(Tl)

NaI is an inorganic ionic crystal like other alkali halides such as NaCl, KCl, and LiF. The lattices of these compounds are generally cubic and many of their properties are strongly affected by Coulomb interactions properties that have been the subject of extensive experimental and theoretical studies in the past [1, 2]. In addition, alkali halides are insulators whose electrons can only occupy discrete energy bands, as can be seen in figure 5.1.

Liberated electrons, released from ionizing radiation, leave behind a charge deficiency that behaves like an electron with a positive charge, i.e. a positive hole. The size of this energy gap is approximately 8 eV and represents a broad region of the spectrum extending into the far ultraviolet where the crystal is normally transparent and photons will not be absorbed. For pure NaI, the return of an electron to the valence band with the emission of an 8 eV photon is not an efficient process and photon detection in this energy range is therefore difficult. Producing an efficient and useful scintillator with emission in the visible region of the spectrum requires the presence of localized energy states in the energy gap. Fortunately, they can be introduced by the incorporation of impurity activators into the crystal lattice. Table 5.1 lists some of the common commercially available scintillation crystals.

Table 5.1 *Properties of common inorganic scintillation crystals (after data from Solon Technologies).*

Material	Efficiency (NaI(Tl) = 100%)	Decay constant $1/e$ (ms)	Emission energy (eV)		Index of refraction	Density (g cm^{-3})	Hygroscopic
			max.	cut-off			
NaI(Tl)	100	0.23	2.99	3.87	1.85	3.67	yes
BaF$_2$	15	0.6	4	5.64	1.49	4.88	no
	2–3	0.0008	5.64	6.89			
B$_4$Ge$_3$O$_{12}$	10–12	0.3	2.58	3.54	2.15	7.13	no
CaF$_2$(Eu)	50	0.94	2.85	3.06	1.44	3.18	no
CdWO$_4$	18	5/20	2.3	2.76	2/2.4	7.9	no
CsF	3–5	0.005	3.18	4.96	1.48	4.64	yes
Cs(Na)	85	0.63	2.95	4.13	1.8	4.51	yes
CsI(Pure)	5–7	0.01	3.94	4.96	1.8	4.51	no
CsI(Tl)	45	1	2.19	3.76	1.8	4.51	no
Li(Eu)	35	1.4	2.56	2.76	1.96	4.08	yes

Table 5.2   *Crystal and ionic radii [2].*

Element	Atomic number	Charge	Radius in crystals	Ionic radius
Na	11	1^+	1.85	0.97
K	19	1^+	2.25	1.33
Tl	81	$1^+, 3^+$	1.67	1.47
Bi	83	$3^+, 5^+$	1.55	0.96
Cl	17	$1^-, 5^+, 7^+$	0.91	1.81
I	53	$1^-, 5^+, 7^+$	1.35	2.16

The scintillation of NaI activated with thallium as an impurity was first reported in 1948 by Hofstadter [3], who grew small crystals in a quartz tube and demonstrated that NaI(Tl) generated appreciably more luminescence than organic phosphors such as naphthalene. The superior light yield of NaI(Tl) makes it the most commonly used scintillator (table 5.1).

When NaI crystals are grown with small amounts of thallium (Tl), the thallium ions randomly replace the Na^+ ions at lattice sites because they are of similar ion size and have the same charge (see table 5.2). The thallium atom has five inner shells (78 electrons) and three outer electrons. For singly ionized thallium there are two allowed transitions. However, when Tl^+ substitutionally replaces Na^+, it strongly interacts with the Coulomb field of the six neighbouring I^- ions and the vibrating crystal lattice causing a shift in its atomic energy levels. This lifts the degeneracies of its excited states.

Measurements at low temperatures of NaI(Tl) show the presence of three primary optical absorption bands (4.25, 5.0 and 5.3 eV) and three emission bands (2.85, 3.75 and 4.45 eV). The significant energy difference between absorption and emission (Stokes shift) can be attributed to a large energy shift between excitation and absorption positions on the energy bands

Experiments have demonstrated that ionizing radiation absorbed in NaI(Tl) also stimulates the same luminescence as the optical stimulation even though the energies are high enough to eliminate the possibility of direct optical absorption [4]. From data such as this, it is inferred that the ionizing radiation is first absorbed by the host lattice and then, by a second process, the deposited energy is transferred to the thallium activator site which emits characteristic light.

The primary radiative emission from the excited thallium ion, $(Tl^+)^*$, occurs at 3 eV above the ground state. The efficiency of the energy transfer process at room temperature has been shown to be approximately 13% [5]. Thus, a 140 keV photon should produce $140 \times 10^3 \times 0.13$ eV/(3.0 eV per photon) $= 6.1 \times 10^3$ photons, and a 511 keV photon 2.2×10^4 photons. Additionally, 20 eV (approximately three times the band gap) is required to produce an electron–hole pair [6]. Thus 7×10^3 pairs are produced from each 140 keV photon (2.6×10^4 for a 511 keV photon). The number of electron–hole pairs and the number of photons are approximately the same, indicating there is a high probability for electron–hole pair recombination at Tl^+ sites and subsequent emission from $(Tl^+)^*$.

The calculation above demonstrates that most of the luminescence (if not all of it) from NaI with 0.1 mol% Tl (1000 ppm) is associated with recombination of charge at the thallium site. Electron charge transport in NaI occurs in the same manner as in semiconductors; the electron is raised into the conduction band and subsequently can freely move through the crystal.

Charge transport occurs rapidly with respect to the lifetimes of thallium emission. The primary decay time of the room-temperature scintillation is 0.23 μs. However, there may be several small components including a 0.15 s. long-lived glow that contributes approximately 9% of the total light yield [7]. In general, for small temperature changes around 22 °C the temperature dependence of the decay time is approximately 5 ns °C^{-1} and the light yield changes 0.1% °C^{-1}. With regard to scintillation light intensity (i.e. light output) as a function of energy, experimental data indicate a linear relationship within the useful nuclear medicine energy range [8]. The detecting efficiency of scintillation light output is a function of many variables such as crystal size, location of interaction in the crystal and differences in photocathode sensitivity. Thus, some variations in energy response should be expected in a nuclear camera system.

Even though it is almost 50 years since the luminescent properties of NaI(Tl) were observed, the scintillator continues to be the primary detector system for clinical use and other counting applications. This is a result of the material's desirable physical properties such as high light output and a short decay time, but also the ability to commercially grow large crystals relatively inexpensively.

NaI(Tl) scintillators are grown in Strockbarge furnaces from a melt of ultra-pure NaI doped with approximately 2000 ppm Tl in a platinum crucible. The total process of purification, growth and annealing can take up to 90 d. The final crystals are typically 30 in diameter and 8–15 in tall with 700–1100 ppm Tl. Since NaI is hygroscopic, preparation of the detectors must be performed in a controlled atmosphere where the crystals are cut, etched, polished etc. The large boules of material are sliced into thin sections for camera crystals with about 20% of the thickness lost during preparation. Any treatment of the fragile crystal must not introduce stress and lattice defects that would reduce transparency and introduce scattering sites and alternative trapping centres. It is also possible, in high-temperature presses, to forge special detector geometries from crystal slices. The final treatment of the detector's surface is crucial for optimization of light output efficiency.

Fortunately the luminescent output is insensitive to variations in thallium iodide content above 0.1 mol% [9]. Typically scintillation light intensity will vary by 3–10% (generally <5%) within a detector system with most of the alterations attributed to the detector optics [10]. Approximately 70% of

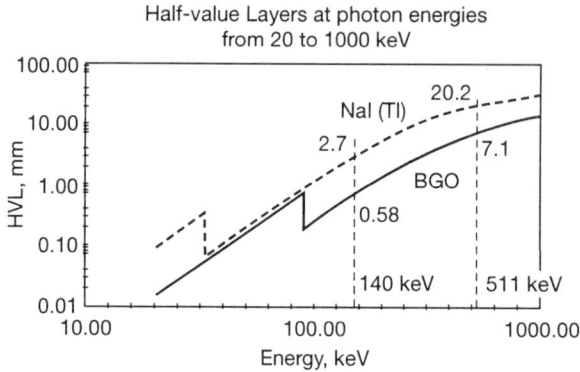

Figure 5.2   *Half-value layers for gamma rays in BGO and NaI(Tl) [12].*

the emitted light reaches the photomultiplier plane, where additional signal variations are introduced by the PMTs.

After treatment, the crystal is sealed in an aluminium can with a special glass window; one surface is bonded to the window with silicon rubber and the other surfaces are surrounded with a diffuse reflector. The final detectors, which may be as large as $\frac{3}{8} \times 14 \times 20$ in, require uniform response with superior optical clarity throughout the crystal.

5.3 SCINTILLATORS: BISMUTH GERMINATE (BGO)

PET systems require large numbers (between 4096 and 18 816 depending on the vendor and model) of small detectors, typically $6.75 \times 6.75 \times 20$ mm^3 in size. A scintillator with stopping power higher than that of NaI(Tl) would also be advantageous. A material with desirable characteristics is bismuth germinate ($B_4G_3O_{12}$) which has been used in most PET systems since it became commercially available in the late 1970s. Like NaI(Tl), it has a cubic structure and is grown from a melt containing in this case only ultra-pure GeO_2 and Bi_2O_3: no dopants are required. At room temperature BGO is stable and not hygroscopic. It also can be machined and no special containers are required. However, BGO is relatively expensive and is not available in very large sizes. It is the bismuth component with its high atomic number (83) that gives BGO a high probability of photoelectric interaction, as shown in figure 5.2. At 511 keV there is almost a factor of three difference in half-value layer (HVL). The differences in total attenuation are almost entirely from the contribution of the photoelectric component.

While scintillation phenomena in alkali halides have been extensively studied, the properties of BGO are relatively unexplored. Based on the studies of Weber and Monchamp [11] and others [12, 13], we can summarize the luminescent characteristics of BGO as follows. A broad fluorescence band at 2.48 eV has been observed from trivalent bismuth ions when BGO is optically excited in the ultraviolet. The same emission band is also stimulated by ionizing radiation excitation and, thus, demonstrates that Bi^{3+} is the primary energy trapping site for energy deposited by gamma and x-rays. The light yield of BGO is about 20–25% of NaI(Tl), however the reduced sensitivity of the PMT below 2.5 eV reduces the photoelectron yield to 10–15% (remember, however, that a 511 keV scintillation will generate 3.6 times the amount of light that a 140 keV event does) [14]. In addition, the energy resolution at 662 keV (Cs^{137}) is 15.5%, compared to 7.7% for NaI(Tl).

5.4 PULSE SHAPE

Emission spectra are an important physical characteristic of scintillators. For BGO, two decay components exist: a fast component of 60 ns with approximately 10% of the light output and the slower primary 300 ns component. In comparison, NaI(Tl) has an initial decay that is much slower, followed by the well known 230 ns component and a long-lifetime phosphorescence (or afterglow) that makes it unsuitable for some applications. Adequate time resolution of the moment of absorption is necessary for coincidence counting in PET scanners and is limited by the statistics of light production in the scintillation. The resolution for 511 keV photons is 1.9 ns (FWHM) for BGO compared to 0.75 ns for NaI(Tl). The ratio to NaI(Tl) is better than expected based on relative light yields since the BGO resolution is strongly influenced by the contribution of the initial fast component. The relatively good time resolution of BGO and the absence of afterglow make BGO a suitable PET scintillator.

5.5 PHOTOMULTIPLIER TUBES (PMTS)

Figure 5.3(*a*) shows the structure of an eight-stage, box-and-grid, head-on PMT with a 60 mm hexagonal window. It employs a potassium–caesium–antimonide (bialkali) photocathode and has a gain of 1.9×10^5 at 800 V. This is typical for a tube of this type since each photoelectron produces three or four secondary electrons at each step in the dynode chain.

This type of tube is frequently chosen for use in scintillation cameras because of its large flat semi-transparent photocathode but overall small size. Figure 5.3(*b*) shows schematically the light conversion of two incident photons. The detector system (crystal + PMT) can be characterized by its total resolution where

$$R_{\text{total intrinsic}} = \sqrt{R_{\text{scintillator}}^2 + R_{\text{PMT}}^2 + R_{\text{other}}^2} \qquad (5.1)$$

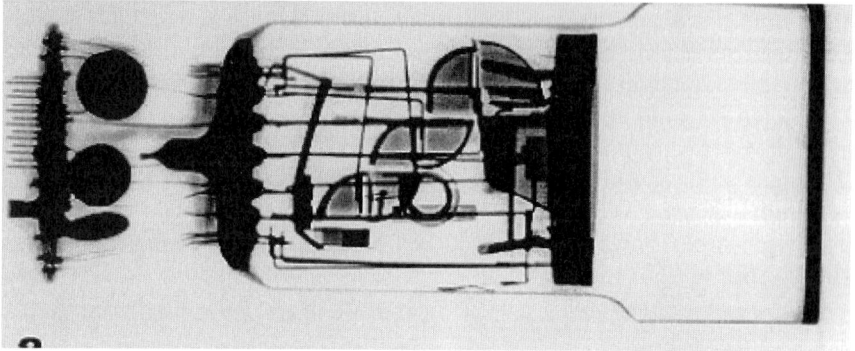

(a)

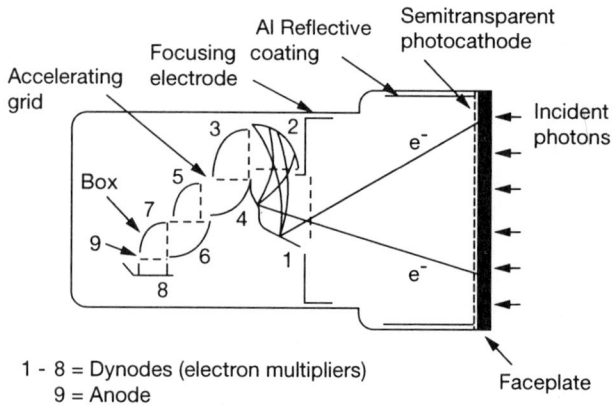

Semitransparent photocathode

Al Reflective coating

Focusing electrode

Accelerating grid

Box

Incident photons

1 - 8 = Dynodes (electron multipliers)
9 = Anode

Faceplate

(b)

Figure 5.3 *A PMT. (a) An x-ray of a Burle S83053 PMT, a type commonly used in nuclear medicine. (b) A schematic diagram of (a) showing the modified box-and-grid multiplier structure.*

and

$$R \equiv \Delta E / E_0. \tag{5.2}$$

E_0 is the energy of a gamma ray that generates a spectrum with a photopeak having a full width at half maximum (FWHM) of ΔE. The first term in the above expression represents the intrinsic resolution associated with light output from scintillation events, which is a function of energy. The second term represents the statistics of the multiplication process at the dynodes within the PMT tube. A third term includes parameters that are not dependent on the number of photons but contribute to the energy resolution.

These include variations in the sensitivity of the photocathode and inhomogeneities of the crystal [10].

The contribution to detector resolution from inherent statistical variations depends on the number of photons. From our previous discussion of NaI(Tl), we know that a 140 keV gamma ray will generate approximately 6100 light photons. The number reaching the photomultiplier plane depends on the crystal dimensions and shape, coupling to the PMTs and other variables. We assume a light loss of 30% in the detector, reducing the number of photons reaching the photocathode to 4270. The photocathode quantum efficiency is approximately 25% and diminishes the signal to 1067 photoelectrons. This is the minimum signal in the detector electronics and is the most sensitive point in the imaging chain for statistical fluctuations. If we assume Poisson statistics govern the processes, the standard deviation is equal to the square root of the mean or 3.1%. To express the result in terms of FWHM, we assume a Gaussian peak shape. Thus, the statistical fluctuation of the detector photons is 2.355 times the standard deviation or 7.3% (if the photocathode loss is not included, the resolution would be 3.6%). Since the energy resolution for modern camera systems is around 9%, the primary contribution to this number (7.3%) is that from the statistical fluctuations of scintillation photons as described above; the contribution (1.7%) contributed by amplification of photocathode electrons and other sources is relatively small.

The photon light yield, as noted earlier, for NaI(Tl) and BGO is proportional to gamma ray energy and, thus, resolution will be a function of energy. From the above definition of resolution,

$$R \equiv \frac{\Delta E}{E_0} = k \frac{\sqrt{E}}{E} = \frac{k}{\sqrt{E}}. \tag{5.3}$$

This demonstrates that detector resolution should be inversely proportional to the square root of the incident gamma ray energy. However, the resolution of NaI(Tl) has been examined experimentally [14] and deviations from this expression have been observed, indicating the presence of nonstatistical sources of peak broadening. A more accurate representation of resolution is

$$R = \frac{\sqrt{\alpha + \beta E}}{E} \tag{5.4}$$

where α and β are constants for a particular scintillator–PMT combination [7].

ACKNOWLEDGMENT

The author thanks Robert Pickar for aid in the preparation of this manuscript.

REFERENCES

[1] Fowler W B (ed) 1968 *Physics of Color Centers* (New York: Academic)
[2] Schulman J H and Compton W D 1962 *Color Centers in Solids* (New York: Pergamon)
[3] Hofstadter H 1948 Alkali halide scintillation counters *Phys. Rev.* **74** 100
[4] Kauffman R G, Hadley W B and Hersh H N 1970 The scintillation mechanism in thallium doped alkali halides *IEEE Trans. Nucl. Sci.* **NS-17** 82
[5] Van Sciever W J and Bogart L 1958 Fundamental studies of scintillation phenomenon in NaI* *IEEE Trans. Nucl. Sci.* **NS-3** 90
[6] Murray R B 1975 Energy transfer in alkali halide scintillators by electron–hole diffusion and capture *IEEE Trans. Nucl. Sci.* **NS-22** 54
[7] Knoll G F 1989 *Radiation Detection and Measurement* (New York: Wiley)
[8] Miyajima M, Sasaki S and Tawara H 1993 Numbers of scintillation photons produced in NaI(Tl) and plastic scintillator by gamma rays *IEEE Trans. Nucl. Sci.* **NS-40** 417
[9] Harshaw J A, Stewart E C and Hay J O 1952 *AEC Report* NYO-1577
[10] Bicron 1995 private communications
[11] Weber M J and Monchamp R R 1973 Luminescence of $Bi_4Ge_3O_{12}$: spectral and decay properties *J. Appl. Phys.* **44** 5495
[12] Blasse G and Brill A 1968 Investigations on Bi^{3+}-activated phosphors *J. Chem. Phys.* **48** 217
[13] Nestor O H and Huang C Y 1975 Bismuth germanate: a high-Z gamma ray and charged particle detector *IEEE Trans. Nucl. Sci.* **NS-22**
[14] Beattie R J D and Byrne J 1972 A Monte Carlo program for evaluating the response of a scintillation counter to monoenergetic gamma rays *Nucl. Instrum. Methods* **104** 63

CHAPTER 6

THE SCINTILLATION CAMERA— BASIC PRINCIPLES

Sven-Erik Strand

6.1 INTRODUCTION

The scintillation camera, also called the Anger camera, was invented by Hal O Anger in 1956/1957 [1, 2]. In 1962, the first commercial camera was presented by Nuclear Chicago, with a NaI(Tl) crystal 25 cm in diameter and thickness 1.25 cm, with 19 photomultiplier (PM) tubes. With the development of techniques for growing larger NaI(Tl) crystals, cameras with wider fields were developed from the mid-1970s. Although ^{99m}Tc was discovered as early as 1938 [3], it was not until the middle of the 1960s that ^{99m}Tc-labelled radiopharmaceuticals became widely available. ^{99m}Tc decays emitting an optimal photon energy of 140.5 keV for scintillation camera imaging and has favourable dosimetric properties. It promoted rapidly growing interest in the scintillation camera technique, with the camera's potential for high sensitivity, large field of view and dynamic studies. During the following years, the camera was widely developed; however, the main principles remain. Today about a dozen larger vendors are manufacturing cameras with about 50 different models [4]. Figure 6.1 shows an example of a single-headed scintillation camera.

The scintillation camera is the most commonly used imaging system in nuclear medicine today for both planar and tomographic studies [5]. The photon energy range covered is from 40–50 keV up to 400 keV normally. Some cameras are shielded and are useful up to 511 keV, making it possible to study positron-emitting radionuclides/pharmaceuticals. Today's system can be characterized as composed of scintillation camera head(s), gantry and computer. The computer is often separated into acquisition and processing stations. The components together make up an integrated system. To

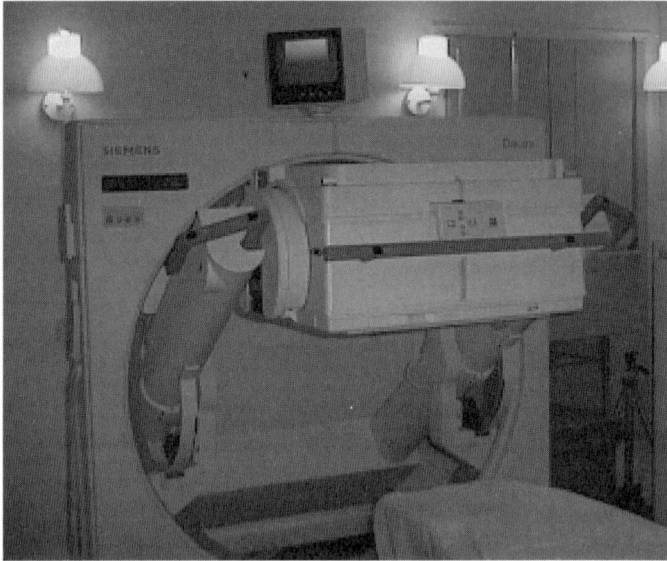

Figure 6.1 *A modern one-headed scintillation camera with SPECT facility.*

increase the sensitivity, multiheaded cameras with two, three or four heads are available. Scintillation camera improvements in recent years are in sensitivity, uniformity, intrinsic spatial resolution, energy resolution and dead time. Also, imaging of positron-emitting radionuclides (511 keV) has become feasible. Details of camera construction and function can be found in several textbooks and review articles [6, 7].

Although many parameters for the camera can be calculated numerically, the Monte Carlo technique is an important tool for optimizing those imaging parameters difficult to calculate, as shown throughout this book. Below is a short summary of important principles for the scintillation camera.

6.2 PRINCIPLES FOR THE SCINTILLATION CAMERA

The scintillation camera can be considered as a photon-detecting and an image-creating detector. Photons emitted from the activity distribution in the patient are projected through the collimator onto the sodium iodine crystal, where they interact and create scintillation light. The light is detected by photomultiplier tubes (PMTs), and transformed into current pulses, proportional to the flux of the incoming photons. These pulses are processed by analogue and/or digital electronic circuits, and computation with suitable algorithms results in output signals (data) representing the centroid of the scintillation light distribution created in the crystal, and the energy deposited in the crystal.

6.3 THE DETECTOR

The detector part of the scintillation camera consists of a NaI(Tl) crystal, optically coupled to a set of PMTs. The crystal can be of different sizes and shapes, and the number of PMTs varies. Circular camera heads are designed with diameters up to 50 cm. The number of PMTs used varies between 37 and about 105 with circular, square or hexagonal cross section. Crystal thickness is mostly 9.5 mm. Rectangular-shaped crystals can be found with dimensions up to 66 cm × 45 cm. Here up to 115 hexagonal or circular PMTs with diameters 1.5–7.5 cm are used.

The primary decay time at room temperature for the scintillation is 0.23 µs for 60% of the light, with 1.15 µs for the rest. However, there exists a 0.15 s long-lived glow that contributes approximately 9% of the total light yield [8, 9]. There is a linear relationship between scintillation light intensity (i.e. light output) and photon energy. The detection efficiency of scintillation light output varies with many factors such as crystal size, location of inter-action in the crystal and variations in photocathode sensitivity, giving some variations in energy response. During the crystal growing process as the crystals are drawn from the molten solution of NaI and Tl, some variations in thallium concentration can be expected along the primary and radial axes. Typically scintillation light intensity will vary less than 5%.

After the NaI(Tl) crystal is grown it is cut, etched and polished, and the surface is treated to get high light-output efficiency. The NaI(Tl) crystal is both delicate and hygroscopic and is hermetically sealed in a thin alumina cover (<1 mm thick), except on the side facing the PMTs where a transpar-ent light guide is applied. The other surfaces are surrounded with a diffuse reflector (alumina oxide). The whole detector assembly is mounted in a lead housing with walls thick enough (about 1–2 cm) for shielding of photon penetration up to at least 500 keV. Thus, the transmission of low energy photons through the cover is small. In front of the detector the cover can be even thicker.

Newer materials for scintillators have been proposed with higher Z and shorter decay time. Such materials are, for example, LSO and YSO to be used for coincidence imaging [10]. None of these are yet available commercially.

6.4 OPTICAL COUPLING—THE LIGHT GUIDE

The crystal and the PMTs are optically coupled via a light guide consisting of a 5–25 mm thick sheet of glass or Perspex. With optical glue, usually silicon oil or grease, the components are optically coupled. The light guide keeps the PMTs further away from the crystal and enables more PMTs to collect scintillation light from the light flash. The light guide helps to scatter light into the PMTs, due to the angle of total reflection between glass and NaI of 53.1°. The geometrical efficiency of the PMT is dependent on the total thickness of crystal and light guide and the area of the PMT.

The PMTs should be mounted so that the difference between collected light of neighbouring tubes should vary linearly with the distance to the scintillation point in the crystal. The larger the difference between these signals the better the ability of the camera to localize the scintillation event. One important factor is the packing density of the PMTs. By using hexagonal PMTs, better packing with less loss of scintillation light is achieved.

To improve the spatial resolution, the thickness of the light guide is reduced. However, then the nonlinearity in positioning is increased. Corrections must be made either mechanically (light guide geometric design, PMT geometry) or electronically. Both energy and spatial resolution in the camera are dependent on good scintillation light–photoelectron conversion efficiency. Improvements have been achieved by matching the spectral sensitivity of the PMT (using PMTs with bialkali photocathodes).

6.5 ELECTRONICS

When a scintillation event occurs in the crystal, scintillation light is collected in the PMTs in proportion to their solid angle from the light emission point. The charge created in the tubes is thus dependent on the distance between scintillation light emission point and PMT. In the array of PMTs, this can be used for calculating the position of the event. The charge in the PMTs is integrated for about 1 µs, to create a pulse including charge from the whole light signal. The total charge registered by all PMTs is summed and is proportional to the energy deposition in the crystal. This signal is designated as the 'energy signal' or the 'Z signal'.

Scintillation camera systems are sometimes called analogue or digital cameras, depending on the degree of digital electronics in the detector. The general meaning is that the circuits doing the position and the energy calculations can be either analogue or digital.

6.5.1 *Analogue circuits*

The pulse shaping occurs in the preamplifiers and the signals are fed through a resistor matrix to summing amplifiers, one for each of four position directions, X_+, X_-, Y_+ and Y_-. All summing, integration, calculation and normalization take place in analogue circuits and position and total energy are calculated from simple algorithms as

$$X = (X_+ - X_-)/Z \tag{6.1}$$

$$Y = (Y_+ - Y_-)/Z \tag{6.2}$$

$$Z = X_+ + X_- + Y_+ + Y_- \tag{6.3}$$

In this simple algorithm the different PMT signals are weighted with a factor dependent on the distance from the centre of the crystal, the origin of the coordinate system. The division of the position signal by the energy signal makes the image coordinates independent of the statistical variation in pulse amplitude within the energy window.

Some electronic operations are slow, and the light decay with the electronics creates a system with a large time constant. To prevent large 'dead times' the pulses are integrated for a limited time and sometimes also truncated. A typical pulse length in the detector electronics is 0.5–3 µs.

6.5.2 Digital circuits

Pulse shaping and sometimes pulse clipping take place in the preamplifiers. An analogue event detector works in real time, detecting a valid event including coarse position and triggering off the digital sampling. The outputs from the preamplifiers are simultaneously digitized in a set of analogue-to-digital converters and all signals or a subset of all signals are processed further. Calculations and normalization for position and energy similar to the analogue circuit are now performed digitally and processor controlled. In camera systems with digital circuits, the processor is often also used for baseline stabilization and automatic PMT/preamplifier gain control.

6.6 PILEUP AND MISPOSITIONING

Dead time and pulse pileup could make absolute quantitation difficult due to the nonlinearity between activity (i.e. photon fluence rate impinging on the crystal) and count rate. Mispositioning could further decrease quantitation ability and deteriorates the image quality [11, 12]. When the crystal is exposed to a high photon fluence rate, the long decay time of the scintillation light creates such a long afterglow at a preceding event site that the PMTs, when registering the scintillation light from an event, still collect light from the former event. Thus, the signal from each PMT is the sum of the total light emission in the crystal and thus the events will be mispositioned. The effects have been described in detail [13]. Remember that it is the photon fluence for the whole energy spectra that is important. All photons interacting in the crystal cause scintillation light and despite their energy they contribute to the signal from the PMTs.

6.7 CORRECTION CIRCUITS

The performance of the scintillation camera detector has improved during a 30-year period. However, there are remaining image quality problems due to errors introduced by the way gamma events are collected and processed. Nonlinearity is one major problem in the detector. Reasons for image degradation are: (i) defects in the NaI(Tl) crystal, (ii) imperfect light coupling between crystal and PMTs, (iii) the limited number of PMTs, (iv) electronic dead time and (v) collimator defects. A number of built-in corrections can be found in a scintillation camera today to reduce degradation in spatial linearity, energy resolution, time resolution and sensitivity.

6.7.1 Corrections in light guide and preamplifier

Stationary mechanical corrections are specially designed light guides with varying light transmission, which are prefabricated with the camera. Preamplifiers with nonlinear amplification can also be used, where an event under the PMT is relatively less amplified compared with an event away from the tube. In a digitally controlled detector the signal from any PMT directly under an event is attenuated

6.7.2 Correction for position nonlinearity

Position errors occur because a discrete array of tubes is used to detect events over a continuous crystal surface. 'Coordinate bunching' causes events that occur between PMTs to be positioned towards the centre of the tubes. A matrix of correction factors can be created with the aid of a suitable mask placed on the camera detector.

6.7.3 Correction for energy nonlinearity

The solution has the same principle, a correction matrix, as for position nonlinearity. A mask over the camera surface defines the matrix points of corrections.

6.7.4 Correction for remaining nonuniformity

In spite of the corrections mentioned above there still is a nonuniformity. This is due to imperfections in these corrections and to varying response in the collimator. Correction for this is made with a flood image with homogenous photon fluence density over the field of view.

6.7.5 Correction for drift

The PMTs and the high voltage can suffer from changes in gain, influencing the signal amplitudes, and can be corrected either for each PMT to be exposed regularly with a stable light flash from a diode, or by monitoring the pulse height distribution in a 'tracking' window, then automatically adjusting the gain or high voltage.

6.7.6 Correction for dead time and pulse pileup

The dead time in modern cameras is about 0.5–3 μs due to processing of signals and corrections. The dead time can be reduced by using variable integration times, double buffering etc. Another cause of dead time is pulse pileup at high photon fluence rates. Pileup in the crystal and in the electronic amplifiers can be corrected for by 'pileup rejection' circuits. The condition for rejection is set up after analysing the event pulses. Two examples of pileup rejection are (i) the integrated value of the pulse is compared at a certain predefined time with the level of the original pulse (a reject condition for nontolerable pileup can be set) and (ii) by analysing the leading edge of the event pulse and matching to a predefined criterion for pileup. Pileup pulses can also, after detection, be restored. In a double event pulse the envelope and value of a single pulse are roughly calculated and subtracted from the pileup pulse. Thus, both pulses are restored, giving fewer system count losses.

6.8 SENSITIVITY AND RESOLUTION

The *system sensitivity* of the scintillation camera is governed by two factors; geometrical efficiency of the collimator, ε_c, defined as the number of photons emitted from the object that passes through the collimator holes and the photo-peak detection efficiency of the crystal, f, defined as the fraction of the photons impinging on the crystal that result in an event, registered in the photo-peak window.

The *system spatial resolution* of the scintillation camera depends on geometrical resolution of the collimator, R_c, the ability of the crystal to transfer the 'absorbed energy image' to a 'scintillation light image' and a scintillation light image to the PMT cathodes and the ability of the electronics to transfer this 'scintillation light image' to a readable image.

The latter two points comprise the intrinsic spatial resolution of the camera, R_i. The position resolution in a scintillation camera is its ability to separate two nearby point sources in the image. The spatial resolution is most accurately described by the point or line spread functions, PSF or LSF, or the modulation transfer function, MTF, which is the absolute value of the Fourier transform of the PSF or LSF. A simplified description is the full width at half maximum or tenth maximum (FWHM or FWTM) of the PSF or LSF. There are two contributions to the detector PSF that can be mathematically described as the convolution of the camera intrinsic PSF with the collimator PSF.

If the crystal is exposed to an infinitesimally narrow beam of photons, the image will be a count rate distribution almost Gaussian in shape. To separate two such line sources these must be separated with a minimal distance R, the resolution distance. The relationship is $R = 0.87 \times \text{FWHM}$ due to the contribution from the LSF tail. The finite width is caused by multiple Compton scattering of photons in the crystal and statistical variation in the production of electrons in the PMT cathode.

The position of the photon interaction point in the crystal will be the centre of gravity of the total emitted scintillation light. The higher the energy and the thicker the crystal the higher the probability for multiple Compton processes, causing deteriorated intrinsic spatial resolution.

The photoelectrons emitted from the photocathode in the PMTs suffer from statistical variations, and the number is proportional to $1/\sqrt{E}$, where E is equal to the energy deposition. Thus, the position resolution in the X- and Y-signals will vary with $1/\sqrt{E}$. To overcome some of that uncertainty, the electron efficiency in the photocathode should be as high as possible.

PMTs with as high a quantum efficiency, $Q(E)$, as possible at the photon energy E are used. The intrinsic resolution relationship then for different photon energies will be

$$R_i(E') = R_i(E)\sqrt{\frac{E}{E'}\frac{Q(E)}{Q(E')}} \qquad (6.4)$$

The intrinsic spatial resolution, R_i, in scintillation cameras has improved since their introduction 30 years ago. Today, R_i is of the order of 3 mm for 140 keV compared to about 12 mm in 1970. Computer simulations taking into account statistical phenomena for the scintillation light, including reflection–refraction–absorption and transmission, have shown that theoretically better than 2 mm will hardly be achievable.

Multiple Compton scatter in the crystal can be decreased using thinner crystals, improving the spatial resolution. However, the sensitivity of the camera is lowered.

The energy resolution, $\Delta E/E$, is proportional to $1/\sqrt{E}$ for the same reason. Energy resolution is however also dependent on a uniform light transmission to the cathodes. The value today is about 8–10% for 140 keV photon energy. The better the energy resolution the narrower the energy window to be used, giving fewer Compton-scattered photons registered in the window. Also dual-radionuclide imaging of nearby photon energies (e.g. ^{99m}Tc and ^{123}I) will be feasible [14].

6.9 PHOTON COLLIMATION

The collimation is a 'mechanical' sorting of the photons impinging on the crystal, preventing nonorthogonal photons from entering the crystal. The dimension of the collimator holes determines the sensitivity and the spatial resolution. The essential factor is the solid angle from the object through the holes that the crystal can see. The parallel hole collimator is the most used, with field of view matched to the useful field of view of the crystal. Focusing collimators such as the fan beam collimator are to some extent used for simple-photon emission computed tomography (SPECT). The pinhole collimator, also applicable for SPECT, can be used when small objects are to be imaged [15, 16]. Specially designed collimators have been made to overcome some collimator limitations [17].

Two parameters characterize collimators; geometric spatial resolution, R_c, and collimator efficiency, ε_c. Overall, both R_c and ε_c are dependent on the ratio between hole diameter and hole length (d/l). R_c varies with d/l and improves when the ratio decreases. ε_c is almost proportional to $(d/l)^2$ and decreases when d/l decreases. Good spatial resolution at large distances needs long small holes whereas high sensitivity requires short and wide holes.

For clinical studies, a compromise must be made due to the opposed dependence of these parameters. One limiting factor is the activity administered, keeping the absorbed dose low to the patient. This will not be the case for applications in radionuclide therapy, where the high photon fluence will make it possible to have specially designed collimators reducing the count rate from the camera below pileup and dead time effects, enabling a better spatial resolution.

Other important parameters are collimator scatter and septum penetration that may broaden the PSF, and thus decrease the spatial resolution and contrast [18, 19]. A rule of thumb is that septum penetration should be below 5%, thus resulting in a septum thickness t, given by

$$t > \frac{6d\mu^{-1}}{1 - 3\mu^{-1}} \qquad (6.5)$$

where μ is the linear attenuation coefficient of lead.

In the choice of collimator, one needs to consider the whole energy spectra for the radionuclide. In studies using ^{123}I it was shown that the best imaging characteristics were not achieved with the collimator optimal for ^{123}I's main photon energy of 159 keV, but a collimator designed for higher energies, due to the septum penetration of the low abundant photons at 440 and 529 keV [20–22].

6.10 SYSTEM SENSITIVITY AND SYSTEM RESOLUTION

The scintillation camera's system sensitivity, S, can be calculated from the collimator geometric sensitivity, ε_c, and the intrinsic efficiency dominated by the photo-peak efficiency of the crystal, f, according to

$$S = n\varepsilon_c\, f \tag{6.6}$$

where n is the number of emitted photons per disintegration. Sensitivities for parallel hole collimators for ^{99m}Tc are between 100 and 200 cps MBq^{-1}. The total spatial resolution for the scintillation camera, R_s, depends on the collimator geometric resolution R_c and the intrinsic spatial resolution R_i.

$$R_s = \sqrt{R_i^2 + R_c^2}. \tag{6.7}$$

The system resolution will also be distorted by Compton-scattered photons collected in the energy window, R_{sc}: then the total spatial resolution for the system will be

$$R_s = \sqrt{R_i^2 + R_c^2 + R_{sc}^2} \tag{6.8}$$

6.11 INTRINSIC LINEARITY

Intrinsic linearity is the ability of the detector (without collimator) to image radioactive line sources as straight lines in the image. Two types of measure are given, intrinsic spatial differential and absolute linearity: (i) differential linearity is the standard deviation of the deviation from that line in the image and (ii) the absolute linearity is the maximum displacement. Typical values of differential linearity for modern cameras are of the order of 0.2–0.5 mm.

6.12 INTRINSIC UNIFORMITY

The ability to image a homogeneous photon fluence rate impinging on the scintillation camera crystal is determined by the nonlinearity in the positioning and the sensitivity variations over the crystal. Position nonlinearity is an effect in the 'camera optics', where light guide, PMTs and electronics cause small shifts in the true coordinates of the event.

Sensitivity variations over the crystal can be due to crystal defects, PMT variations, optical coupling etc. One direct effect is variation in the pulse-height distribution, where the full energy peak will shift between different channels. Values calculated before any uniformity correction are of the order of 2–6% and 2–4% for integral and differential uniformity respectively.

6.13 SPECT

When rotating the scintillation camera for tomographic imaging (SPECT), factors to consider are the reconstruction algorithm, attenuation and scatter problems, mechanical stability and functional variations with rotation. One important parameter is the centre of rotation (COR) where the mechanical rotation axis, projected to the scintillation crystal face, should coincide with the centre of the image matrix. Also it is very important that the camera face is parallel to the rotation axis. The coordinate system for the image matrix must also be parallel to the rotation axis.

During rotation the functional parameters of the camera can change and magnify nonuniformity. Causes might be mechanical in coupling between crystal, light guide and PMTs due to gravitation; the gain from the PMTs can vary with the changes in the Earth's magnetic field (which mostly can be overcome with magnetic shielding around the tubes) and thermal changes within the camera head.

6.14 511 keV IMAGING

Anger in his earlier work proposed the use of coincidence techniques for scintillation cameras for imaging positron emitters [23]. However, the early development was towards dedicated ring systems nowadays used for positron emission tomography (PET) imaging. The subsequent developments of specific PET radiopharmaceuticals has renewed the interest in imaging them with less expensive scintillation cameras. One such radiopharmaceutical, [18]FDG, of special interest in neurology, cardiology and oncology, has promoted the scintillation camera technique towards 511 keV imaging [24]. The simplest approach is to equip the camera with a super-high-energy collimator. These collimators are very heavy, 100–200 kg, have thick septa, 2–4 mm, and long holes, 60–80 mm. Their sensitivity is about 50 cps MBq^{-1} for ^{18}F with a spatial resolution of 10–15 mm at 10 cm from the collimator face [25]. Both planar imaging and SPECT can be performed at 511 keV.

A possible technique to improve the use of a single energy window over the 511 keV photo-peak is to complement with an additional energy window around 320 keV, just below the energy for the Compton edge of 341 keV. The method increases the sensitivity and in fact also gives better spatial resolution. Also scatter correction can be implemented for quantification [26, 27].

Another approach is to use two anti-parallel uncollimated scintillation camera heads in coincidence mode registering both annihilation quanta. Then transverse tomography—limited angle tomography—is possible [28, 29]. When rotating the cameras, a complete tomographic data set is obtained, and transaxial images can be reconstructed [30, 31]. A hybrid camera is the PENN–PET consisting of large-area, position-sensitive NaI(Tl) scintillation crystals coupled to 30 PMTs in a hexagonal array. The spatial resolution for uncollimated cameras is 4–7 mm.

6.15 MONTE CARLO SIMULATIONS FOR SCINTILLATION CAMERAS

The scintillation camera has become the 'work horse' in most nuclear medicine imaging procedures and will be for a long time into the future. Many of the camera's parameters have been improved during the years. Many steps in the process, from the emission of the photon at the decay point in the patient until the image has been created, still need to be optimized.

From the above presentation of the characteristics for different parameters of the scintillation camera, many parameters are obviously measurable; others cannot be measured or depend on factors difficult to measure. Then the technique with Monte Carlo (MC) simulations becomes an important tool, where the photon is followed from the emission point until it is totally absorbed. When the absorption happens in the crystal, the creation of the image information can be investigated in detail. In this way multiple scattering of photons can be evaluated, attenuation properties investigated, crystal interaction points determined and so on.

The earliest MC studies of scintillation camera parameters were made by Anger who calculated intrinsic efficiency and intrinsic spatial resolution for NaI(Tl) crystals. Also, collimator influence on image creation was later studied by the MC technique. In one of our group's earlier studies on the pileup effect and mispositioning at high photon fluence rates, the MC method was used to study this problem, which is almost impossible to evaluate analytically [13]. Other good examples where the MC technique is an excellent tool are the optimization of collimators (Chapters 9, 10, 14), temporal resolution (Chapter 11) and scatter correction in SPECT (Chapter 12). Using the powerful tool of MC simulation with fast computers, one can expect more individual optimization of each investigation procedure, such as the development of patient-related dose planning in radionuclide therapy [32–34].

ACKNOWLEDGMENTS

The author would like to thank Ingemar Larsson and Kaj Jönsson, Lund University Hospital, for their valuable contributions and suggestions. Special thanks to Ingemar, recently retired, who guided the author into the field of scintillation camera imaging, acting as a real tutor. This work has been supported by grants from the Swedish Cancer Foundation (grant 2353–B95–09XAB), the Gunnar, Arvid and Elisabeth Nilsson Foundation, the Mrs Berta Kamprad Foundation and the John and Augusta Persson Foundation.

REFERENCES

[1] Anger H O 1957 A new instrument for mapping gamma-ray emitters *Biol. Med. Q. Rep. U. Cal. Res. Lab.* **38** 3653

[2] Anger H O 1958 Scintillation camera *Rev. Sci. Instrum.* **29** 27–33

[3] Segré E and Seaborg G T 1938 Nuclear isomerism in element 43 *Phys. Rev.* **54** 772

[4] 1996 Stationary gamma cameras: a product vomparison chart *Medical Electronics and Equipment News* **36** (Reilly)

[5] Ott R J, Flower M A, Babich J W and Marsden P K 1988 The physics of radioisotope imaging *The Physics of Medical Imaging* vol 6 ed S Webb (Bristol: Hilger) pp 142–318

[6] Rosenthal M S, Cullom J, Hawkins W, Moore S C, Tsui B M W and Yester M 1995 Quantitative SPECT Imaging: a review and recommendations by the focus committee of the Society of Nuclear Medicine Computer and Instrumentation Council *J. Nucl. Med.* **36** 1489–513

[7] Sorensson J A and Phelps M E 1987 *Physics in Nuclear Medicine* (Philadelphia, PA: Saunders)

[8] Birks J B 1964 *The Theory and Practice of Scintillation Counting* (Oxford: Academic)

[9] Hine G J 1967 *Instrumentation in Nuclear Medicine* (New York: Academic)

[10] Dahlbom M, MacDonald L R, Eriksson L, Paulus M, Andreaco M, Casey M E and Moyers C 1997 Performance of a YSO/LSO detector block for use in a PET/SPECT system *IEEE Trans. Nucl. Sci.* **44** 1114–9

[11] Ceberg C, Larsson I and Strand S-E 1991 A new method for quantification of image distortion due to pile-up in scintillation cameras *Eur. J. Nucl. Med.* **18** 959–63

[12] Strand S-E and Larsson I 1978 Image artifacts at high photon fluence rates in single-crystal NaI(Tl) scintillation cameras *J. Nucl. Med.* **19** 407–13

[13] Strand S-E and Lamm I-L 1980 Theoretical studies of image artifacts and counting losses for different photon fluence rates and pulse-height distributions in single-crystal NaI(Tl) scintillation cameras *J. Nucl. Med.* **21** 264–75

[14] Ivanovic M and Weber D A 1994 Feasibility of dual radionuclide brain imaging with ^{123}I and ^{99m}Tc *Med. Phys.* **21** 667–74

[15] Weber D A, Ivanovic M, Franceschi D, Strand S-E, Erlandsson K, Franceschi M, Atkins H L, Coderre J A and Ljunggren K 1994 Pinhole SPECT: an approach to *in vivo* high resolution SPECT imaging in small laboratory animals *J. Nucl. Med.* **35** 342–8

[16] Wanet P M, Sand A and Abramovici J 1996 Physical and clinical evaluation of high-resolution thyroid pinhole tomograhy *J. Nucl. Med.* **37** 2017–20

[17] Kimiaei S, Larsson S A and Jacobsson H 1996 Collimator design for improved spatial resolution in SPECT and planar scintigraphy *J. Nucl. Med.* **37** 1417–21

[18] Kibby P M 1969 The design of multichannel collimators for radioisotope cameras *Br. J. Radiol.* **42** 91–101

[19] deVries D J, Moore S C, Zimmerman R E, Mueller S P, Friedland B and Lanza R C 1990 Development and validation of a Monte Carlo simulation of photon transport in an Anger camera *IEEE Trans. Med. Imaging* **MI-9** 430–8

[20] Bolmsjö M, Strand S-E and Persson B R R 1977 Imaging ^{123}I with a scintillation camera: a study of detection performance and quality factor concepts *Phys. Med. Biol.* **22** 266–77

[21] Macey D J, DeNardo G L, DeNardo S J and Hines H H 1986 Comparison of low- and medium-collimators for SPECT imaging with ^{123}I-labelled antibodies *J. Nucl. Med.* **27** 1467–75

[22] DeGeeter F, Franken P R, Defrise M, Andries H, Saelens E and Bossuyt A 1996 Optimal collimator choice for sequential ^{123}I and ^{99m}Tc imaging *Eur. J. Nucl. Med.* **23** 768–74

[23] Anger H O 1959 Scintillation and positron camera *U. Cal. Res. Lab.* 9640

[24] Kalff V, Berlangier U, Every B V, Rowe J L, Lambrecht R M, Tochon-Dnaguy H J, Egan G F, McKay W J and Kelly M J 1995 Is planar thallium-201/fluorine-18 fluorodeoxyglucose imaging a reasonable clinical alternative to positron emission tomographic myocardial viability scanning? *Eur. J. Nucl. Med.* **22** 625–32

[25] Leichner P K, Morgan H T, Holdman K P, Harrison K A, Valentino F, Lexa R, Kelly R F, Hawkins W G and Dalrymple G V 1995 SPECT imaging of fluorine-18 *J. Nucl. Med.* **36** 1472–5

[26] Ljungberg M, Ohlsson T, Sandell A and Strand S 1996 Scintillation camera imaging of positron-emitting radionuclides in the Compton region *Conf. Records IEEE Medical Imaging Conf. (San Francisco, CA, 1995)* vol 2 (Piscataway, NJ: IEEE) pp 977–81

[27] Ljungberg M, Danfelter M, Strand S, King M A and Brill B A 1997 Scatter correction in scintillation camera imaging of positron-emitting radionuclides *Conf. Records IEEE Medical Imaging Conf. (Anaheim, CA, 1996)* vol 3 (Piscataway, NJ: IEEEE) pp 1532–6

[28] Muehllehner G, Buchin M P and Dudek J H 1976 Performance parameters of a positron imaging camera *IEEE Trans. Nucl. Sci.* **NS-23** 528–37

[29] Muehllehner G, Atkins F B and Harper P V 1977 Positron camera with longitudinal and transverse tomographic capability *Medical Radionuclide Imaging* (Vienna: IAEA) pp 291–307

[30] Paans A M J, Vaalburg W and Woldering M G 1985 A rotating double-headed positron camera *J. Nucl. Med.* **26** 1466–71

[31] Sandell A, Ohlsson T, Erlandsson K, Hellborg R and Strand S-E 1992 A PET system based on 2-18FDG production with a low energy electrostatic proton accelerator and a dual headed PET scanner *Acta Oncol.* **31** 771–6

[32] Erdi A K, Erdi Y E, Yorke E D and Wessels B W 1966 Treatment planning for radio-immunotherapy *Phys. Med. Biol.* **41** 2009–26

[33] Tagesson M, Ljungberg M and Strand S-E 1994 Transformation of activity distribution in quantitative SPECT to absorbed dose distribution in a radionuclide treatment planning system *J. Nucl. Med.* **35** 123P (abstract)

[34] Kolbert S K, Sgouros G, Scott A M, Bronstein J E, Malane R A, Zhang J, Kalaigian H, McNamara S, Schwartz L and Larson S M 1997 Implementation and evaluation of patient-specific three-dimensional internal dosimetry *J. Nucl. Med.* **38** 301–8

CHAPTER 7

THE SIMSET PROGRAM

Tom K Lewellen, Robert L Harrison and Steven Vannoy

7.1 INTRODUCTION

Our laboratory is developing a software simulation system for emission tomography (SimSET) using modern software engineering techniques [1, 2]. The code is designed for both single-photon emitters (single-photon emission computed tomography (SPECT) and planar systems) and positron emitters (positron emission tomography (PET) systems). As modules in the system are completed we are placing them in the public domain. The package in release as of February 1996 includes the photon history generator (PHG), the object editor, a collimator module and detection and binning modules [3, 4].

The PHG tracks photons through the tomograph field of view (FOV), creating a photon history list containing the photons that reach the tomograph surface within user-specified limits. The object editor helps the user define voxelized activity and attenuation objects for the PHG. The current collimator module includes Monte Carlo simulation of PET collimators based on PETSIM [5] and simulation of SPECT collimators based on geometric transfer functions [6, 7]. The detector and binning modules allow for Gaussian blurring of the energy, and binning by any combination of number of scatters, axial position, transaxial distance and angle and photon energy. Many extensions and additions to these modules are planned over the coming years, including Monte Carlo simulation of detectors, Monte Carlo simulation of SPECT collimators, improved importance sampling and better modelling of coherent scatter, positron range and other physical effects.

The code development was undertaken to accomplish several goals: (i) to produce a software system that was highly modular, portable and efficient;

(ii) to optimize the code for simulating heterogeneous objects in three dimensions and (iii) to produce an easy-to-use code without unnecessary complexities. Early in the development process, many of the existing Monte Carlo codes were carefully investigated (e.g. EGS4, GEANT and MCNP) and a survey was made of many potential Monte Carlo users at laboratories throughout the USA. As a result, the programming language selected was C, reflecting the popularity and support of that language on modern workstations and desktop computers. Instead of the complex geometric object descriptions used in the larger Monte Carlo implementations, SimSET utilizes simple geometric and voxelized representations of the object. The SimSET package does implement techniques to improve efficiency, as discussed briefly below, but does not include many options in the more complex packages that are often not relevant to nuclear medicine applications (e.g. pair production).

The distribution of the code is via the Internet. The package includes all of the source codes, manuals in Postscript and Microsoft Word format files, installation scripts and test data sets for verifying successful installation of the code. Installation scripts that create directories, compile code etc are included for Data General AViiON, DEC Alpha, DEC VAX, IBM R6000, SGI, SUN, HP and generic Unix systems. The code has also been successfully installed and operated on a variety of Macintosh computer systems. To gain access to the code, and to be added to a list of users for update notices, interested individuals should send electronic mail to simset@u. washington.edu.

7.2 SOFTWARE DEVELOPMENT

To make the SimSET modules easy to maintain, extend and transport, we developed them using formal software engineering techniques. We split our development into five stages: analysis, design, implementation, validation and maintenance. In the analysis stage we elicited requirements from likely users of the software both within our own group and from groups doing similar work at other institutions. We reduced the users' input to a series of functional requirements and developed data flow diagrams from them. The design process was started by defining functional modules which would be needed for the particular software task. We then developed control flow diagrams of the modules and pseudo-English definitions for each of the functions.

The careful analysis and design focused the effort of the scientific research staff on the functional level, leaving the implementation to be carried out by our programming staff. This division facilitated our goal of producing functionally hierarchical software. The base of the hierarchy is a layer of functionally cohesive routines for both general purpose and simulation-specific tasks. All operating system dependences are restricted to these low-level functions.

Our validation strategy is composed of functional testing at each layer of the hierarchy, comparison of simulation results with analytic predictions, statistical analysis, structured code inspection, and independent beta testing. A modular approach to development allows the process from design to source code to occur in independent steps. For example, one step in the photon transport process is the calculation of a scatter angle. Starting with the Kahn algorithm [8] for sampling from the Klein–Nishina distribution, our programmers wrote a module to compute scatter angle for modelling Compton scatter. The code was inspected line by line with a member of the scientific staff. An independent test shell was created to exercise the scatter module. The resulting distribution of scatter angles was compared to an analytically derived scatter angle distribution. Finally, when enough modules were completed to perform a simple simulation, the results from geometri-cally simple situations were compared with analytic predictions.

Validation of importance sampling techniques provided a significant chal-lenge. These techniques add complexity to the source code, exacerbating the difficulty of validating both the algorithm and its implementation. We attacked this problem from two angles; first we developed a statistical test for validating that there was no bias introduced by the use of importance sampling, and then we developed a series of simulations that would highlight potential sources of bias.

Our statistical test method was implemented using Microsoft Excel spread-sheet software. Whenever possible we have used existing third-party applica-tions for validation. This reduces concern about implementation errors and correlation between code that creates the data and code that analyses it. We performed each test simulation with all combinations of importance sam-pling techniques turned on and off. The results were analysed using Excel spreadsheets and reviewed by the scientific staff.

The final stage of validation consists of independent beta testing by investi-gators at other institutions. No amount of in-house testing can substitute for the rigours of this process. In the beta test of our first version of the PHG, eleven 'bugs' were discovered in 11 months of testing. These ranged from innocuous to data corrupting in some circumstances. Not surprisingly, the majority of the bugs occurred in the most recently developed portions of the code, and in areas where new features were added during the testing process. No further bugs have been discovered in the original software. In the current release, available since February 1996, only two bugs have been discovered, both in the newest sections of the software.

SimSET is an essential tool for many researchers. As the needs of our users evolve, the capabilities of SimSET are being expanded. The care taken in design and implementation has minimized the effort necessary to maintain

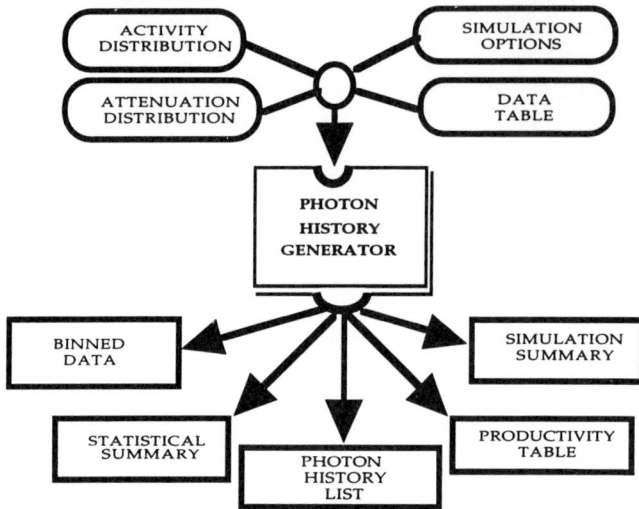

Figure 7.1   *The input/output data flow of the PHG module.*

and extend the software. This effort also eases the adaptation of SimSET to research at other institutions.

7.3 PHOTON HISTORY GENERATOR

The photon history generator (PHG) is the module that generates and tracks photons within the FOV of the tomograph being simulated. The attenuation and isotope distributions and overall simulation instructions are read by the PHG module at run time. Figure 7.1 illustrates the input/output of the PHG module.

The object editor is used to define the attenuation and isotope distributions used by the PHG. To allow for maximum flexibility, the attenuation and activity distributions are modelled as voxelized objects in three dimensions (see figure 7.2). Voxel dimensions can vary in each slice so that an object can be finely described in the slice(s) of interest, yet only requires a coarse description of the rest of the object. Arbitrary heterogeneous attenuation and activity distributions can be simulated. The object editor provides the choice of initializing each slice with a constant value or from voxel-based physiologically realistic phantoms, e.g. the digitized anthropomorphic phantom of Zubal *et al* [9, 10]. After initializing the object with constants or from an anthropomorphic phantom, the users may add geometric elements (e.g. cylinders, ellipsoids) to the object. Tables are used to translate the numbers in the voxelized phantom into actual activity or attenuation values

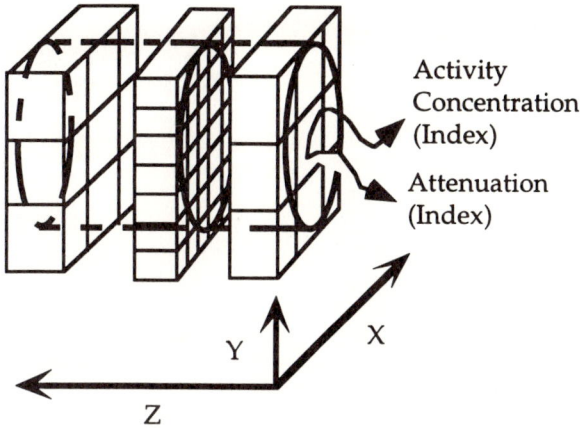

Figure 7.2 *The object to be imaged is defined as a series of contiguous slices (the gaps shown here are only an aid to visualization). The slices can have different thicknesses and resolutions. Each voxel contains indices to a table of activity concentrations and a table of attenuation properties. Voxels outside the largest cylinder contained within the bounding rectangle are automatically given zero activity and attenuation.*

so that the users may change their distributions without changing values in the digitized phantoms (see figure 7.3). When the PHG processes the object files, it produces three-dimensional integer arrays defining the attenuation and isotope distributions. These may be viewed using most image display

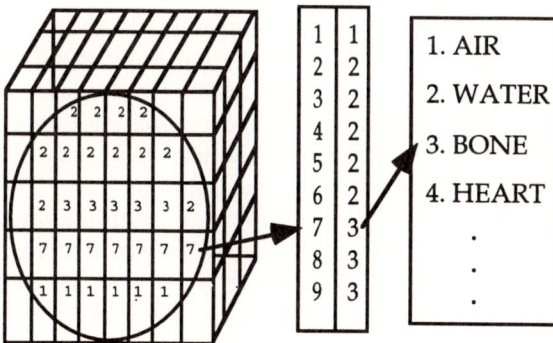

Figure 7.3 *An attenuation index in the voxelized object does not refer directly to the type of attenuation material in the attenuation table, but rather to an attenuation translation table. This allows digitized phantoms with different indexing schemes to be used without altering the attenuation table. The activity indices are handled in the same manner.*

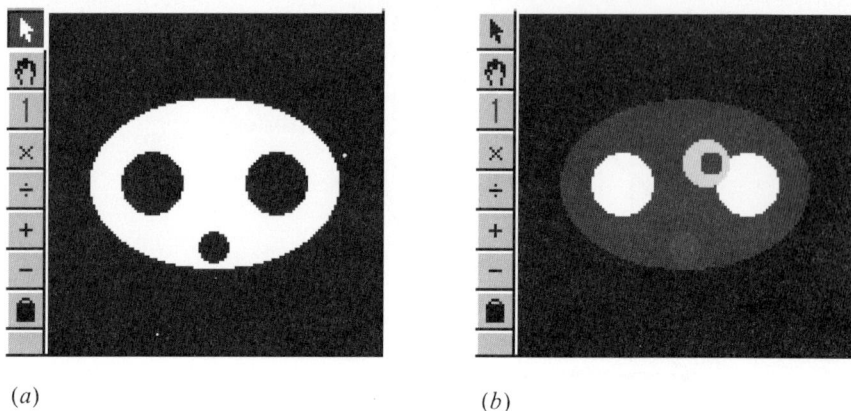

(a)   (b)

Figure 7.4 *Examples of viewing an attenuation object (a) and an emission object (b) produced by the PHG module using a display program on a Macintosh computer. These images are from the PHG tutorial, included in the SimSET software package.*

programs. For example, figure 7.4 depicts an attenuation and emission object used for a SPECT simulation used in the PHG tutorial (part of the SimSET distribution package).

Figure 7.5 illustrates the basic photon tracking algorithm used in the PHG. Each photon is tracked from its starting location to the surface of the 'object cylinder' (a cylinder specified by the user that encloses the object). From there, the photon is projected to a bounding cylinder referred to as the 'target cylinder'. The radius and axial extent of the target cylinder are user specified; they will usually be determined by the inner radius and axial dimensions of the detection/collimation system to be simulated. The space between the object cylinder and target cylinder is modelled as a vacuum. All photons that reach the target cylinder within a user-specified axial 'acceptance angle' will be included in the photon history list. Figure 7.5 includes a 'collimate photon' and a 'detect photon' step. The PHG offers the user the option to (i) save each photon reaching the target cylinder in the history file or (ii) collimate, detect and bin the data event by event during the simulation. If a history file is created, it can later be run through the collimation, detection and binning modules. These modules are discussed below.

A simplified conceptual model of the PHG tracking algorithm is presented below.

(i) Sample an initial direction vector for the photon.
(ii) Sample the number of free paths to travel before an interaction occurs.

Tracking Photons

```
                           ┌─────────────────────┐
                    ┌──────│   Generate Decay    │
  SPECT             │      └─────────────────────┘
                    │                 ↓
                    │        ┌──────────────────┐
                    │        │ Produce Photon(s)│
                    │        └──────────────────┘
                    │                 ↓
                    │   ┌─────────────────────────────┐
           PET      │   │  Project to Target Cylinder │
        ──────────  │   │   "Initial Forced Detection"│
         ◇──────────┘   └─────────────────────────────┘
         2nd Photon              ↓
            ┌──────────────────────────────────────────────┐
            │ Force Scatter & Project: "Forced Detection"   │←──
            └──────────────────────────────────────────────┘   │
                              ↓                                 │
            ┌──────────────────────────────────────┐            │
            │   Track Until Interaction or Escape   │           │
            └──────────────────────────────────────┘            │
```

Escaped &   Escaped &
Detected   Discarded

Update Statistics

Collimate Photon

Detect Photon

Compute
Scatter Angle

Interact

Bin Photon

Store In History

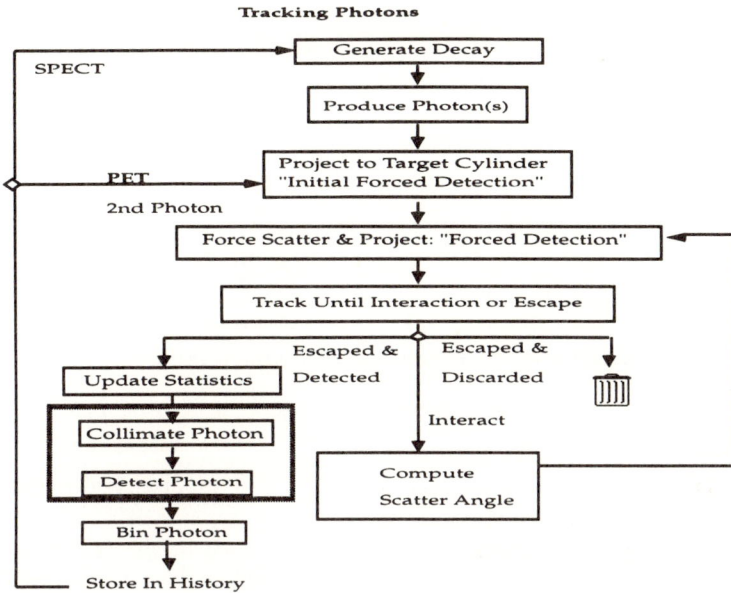

Figure 7.5 *The basic photon tracking logic in the PHG module.*

(iii) Calculate the travel distance corresponding to the selected free paths based on the user-supplied attenuation distribution.

(iv) If the photon interacts before reaching the object cylinder, choose a scatter angle and corresponding energy value, and continue tracking with step (ii). Otherwise, if the photon reaches the object cylinder, project it to the target cylinder. If it hits the target cylinder within the specified axial limits and acceptance angle add it to the photon history list or perform the user-specified collimator, detector and binning operations.

The main factors that determine the computational efficiency of a simulation using the PHG are the geometry of the tomograph being simulated, the importance sampling options used and the voxelization of the attenuation object. The portion of the object being imaged that falls within the tomograph FOV and the solid angle subtended by the tomograph limit the efficiency of conventional simulations. Events occurring or scattering outside the FOV are simulated because of the possibility they will subsequently scatter into the FOV. Similarly events within the FOV that start or scatter in directions that cannot be detected are tracked.

The computational cost of these obstacles can be reduced using importance sampling techniques. The techniques we have adapted include (i) forced detection, (ii) stratification and (iii) weight windows [8, 11–13]. The details

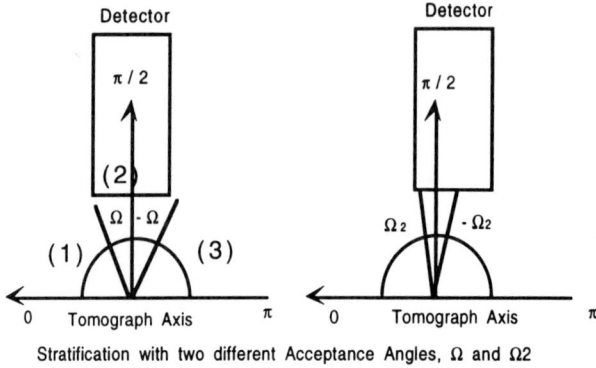

Stratification with two different Acceptance Angles, Ω and Ω2

Figure 7.6   *An example of stratification geometry used by the PHG.*

of our approach have been published [14, 15], and are discussed in Chapter 2, 'Variance Reduction and Stratification'.

The target cylinder represents the innermost face of the collimator for PET, or the cylinder defined by the rotation of the collimator for SPECT. For an unscattered photon to be detected the ray defined by its starting location and direction must intersect the target cylinder within the tomograph or collimator's acceptance angle (though this does not guarantee the photon will be detectable). Similar conditions for detectability can be set at a scattered photon's last interaction point. To avoid wasting computer time simulating events that will not be detected, the simulated decay locations and initial photon direction cosines are stratified by object slice and axial direction cosine: more photons are simulated in slices and angles that are likely to produce detections. The PHG uses a 'productivity table' to tell where the greatest number of 'starts' should take place. The productivity table stratifies the possible slices and axial angles and assigns 'productivities' for each slice/angle bin. Different productivities are computed for scattered and unscattered events. Figure 7.6 illustrates the stratification of the emission angles.

For a given acceptance angle Ω, stratification divides the potential starting angles into three areas: (i) from zero to $\pi/2-\Omega$, (ii) from $\pi/2-\Omega$ to $\pi/2+\Omega$ and (iii) from $\pi/2+\Omega$ to π. We then subdivide these areas into potential starting angles. The nonproductive areas (i) and (iii) are subdivided into six angles each. The productive area, (ii), is divided into 12 potential angle ranges. A short simulation with stratification disabled is used to compute the productivity of each of these ranges.

The principle of forced detection is illustrated in figure 7.7. A photon that would otherwise not 'hit' the target cylinder within the acceptance angle is forced to scatter in such a way that it will. The technique is implemented in the PHG with the use of pre-computed probabilities that a photon with a

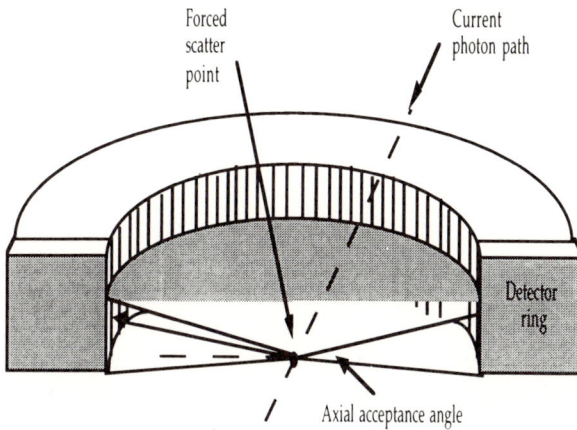

Figure 7.7 *The basic concept of forced detection—force a photon to scatter and hit the target cylinder within the acceptance angle.*

specific direction and angle will scatter into a selectable range of outgoing angles. In particular, the forced detection tables convert from the coordinates used by the Klein–Nishina formula (incoming energy and scatter angle) to the coordinates needed to implement forced detection (incoming energy, outgoing z-direction cosine and sine of outgoing azimuthal angle).

These techniques (stratification and forced detection) require that each photon be assigned a starting weight that indicates how many photons it represents in the real world. The weight is adjusted at every importance sampling step. A list of N detected events with varying weights is worth less (i.e. statistics computed from them will have a greater variance) than a list of N uniformly weighted detected events by a multiplicative factor between zero and unity. This is called the quality factor (QF) and is used in the calculation of the computational figure of merit (CFOM) [15]

$$\text{CFOM} = (N \times \text{QF})/(\text{CPU seconds}) \qquad (7.1)$$

For conventional simulations CFOM gives the number of events produced per central processing unit (CPU) second. For simulations using importance sampling it gives a similar figure adjusted by the quality factor. In general, the CFOM improves significantly with the use of stratification and forced detection.

Increasing the number of voxels in the attenuation object usually increases the time spent tracking each photon: every time a photon crosses into a new voxel the voxel attenuation value must be looked up and the distance and free-path lengths to the next voxel calculated. An exception to this occurs when using stratification: increasing the number of slices may actually improve the efficiency of the stratification and thus of the simulation as a

Table 7.1 *Efficiency of the PHG for several simulations. All simulations were performed on a 25 MIPS computer.*

Tomograph	Voxelization	CFOM without IS (events/CPU s)	CFOM with IS (events/CPU s)
SPECT	$1 \times 1 \times 1$	82	213
SPECT	$1 \times 128 \times 128$	48	108
SPECT	$5 \times 128 \times 128$	48	92
PET	$1 \times 1 \times 1$	58	83
PET	$1 \times 128 \times 128$	29	85
PET	$5 \times 128 \times 128$	28	102

whole. However, once the voxelization has been chosen, the heterogeneity of activity and attenuation within the voxelized object has no effect on computation time.

Efficiency results for a cylindrical phantom with three different voxelizations and in two different imaging situations are given above both with and without importance sampling (IS) (table 7.1). Photon history lists were prepared to give incident photon fluxes for two tomographs: a typical SPECT tomograph (denoted SPECT) and a volume PET tomograph (denoted PET). The phantom simulated is a circular cylinder of water with uniform activity 20 cm in diameter and 20 cm in length. It is simulated using one voxel ($1 \times 1 \times 1$—the activity and attenuation are truncated to the object cylinder), using one 20 cm thick slice with 128×128 voxels ($1 \times 128 \times 128$), and using five 4 cm thick slices each with 128×128 voxels ($5 \times 128 \times 128$).

The IS results used stratification and forced detection. These results are from the first release version of the software. Previously reported results [15], which were from prototype software, showed a larger efficiency improvement when using IS. We are currently investigating whether similar speedups can be achieved without sacrificing the maintainability and extensibility of the software.

7.4 THE BINNING MODULE

The SimSET software automatically reports a variety of statistics on the output photons to provide a general description of the simulation results. Most users will want to further analyse the data in ways specific to their research goals. The binning module can be used to create multidimensional histograms of statistics. Binning can be performed using data from the PHG, the collimator module or the detector module, either on an event-by-event basis while the other modules are running or afterwards from a photon history list. The user controls the binning using the binning parameters file. Histograms can be formed over any combination of photons' axial position,

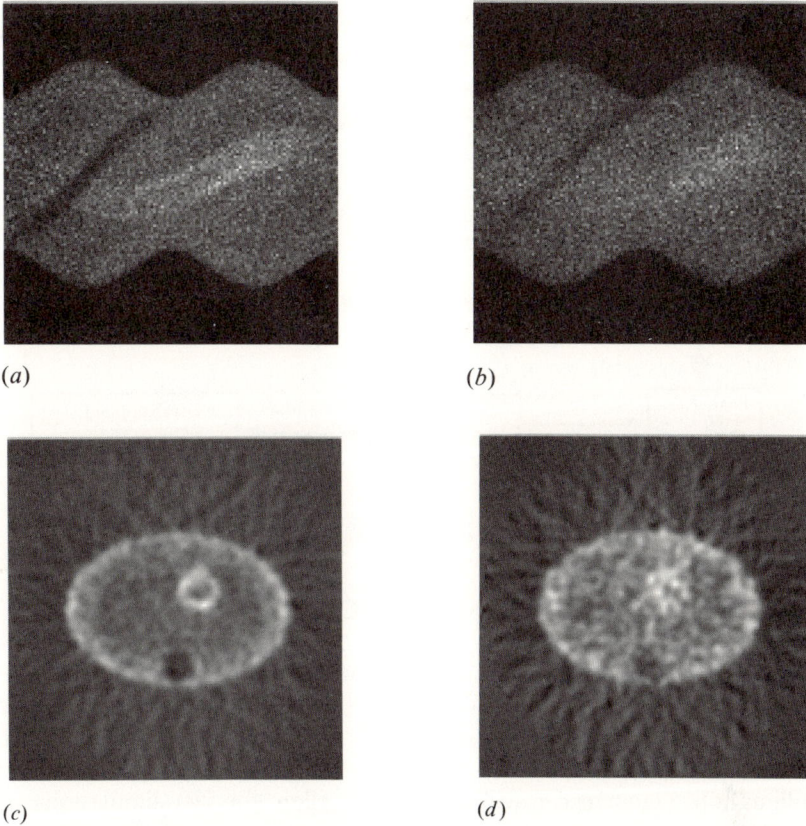

(a)

(b)

(c)

(d)

Figure 7.8 *Examples of binned sinograms for unscattered (a) and scattered (b) photons in a simulation study. (c) A reconstruction of the unscattered photons; (d) a reconstruction of the scattered photons using the Donner reconstruction library. These images are from the PHG tutorial, included in the SimSET software package.*

transaxial distance and angle, number of scatters and final energy. The ranges and number of bins for each dimension are set in the binning parameters file. Examples of a set of sinograms produced by the binning module and the resulting reconstructed images (using the Donner reconstruction library [16]) are shown in figure 7.8.

Figure 7.9 illustrates the easiest method for users to add their own binning options to the PHG modules. The structure of the code supports the incorporation of user modules that conform to the data structures as defined in the PHG manuals.

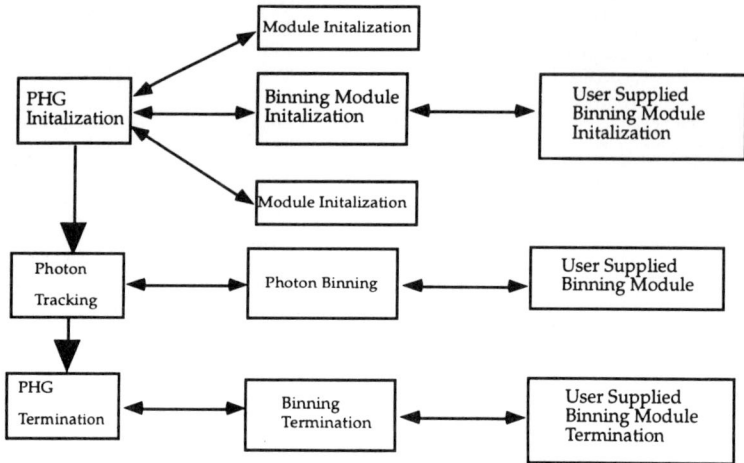

Figure 7.9   *PHG data flow and interaction with user-written modules.*

7.5 THE COLLIMATOR AND DETECTOR MODULES

Both SPECT and PET collimator simulations are provided with SimSET. For SPECT we have implemented, in collaboration with researchers from the University of North Carolina, a collimator simulation using geometric transfer functions [6]. There are options for parallel, fan and cone beam collimation. Geometric transfer functions allow for fast simulations with realistic geometric effects, but do not address issues such as penetration, scatter in the collimator and lead x-rays. A photon tracking version of the collimator, discussed below under 'future development', will be required to study such effects.

For PET collimator simulations we have adapted the model used in PETSIM of Thompson *et al* [5]. This model allows for multiple layers of septa, both axially and radially, and for radially tapered septa (figure 7.10). Different materials can be used for different parts of the collimator. Photons are projected from the target cylinder (the cylindrical surface to which the PHG projects photons) to the innermost surface of the collimator. From there the photons are tracked until they either (i) exit the outermost radial boundary of the collimator and are saved to the photon history list or passed to the detector/binning module for further processing, (ii) exit the innermost radial, topmost axial or bottommost axial boundaries and are discarded or (iii) fall below a user-specified minimum energy and are discarded.

The current SimSET detector module only performs a Gaussian energy blur on each photon. However, photon tracking detector simulations are an

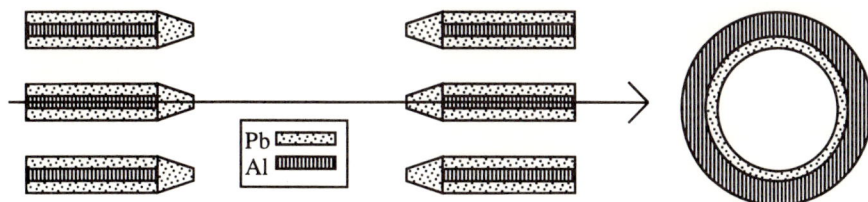

Figure 7.10   *Axial and radial cross sections through a PET collimator with multiple axial and radial layers. The innermost septum is tapered.*

active area of development and are discussed under 'future development' below.

7.6 UTILITIES

A variety of utility tools and special options are also provided for the user. In addition to the many error-checking routines in the software, the user can also turn on a 'debug' mode where many additional files are created to capture critical information about the simulation process as it occurs. This mode is very useful in debugging new binning code added by the user. Additional utilities are included for a variety of data file conversions to assist in moving files from different processors (e.g. from a DEC Alpha to a HP 700 series workstation) as well as preparing files for processing by other programs (e.g. a display program to view sinograms or binned planar images). Other options include the ability to run multiple simulations and add the results together, with the proper adjustment of the photon weights—particularly useful for simulating dynamic scans.

Several programs and data sets that complement or extend the functionality of the SimSET software can be downloaded from the SimSET ftp site. These include Michael Ljungberg's SIMIND, a simulation of SPECT imaging [17], the anthropomorphic digitized phantom of Zubal *et al* [9, 10] and a digitized version of the 3D Hoffman brain phantom [18], which can be used in conjunction with SimSET to simulate realistic clinical imaging situations; and the version of DETECT extended by Levin and Moisan [19], a simulation of scintillation light in detector crystals.

7.7 FUTURE DEVELOPMENT

We are currently preparing a new version of the SimSET software for release in the summer of 1997. We are adding significant new capabilities to the detector and binning modules, adding support for simulations of coincidence imaging using dual-headed gamma cameras, increasing the number of physical effects modelled by the photon tracking algorithms and making several modifications to simplify use.

Photon scatter in detectors significantly affects both spatial and energy resolution in modern tomographs. To facilitate the study of these effects, our next release will include photon tracking detector modules for SPECT and PET. The SPECT detector module will model conventional planar gamma cameras. It will allow for multiple layer detectors and graded absorbers (e.g. to reduce low-energy photon counts [20]). As each photon is tracked through the detector, a list of its interaction locations and the energy deposited at each location will be maintained. The total deposited energy and energy-weighted centroid of the interactions will be computed and may be stored in a photon history list or passed directly to the binning module.

Two PET detector modules will be included in the next release. One will use two planar SPECT detectors, as described in the paragraph above, 180° apart to simulate coincidence imaging with dual-headed gamma cameras [21, 22]. The other module will model the PET detection system as a cylindrical crystal. As above, a list of each photon's interaction locations and the energy deposited at each location will be maintained. This list will be used to compute the total deposited energy and energy-weighted centroid of the interactions.

Two major changes are being made to the binning module. Previous versions of SimSET have used a fixed order of precedence to determine which binning index will vary the most quickly in the output data. The new software will allow the user to set the order of precedence for each simulation. The new version will also include new binning parameters for 3D PET reconstruction algorithms.

The new release will include models for positron range, annihilation photon noncollinearity and coherent scatter. Positron range and annihilation photon noncollinearity are becoming important factors limiting the resolution of some new tomograph designs. To track positrons from decay to annihilation would require a great deal of computer time. Therefore we will model positron range by randomly selecting a distance from the in-water range distribution for the appropriate isotope [23]. A random direction from the decay location will be chosen, and the selected positron range will be adjusted to account for the electron density of materials in the object in that direction. Annihilation photon noncollinearity will be randomly selected based on [24].

Coherent scatter can have a significant impact on the distribution of scatter in emission tomography [25, 26]. Our first model will use the independent-atom approximation in water to define the coherent scatter distribution. However our implementation will allow for different materials and interference effects to be modelled by simply modifying a table to give the appropriate distribution [27].

We are continuing to work on the PHG to improve its ease of use and documentation. The next release will include an expanded header, more informative screen output and a series of new utilities to help users manipulate multidimensional data, perform attenuation corrections and statistically analyse simulation data. We are also revamping the manual, tutorials and sample runs provided with SimSET.

ACKNOWLEDGMENT

Work by the author presented in this chapter was supported by PHS grant CA42593.

REFERENCES

[1] Lewellen T K, Anson C P, Haynor D R, Harrison R L, Bice A N, Schubert S F, Miyaoka R S, Gillispie S B and Zhu J B 1988 Design of a simulation system for emission tomographs *J. Nucl. Med.* **29** 871

[2] Anson C P, Harrison R L, Lewellen T K, Gillispie S B, Pollard K P, Bice A N, Miyaoka R S, Haynor D H and Zhu J B 1989 On using formal software engineering techniques in an academic environment *IEEE Eng. Med. Biol. Soc.* **11** 2011–12

[3] Harrison R L, Haynor D R, Gillispie S B, Vannoy S D, Kaplan M S and Lewellen T K 1993 A public-domain simulation system for emission tomography: photon tracking through heterogeneous attenuation using importance sampling *J. Nucl. Med.* **34** 60P

[4] Harrison R L, Vannoy S D, Haynor D R, Gillispie S B, Kaplan M S and Lewellen T K 1993 Preliminary experience with the photon history generator module of a public-domain simulation system for emission tomography *IEEE NSS-MIC Conf. Record* **2** 1154–8

[5] Thompson C J, Moreno C J and Picard Y 1992 PETSIM: Monte Carlo simulation of all sensitivity and resolution parameters of cylindrical positron imaging systems *Phys. Med. Biol.* **37** 731–49

[6] Tsui B M W and Gullberg G T 1990 The geometric transfer function for cone and fan beam collimators *Phys. Med. Biol.* **35** 81–93

[7] Metz C, Atkins F and Beck R 1980 The geometric transfer function component for scintillation camera collimators with straight parallel holes *Phys. Med. Biol.* **25** 1059–70

[8] Kahn H 1956 *Use of Different Monte Carlo Sampling Techniques* (New York: Wiley) pp 146–90

[9] Zubal I, Gindi G, Smith E, Harrell C and Hoffer P 1990 High resolution anthropomorphic phantom for Monte Carlo simulation of nuclear medicine images *J. Nucl. Med.* **31** 729

[10] Zubal I and Harrell C 1994 Computerized three-dimensional segmented human anatomy *Med. Phys.* **21** 299–302

[11] Booth T E and Hendricks J S 1984 Importance estimation in forward Monte Carlo calculations *Nucl. Tech. Fus.* **5** 90–100

[12] Brown R S and Hendricks J S 1987 Implementation of stratified sampling for Monte Carlo applications *Nucl. Sci. Eng.* **97** 245–8

[13] Cramer S N, Gonnord J and Hendricks J S 1986 Monte Carlo techniques for analyzing deep-penetration problems *Nucl. Sci. Eng.* **92** 280–8

[14] Haynor D R, Harrison R L, Lewellen T K, Bice A N, Anson C P, Gillispie S B, Miyaoka R S, Pollard K R and Zhu J B 1990 Improving the efficiency of emission tomography simulations using variance reduction techniques *IEEE Trans. Nucl. Sci.* **NS-37** 749–53

[15] Haynor D R, Harrison R L and Lewellen T K 1991 The use of importance sampling techniques to improve the efficiency of photon tracking in emission tomography simulations *Med. Phys.* **18** 990–1001

[16] Huesman R, Gullberg G, Greenberg W and Budinger T 1977 *Users Manual: Donner Algorithms for Reconstruction Tomography* (Berkeley, CA: Lawrence Berkeley Laboratory)

[17] Ljungberg M and Strand S-E 1989 A Monte Carlo program for the simulation of scintillation camera characteristics *Comput. Methods Biomed.* **29** 257–72

[18] Hoffman E J, Cutler P D, Digby W M and Mazziota J C 1990 3D phantom to simulate cerebral blood flow and metabolic images for PET *IEEE Trans. Nucl. Sci.* **NS-37** 616–20

[19] Levin A and Moisan C 1996 A more physical approach to model the surface treatment of scintillation counters and its implementation into DETECT *Proc. Nucl. Science Symp. Medical Imaging Conf. (Anaheim, CA)* (Piscataway, NJ: IEEE) pp 702–6

[20] Muehllehner G, Jaszczak R and Beck R 1974 The reduction of coincidence loss in radionuclide imaging cameras through the use of composite filters *Phys. Med. Biol.* **19** 504–10

[21] Nellemann P, Hines H, Braymer W, Muehllehner G and Geagan M 1995 Performance characteristics of a dual head SPECT scanner with PET capability *Proc. Nucl. Science Symp. Medical Imaging Conf. (San Francisco, CA)* (Piscataway, NJ: IEEE) pp 1751–5

[22] Miyaoka R, Costa W, Lewellen T, Kohlmyer S, Kaplan M, Jansen F and Stearns C 1996 Coincidence imaging using a standard dual headed gamma camera *Proc. Nucl. Science Symp. Medical Imaging Conf. (Anaheim, CA)* (Piscataway, NJ: IEEE) pp 1127–9

[23] Derenzo S E 1986 Mathematical removal of positron range blurring in high resolution tomography *IEEE Trans. Nucl. Sci.* **NS-33** 565–9

[24] Derenzo S E, Budinger T F and Huesman R H 1985 *Detectors for High Resolution Dynamic Positron Emission Tomography* (New York: Raven) pp 21–31

[25] Johns P C and Yaffe M J 1983 Coherent scatter in diagnostic radiology *Med. Phys.* **10** 40–50

[26] Leliveld C J, Maas J G, Bom V R and Eijk C W E 1996 Monte Carlo modeling of coherent scatter: influence of interference *IEEE Trans. Nucl. Sci.* **NS-43** 3315–21

[27] Brookhaven National Laboratory (USA Department of Energy) 1983 National Nuclear Data Center, online data service

CHAPTER 8

VECTORIZED MONTE CARLO CODE FOR MODELLING PHOTON TRANSPORT IN NUCLEAR MEDICINE

Mark F Smith

8.1 INTRODUCTION

Monte Carlo modelling of photon transport and detection is an important tool in nuclear medicine. It has been used to study gamma camera and detector design for planar and single-photon emission computed tomography (SPECT) imaging and to simulate projection data acquisition to test quantitative SPECT reconstruction algorithms. For example, simulations have been used to model the resolution and detection efficiency of sodium iodide scintillators for gamma cameras [1], to design detector modules for a cylindrical SPECT detector [2], to model collimator transfer functions [3], to study photon scatter within a collimator [4] and to simulate gamma ray detection by a surgical probe [5]. Projection data acquisition using parallel beam [6], cone beam [7] and rotating slit [8] collimators have been simulated and studies of photon scatter [9–12] and how to build system matrices for matrix-based image reconstruction [13–15] have been performed using Monte Carlo methods. Codes for nuclear medicine application have become more sophisticated and the current generation of codes is capable of modelling photon transport in heterogeneous media [16–21]. The applications in nuclear medicine have been extensive and the preceding references are representative rather than exhaustive. Other applications may be found in the review papers by Raeside [22] and Andreo [23] and in the other chapters of this book.

Simulations of nuclear medicine applications are computationally demanding, even with variance reduction techniques such as forced detection [6] and stratified sampling [24] to increase the effective speed of the modelling. An additional strategy to increase simulation speed is to structure the codes to take advantage of vector processors possessed by some computers. The purpose of this chapter is to explain how this is accomplished and to describe the characteristics, performance and applications of a vectorizable Monte Carlo code to model photon transport in a uniform cylinder and a more general vectorizable code to model photon transport in heterogeneous media.

8.2 MONTE CARLO CODE STRUCTURES

Codes simulating photon transport are usually written using a history-based algorithm, in which photon histories are computed sequentially. The general structure of a history-based code is to have a large outer loop over photon histories. The photon history data are usually stored in scalar variables, which take on new values for subsequent photon histories (figure 8.1A). A history-based approach to computing photon histories is illustrated in figure 8.2. In some numerical implementations, photons may share a part of their history if forced detection is performed at each scattering location when simulating multiple-order scatter [6] or if the variance reduction technique of splitting is used [24].

The key to exploiting vector processors for Monte Carlo modelling is to use an event-based algorithm, in which photon histories are processed in groups. Computations for a given event in the photon history are performed for each photon in the group before computations for the next event are made. For example, the emission locations for all photons in a group are randomly chosen before raypath lengths to the boundary of the object or the attenuation contribution to the photon history weights are computed. An event-based approach is illustrated in figure 8.3.

The photon history information is stored in array variables and computations are performed in many small, tightly coded loops (figure 8.1B). Since computer memory restrictions will place limits on the lengths of the arrays, multiple groups of photon histories may need to be generated in order to simulate the desired number of photon histories. A more complete discussion of a methodology for implementing an event-based code for single-photon imaging may be found in an article by Smith *et al* [25]. Event-based vectorizable Monte Carlo codes have also been used to model particle transport in high-energy physics [26, 27]. In nuclear medicine de Vries *et al* [4] have employed an event-based code structure to utilize an array processor for modelling photon scatter within a collimator.

```
c        A: history-based code example
c
         do 100 i=1,nvoxel  ! loop over source voxels
         do 110 j=1,nhist   ! loop over photon histories

c(source code for photon emission, transport of unscattered photons)

c hwt:   photon history weight
c xmu:   linear attenuation coefficient
c d:     distance from emission point to the edge of the cylinder
c        for forced detection
c exp(-xmu*d): attenuation contribution to photon history weight
c
         hwt=hwt*exp(-xmu*d)      ! update weight with attenuation factor

c  (source code for detection of unscattered photons)

         if(iflag.eq.1) then   ! model scatter if desired
         do 120 k=1,nscat          ! loop over desired no. of scattering events

c  (source code for transport and detection of scattered photons)
c  (exit loop and end photon history if photon energy is too low)

120      continue
         endif
110      continue
100      continue

c        B: event-based code example
c
         do 100 i=1,nvoxel  !  loop over source voxels
         ngroup=nhist/nhistg !  determine number of groups
c                                      with nhistg histories/group
         do 110 ii=1,ngroup !  loop over the groups

c  (source code for photon emission, transport of unscattered photons)

c hwt(j):     photon history weight
c xmu(j):     linear attenuation coefficient
c d(j):       distance from emission point to the edge of the cylinder
c             for forced detection
c exp(-xmu(j)*d(j)): attenuation contribution to photon history weight
c
         do 212 j=1,nhistg    ! loop over photon histories
212      hwt(j)=hwt(j)*exp(-xmu(j)*d(j))  ! update weight with atten. factor

c  (source code for detection of unscattered photons)

         if(iflag.eq.1) then   ! model scatter if desired
         do 120 k=1,nscat          ! loop over desired no. of scattering events

c  (source code for transport and forced detection of scattered photons)
c  (cull photons with low energies and loop over fewer histories)

120      continue
         endif
110      continue
100      continue
```

Figure 8.1   *Source code segments showing the overall code structure and update of the photon history weight with an attenuation factor for (A) history-based and (B) event-based Monte Carlo codes.*

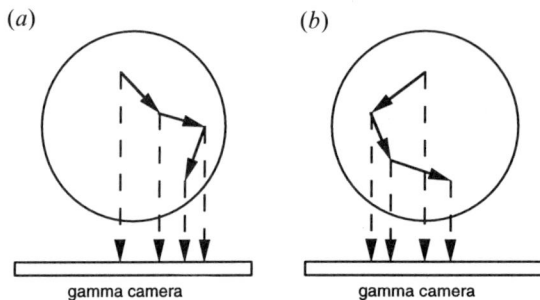

Figure 8.2 *An illustration of a history-based Monte Carlo algorithm. Calculations for the history of one photon (a) are completed before the transport of another photon (b) is simulated. Raypaths for forced detection into the gamma camera are shown from the emission point and from three scattering points.*

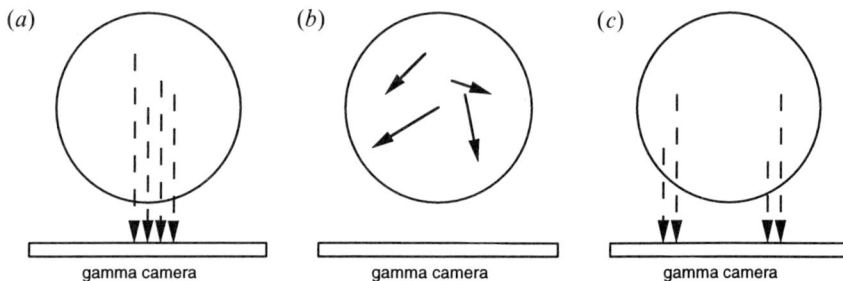

Figure 8.3 *An illustration of an event-based Monte Carlo algorithm, in which photon histories are computed in groups. (a) Forced detection from the emission points, (b) paths to the first scattering points and (c) forced detection from the scattering points.*

When source code is structured using an event-based algorithm, computers with vector processors can pipeline the computations to achieve significant speed increases [28]. Vector arithmetic computations are generally more efficient than performing an equivalent number of operations with scalar variables, since a new arithmetic operation can be performed with each clock cycle once data have been loaded into vector registers. The number of photons in each group must be large enough to compensate for the overhead in setting up the pipeline, of course.

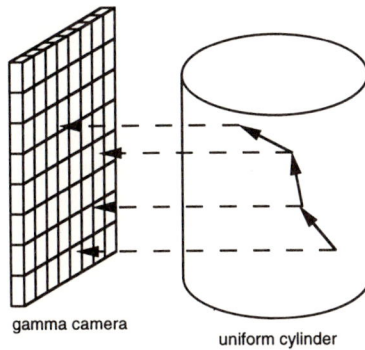

gamma camera

uniform cylinder

Figure 8.4 *Diagram of the circular cylinder and gamma camera for a simulation with the code MCMATV. The path of a photon is shown, with forced detection at the emission and scattering points.*

8.3 MCMATV: A VECTORIZED MONTE CARLO CODE FOR PHOTON TRANSPORT IN A UNIFORM CYLINDER

8.3.1 Description of the code

The program MCMATV (Monte Carlo Matrix Vectorized) is a code which models photon transport within a uniform circular cylinder and photon detection by a gamma camera (figure 8.4). It is used to calculate elements of the system matrix **A** in the equation

$$p = As \qquad (8.1)$$

where p is the projection data vector and s is the discretized source activity distribution. The matrix element A_{ij} represents the probability that a photon emitted in source voxel j is detected in projection pixel i. This system matrix can be used to generate projection data for different source models or to estimate source activity from projection data in a matrix-based image reconstruction. MCMATV is written in FORTRAN77 and was adapted from the history-based Monte Carlo code developed by Beck *et al* [6] and modified by Floyd *et al* [9, 13]. Some of the features of MCMATV have been discussed in these publications and by Smith *et al* [25] in describing conversion of the code from a history-based to an event-based structure. This section provides an overview of the code.

Photon transport is modelled within a circular cylinder whose radius and length are set by the user. The axis of the cylinder is parallel to the face of the gamma camera and its distance to the camera face is user specified. A cross section of the cylinder is divided into circular rings, which are subdivided into small sectors of approximately equal area (figure 8.5(*a*)). The source voxels for the Monte Carlo modelling have cross sections which are

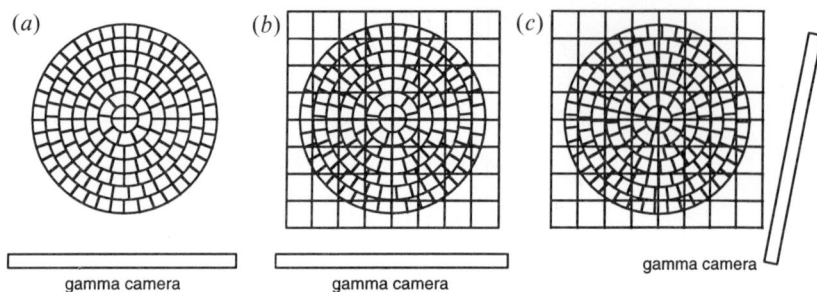

Figure 8.5 *(a) A cross section of the cylinder showing the circular rings and approximately equal area sectors. Usually 100 rings are used for the modelling, for which there are 31 465 sectors. (b) System matrix elements for a square grid are formed from area-weighted averages of the matrix elements for the circular sectors. Normally the square grid is 64 × 64 or 128 × 128. (c) The circular grid is rotated with respect to the square grid, which remains fixed in space, to compute the system matrix elements for the gamma camera at a different angle. The source activity distribution is defined with respect to the square grid.*

these sectors and an axial range which is set by the user. The photon detection probabilities are output by MCMATV on this circular grid. Detection probabilities for a conventional square grid are computed in a separate program by taking area-weighted averages of the sectors which overlay the square pixels (figure 8.5(*b*)) [29]. The use of a fine circular grid allows a system matrix characterizing SPECT projection data acquisition at many angles to be constructed from a Monte Carlo simulation at only one angle (figure 8.5(*c*)). This reduces the time to model the system matrix for a uniform cylinder whose axis is coincident with the axis of rotation of the gamma camera.

Line source response functions can be modelled by having the source voxels extend the length of the cylinder (figure 8.6(*a*)). The source voxels are indexed so that the starting and ending voxels in a simulation can be chosen. This enables a simulation to be restarted in the event of a system crash and also allows a large Monte Carlo simulation to be partitioned between different computers.

The circular cylinder is modelled to be uniformly attenuating with the material properties of water, though there is a program option to turn off attenuation. A lookup table is used to determine the energy-dependent attenuation coefficient, which includes the effect of photoelectric and Compton scatter. There are options to model the transport of only unscattered photons or of unscattered and scattered photons. The maximum number of scattering events in a photon history is set by the user. Only Compton-scattered photons are followed in the program, which is a reasonable

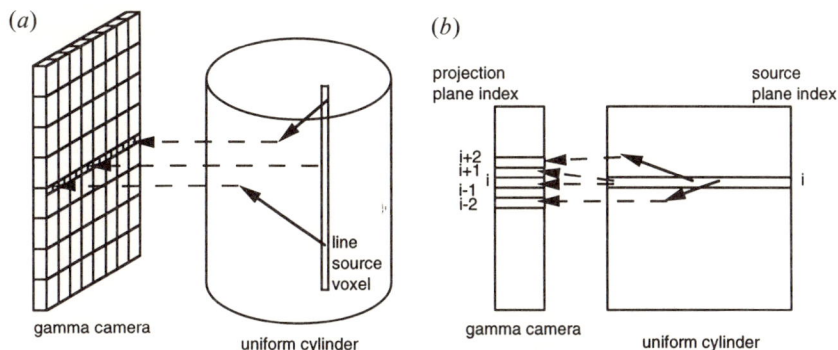

Figure 8.6 *(a) Line source response functions are modelled by extending the source voxel the length of the cylinder and detecting photons in a central plane. (b) Cross-plane effects due to depth-dependent detector response and photon scatter are studied by computing system matrices with probabilities of detection in different transverse planes from emissions in a single source plane. The source voxels have a height equal to the dimension of a projection pixel.*

approximation for photon energies near 140 keV, the energy of gamma ray emissions by ^{99m}Tc. (The Compton scatter cross section in water at 150 keV is 98% of the total photon interaction cross section and at 80 keV it is 94% of the total cross section [30].)

The program models a gamma camera with a parallel hole collimator. Hole length, average hole diameter and septal thickness are user-specified parameters. The collimator specifications are used to model depth-dependent detector response. Sensitivity factors are calculated using the method of Anger [31]. Neither collimator penetration nor photon scatter within the collimator are modelled. The energy-dependent sensitivity of a $\frac{3}{8}$ in NaI(Tl) scintillation crystal is modelled using a lookup table [1]. The gamma camera energy resolution and intrinsic spatial resolution of the scintillation crystal are modelled by a sampling from Gaussian distributions whose full widths at half maximum (FWHMs) are set by the user.

The detector array is square with dimensions and pixel sizes set by the user. Photons are detected in specified transverse projection planes and the history weights from adjacent planes may be summed if desired. When modelling line source response functions, photons are detected in a single projection plane near the centre of the cylinder (figure 8.6(*a*)). Photons can be emitted in one plane and detected in another plane due to depth-dependent collimator response and photon scatter. These effects can be studied by restricting the axial range of photon emission to a single projection plane and by detecting photons separately in several different projection planes (figure 8.6(*b*)). Photons may be detected in one or two energy windows with user-defined energy ranges.

The energy of the emitted photons and the number of photon histories per source voxel are input parameters to the program. Groups of photon histories are simulated for each source voxel in turn. The code uses the variance reduction technique of forced detection and photon history weights are updated to account for attenuation along the raypath into the detector. When modelling the transport of scattered photons, detection is forced at each scatter event in a photon history (figure 8.4). Photons are culled from the group after each scattering event if their energy falls below a threshold, which is based on the gamma camera energy window settings and the gamma camera energy resolution. The detection probabilities for scattered and unscattered photons may be summed before output or written separately to the output system matrix file. The sums of the history weights of detected unscattered photons and scattered photons (as a function of scatter order) are tabulated and output for each energy window. An energy spectrum is also output.

The code has an option to use the portable random-number generator of Marsaglia *et al* [32] as modified by James [33] or to use library calls to machine-specific random-number generators. The former is useful for comparing the performance of the code on different workstations, while the latter is useful for obtaining optimal performance on a given hardware platform. MCMATV has been compiled and executed on Stardent GS1000 computers, Sun SPARCstation 2 and SPARCstation 10 workstations, a Hewlett–Packard Apollo series 700 model 720 workstation, a Cray-2 supercomputer and a Digital Equipment Corporation 300 model 400 workstation, though not all of these machines have a vector processor.

8.3.2 *Program performance and applications*

MCMATV has been validated by comparison with experimental projection data measurements from a clinical scanner, with scatter fractions derived from experimental measurements and with the history-based code from which it was adapted [25]. Performance gains due to vectorization were benchmarked on a Stellar GS1000 computer (Stellar, Newton, MA), which has a vector floating point processor. The benchmark calculated the system matrix elements for line source response functions for emissions by ^{99m}Tc in a 11.3 cm radius, 22.0 cm long water-filled cylinder. The source code was compiled with and without invoking the vector processor. When only unscattered photons were modelled, the Monte Carlo simulation speed increased by a factor of 5.1 with vectorization [25]. When the detection of both scattered and unscattered photons was modelled, the speed increase was a factor of 2.9. The smaller increase for the scatter case is due in part to the rejection sampling method for randomly selecting a Compton scattering angle [34], which was not vectorized, and to the unavoidable use of additional 'IF' statements in the scatter portion of the code.

Monte Carlo modelling with MCMATV has been instrumental in several projects. It has been used to generate system matrices for the development of generalized matrix inverse reconstruction methods [15] and for the evaluation of projection pixel-dependent scatter subtraction factors [35]. The three-dimensional effects of photon scatter and detector response have been studied using the code [36]. MCMATV has also been used for the development and evaluation of an image reconstruction method which uses primary- and secondary-energy-window projection data simultaneously to constrain source activity estimates [37]. Though the code does not have the flexibility to model nonuniform attenuation, its speed and the capability to generate a SPECT system matrix for multiple-angle acquisition from a Monte Carlo simulation at one angle make it a useful tool for research projects for which a uniformly attenuating cylinder is a suitable model.

8.4 MCMATV3D: A VECTORIZED MONTE CARLO CODE FOR PHOTON TRANSPORT IN HETEROGENEOUS MEDIA

8.4.1 Description of the code

The program MCMATV3D (Monte Carlo Matrix Vectorized 3D) is a generalization of MCMATV which models photon transport in heterogeneous media [21]. MCMATV3D is perhaps unique among photon transport codes for nuclear medicine in that it has the capability to model both projection data for simulated SPECT studies and system matrices for use in matrix-based image reconstruction. The additional features of MCMATV3D and its differences from MCMATV are described in this section.

Photon transport is simulated in a three-dimensional (3D) region built from geometric primitives, which consist of elliptical cylinders and ellipsoids. The size and spatial location of these primitives are specified by the user. For each geometric primitive, a density and an index to energy-dependent scatter cross sections are input; there are options to model the attenuation properties of air, water, bone or muscle using the tables of Hubbell [30]. The geometric primitives are indexed in the order in which they are input to the program. Higher-indexed regions overlay lower-indexed regions, enabling complex heterogeneous 3D regions to be constructed.

If projection data are simulated, a 3D source activity distribution is built with a second independent set of geometric primitives. The activity concentration within each primitive is input by the user. Since these source activity primitives overlay each other in the same fashion as the primitives describing the material properties of the model, complex nonuniform activity distributions can be formed (figure 8.7). There are also options within the code for specifying activity distributions for MIRD-style heart [38] and liver [39] models. The source primitive within which a photon is emitted is saved as

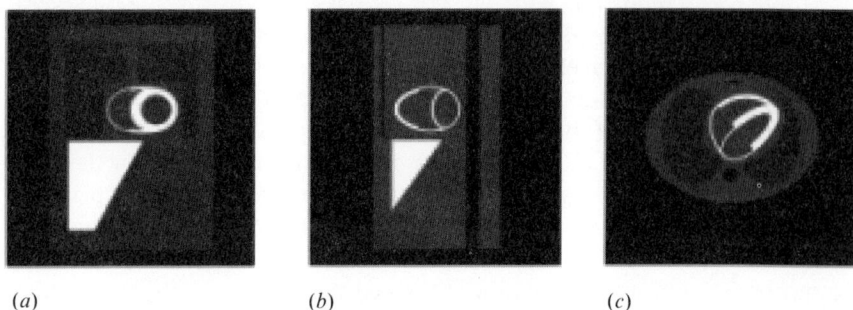

(a)   (b)   (c)

Figure 8.7 *Coronal (a), sagittal (b) and transverse (c) slices through a 3D source activity volume showing the activity distribution for a simulated myocardial perfusion study. The model is contained within a 40 cm high elliptical cylinder with major and minor axis lengths of 31.8 and 23.4 cm, respectively.*

part of the photon history, and there is a program option to generate and output projection data from different geometric primitives. This enables projection data from different organs to be modelled in a single pass of the code.

In contrast with MCMATV, which uses a circular grid, the source region for MCMATV3D is subdivided into cubic voxels with dimensions set by the program user. When modelling system matrices the matrix elements are computed at each gamma camera position, since the attenuating region is generally not symmetric about the axis of rotation of the gamma camera. When modelling projection data acquisition or computing system matrix elements, the gamma camera orbit must be circular and the angular increment of the gamma camera must be a constant. The starting and ending angles are set by the user.

8.4.2 Program performance and applications

Monte Carlo simulations from MCMATV3D were compared with projection data acquired on a clinical triple headed gamma camera (Triad, Trionix Research Laboratory, Twinsburg, OH) [21]. A ^{99m}Tc line source was placed in the centre of a nonuniform thorax phantom (figure 8.8). The gamma camera was equipped with a low-energy ultra-high-resolution (LEUH) collimator and a 20% energy window was centred on the 140 keV photopeak. There is good agreement between projection profiles at different gamma camera angles, including the shape of the scatter tails (figure 8.9). The angular dependence of peak width measurements is reproduced by the code (figure 8.10).

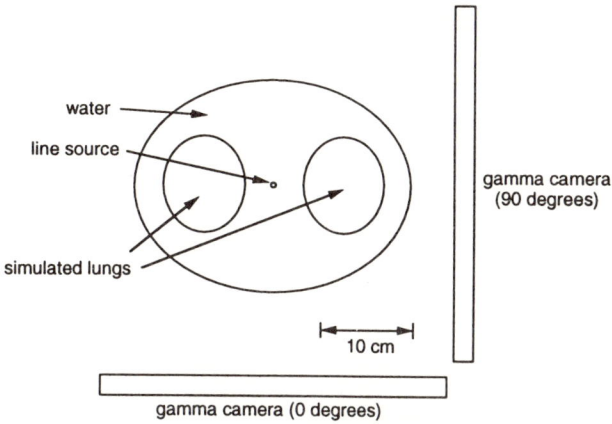

Figure 8.8 *A cross section of the thorax phantom for an experimental projection data acquisition. The phantom is filled with water and has two balsa wood inserts to simulate lungs. (Reprinted with permission from Smith [21].)*

The performance of MCMATV3D on a Stellar GS1000 computer with a vector processor has been evaluated [21]. For a uniform water-filled cylinder, vectorization increases the speed of the code by a factor of 2.1 when modelling the transport of unscattered photons and by a factor of 1.8 when modelling the transport of unscattered and scattered photons. These speed increases are not as great as those achieved by MCMATV for this same test (factors of 4.0 and 2.7, respectively). MCMATV3D also is slower than MCMATV by a factor of 2.4 when modelling the transport of unscattered photons and by a factor of 1.9 when modelling the transport of unscattered and scattered photons. These results illustrate that a photon transport code for a very simple geometry with uniform attenuation can be written to execute faster and to vectorize more efficiently than a more general code. The more modest performance gains with vectorization for MCMATV3D also indicate that further tuning of the code might be possible in order to achieve better performance gains with vectorization.

MCMATV3D was also benchmarked on a Stellar GS1000 computer for simulations of photon transport within models of the thorax containing from one to six geometric primitives. Central processing unit (CPU) time increases approximately linearly with the number of primitives used to model nonuniform attenuation [21] and CPU time for a five-subregion thorax model is approximately twice that for a uniform-attenuation model. The speed increase due to vectorization decreases as the number of subregions increases, probably due to the fact that the part of the code which computes raypath intersections with the geometric primitives does not vectorize well.

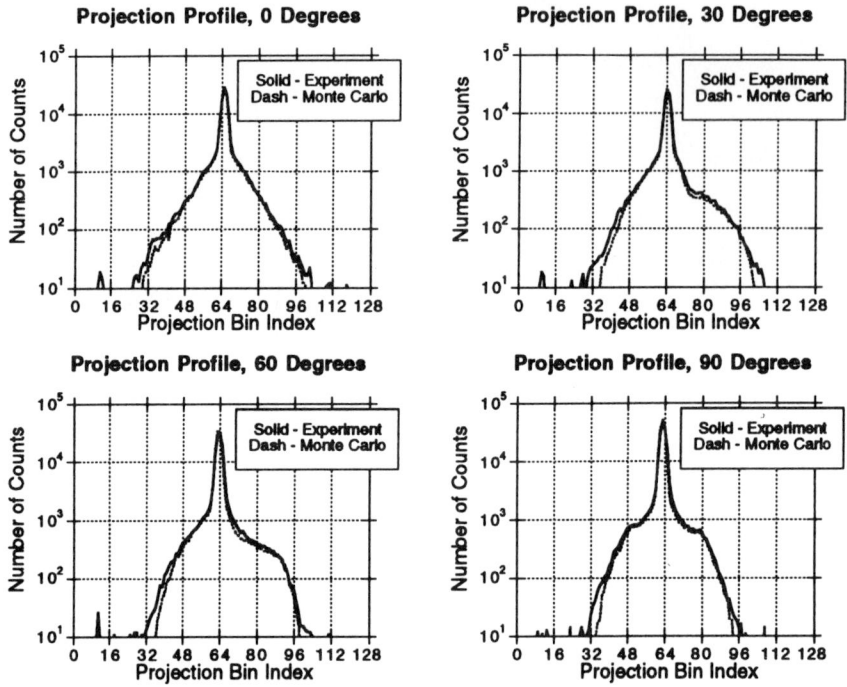

Figure 8.9 *Experimentally measured (solid) and Monte Carlo-simulated profiles (dashed) for different gamma camera positions. The position of the gamma camera is measured counterclockwise from a posterior position (figure 8.8). (Reprinted with permission from Smith [21].)*

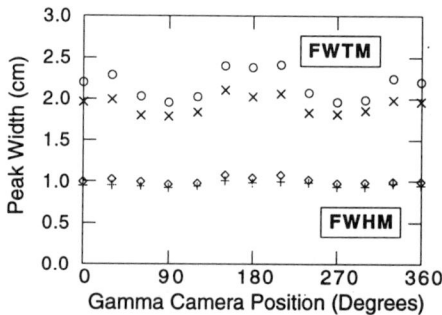

Figure 8.10 *Peak width measurements (FWHM and full width at tenth maximum (FWTM)) for a line source in a nonuniform thorax phantom. The gamma camera position is measured counterclockwise from a posterior position. ◇, FWHM, experiment; +, FWHM, Monte Carlo; ○, FWTM, experiment; × FWTM, Monte Carlo. (Adapted from Smith [21].)*

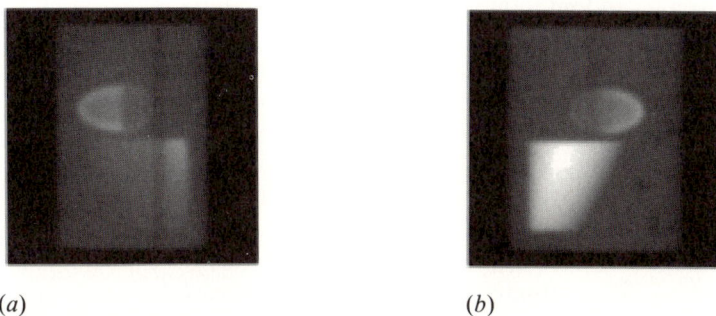

(a)   (b)

Figure 8.11 *Monte Carlo-simulated projection data for the source activity model of figure 8.7. The attenuation model is nonuniform, with different scattering properties within the simulated soft tissue, lungs and bone. (a) The gamma camera rotated 45° anti-clockwise from a posterior position and (b) an anterior view.*

With six subregions modelling the two lungs, cardiac region, spine, sternum and the surrounding soft tissue, the speed increase due to vectorization is about 1.6 both when the transport of only unscattered photons is modelled and when the transport of unscattered and scattered photons is modelled.

MCMATV3D has been used to study nonuniform attenuation compensation for simulated SPECT myocardial perfusion studies [21] and to evaluate improved scatter subtraction methods in nonuniform media for myocardial perfusion imaging [40]. It has also been used to study the effect of photon scatter from the liver on myocardial perfusion quantitation, using fully 3D modelling of photon transport in a 3D thorax model [41]. An example of Monte Carlo-simulated projection data for the 3D source activity model of figure 8.7 is shown in figure 8.11. A 20% energy window was centred on the 140 keV photopeak for emissions by ^{99m}Tc. The transport of unscattered and scattered photons was modelled. MCMATV3D has been compiled and executed on Stardent GS1000 computers and on Sun SPARCstation 2 and 10 workstations.

8.4.3 Additional development of MCMATV3D

The development of MCMATV3D is continuing. The code has recently been modified so that the modelling of scattered photons and unscattered photons is completely independent; the user can choose to model unscattered photons only, scattered photons only, or both scattered and unscattered photons. A faster alternative to using a rejection sampling method [34] for randomly selecting a Compton scattering angle is now offered. With this option, a table with cumulative scatter probability as a function of scattering angle is pre-computed using the Klein–Nishina formula [42]; a simple lookup operation using a random number between zero and unity is then used to select a scattering angle.

The variance reduction technique of stratified sampling [24] has been recently incorporated in the code [41]. Photon emission is stratified according to location when modelling unscattered photons and according to location and emission angle (with respect to a normal to the gamma camera surface) when modelling scattered photons. The physical dimensions of a stratification cell can be set to be a multiple of the dimension of a source voxel, in order to save memory and computing time. At each gamma camera position an initial Monte Carlo simulation is performed to determine the stratification weights for that angle. The user sets the number of photon histories per source voxel for the initial pass and the average number of photon histories per source voxel for the second modelling pass. With stratification the effective speed of the code for modelling unscattered and scattered photons is increased by factors of about 4.3 and 2.3, respectively, for the simulation of a myocardial perfusion study.

MCMATV3D is designed so that options such as converging beam collimation and noncircular gamma camera orbits can be added in a straightforward manner. Implementation of the additional variance reduction techniques of Russian roulette and splitting is planned, as is the capability to input voxel-based source activity and attenuation maps. The next major challenge in improving the performance of MCMATV3D is to structure the code to take advantage of multiple processors. Parallel processing has been supported on large mainframe computers (e.g. Cray-2 and IBM 3090) and is now supported on lower-cost workstations (e.g. Sun SPARCstation 10). Parallel processing has the potential to significantly reduce Monte Carlo simulation times [23] and to make the modelling of photon transport for complex source and attenuation models easier to achieve.

ACKNOWLEDGMENTS

This work has been supported in part by a grant from The Whitaker Foundation, by the Duke-North Carolina National Science Foundation Engineering Research Center for Emerging Cardiovascular Technologies (CDR-8622201), by training grant No HL-07063-15 awarded by the National Institutes of Health, by PHS grant No R01-CA33541 awarded by the National Cancer Institute and by grant No DE-FG05-91ER60894 awarded by the Department of Energy.

REFERENCES

[1] Anger H O and Davis D H 1964 Gamma-ray detection efficiency and image resolution in sodium iodide *Rev. Sci. Instrum.* **35** 693–7

[2] Bradshaw J, Burnham C and Correia J 1985 Application of Monte-Carlo methods to the design of SPECT detector systems *IEEE Trans. Nucl. Sci.* **NS-32** 753–5

[3] Metz C E, Atkins F B and Beck R N 1980 The geometric transfer function component for scintillation camera collimators with straight parallel holes *Phys. Med. Biol.* **25** 1059–70

[4] de Vries D J, Moore S C, Zimmerman R E, Mueller S P, Friedland B and Lanza R C 1990 Development and validation of a Monte Carlo simulation of photon transport in an Anger camera *IEEE Trans. Med. Imaging* **MI-9** 430–8

[5] Saffer J R, Barrett H H, Barber H B and Woolfenden J M 1991 Surgical probe design for a coincidence imaging system without a collimator *Information Processing in Medical Imaging* ed A C F Colchester and D J Hawkes (Berlin: Springer) pp 8–22

[6] Beck J W, Jaszczak R J, Coleman R E, Starmer C F and Nolte L W 1982 Analysis of SPECT including scatter and attenuation using sophisticated Monte Carlo modeling methods *IEEE Trans. Nucl. Sci.* **NS-29** 506–11

[7] Jaszczak R J, Floyd C E Jr, Manglos S H, Greer K L and Coleman R E 1986 Cone beam collimation for single-photon emission computed tomography analysis simulation and image reconstruction using filtered backprojection *Med. Phys.* **13** 484–9

[8] Webb S, Binnie D M, Flower M A and Ott R J 1992 Monte Carlo modeling of the performance of a rotating slit-collimator for improved planar gamma-camera imaging *Phys. Med. Biol.* **37** 1095–108

[9] Floyd C E, Jaszczak R J, Harris C C and Coleman R E 1984 Energy and spatial distribution of multiple order Compton scatter in SPECT: a Monte Carlo investigation *Phys. Med. Biol.* **29** 1217–30

[10] Frey E C and Tsui B M W 1990 Parameterization of the scatter response function in SPECT imaging using Monte Carlo simulation *IEEE Trans. Nucl. Sci.* **NS-37** 1308–15

[11] Rosenthal M S and Henry L J 1990 Scattering in uniform media *Phys. Med. Biol.* **35** 265–74

[12] Ljungberg M and Strand S-E 1990 Scatter and attenuation correction in SPECT using density maps and Monte Carlo simulated scatter functions *J. Nucl. Med.* **31** 1560–7

[13] Floyd C E Jr, Jaszczak R J, Greer K L and Coleman R E 1986 Inverse Monte Carlo as a unified reconstruction algorithm for ECT *J. Nucl. Med.* **27** 1577–85

[14] Bowsher J E and Floyd C E Jr 1991 Treatment of Compton scattering in maximum-likelihood expectation-maximization reconstructions of SPECT images *J. Nucl. Med.* **32** 1285–91

[15] Smith M F, Floyd C E Jr, Jaszczak R J and Coleman R E 1992 Reconstruction of SPECT images using generalized matrix inverses *IEEE Trans. Med. Imaging* **MI-11** 165–75

[16] Ljungberg M and Strand S-E 1989 A Monte Carlo program for the simulation of scintillation camera characteristics *Comput. Programs Biomed.* **29** 257–72

[17] Haynor D R, Harrison R L, Lewellen T K, Bice A N, Anson C P, Gillispie S B, Miyaoka R S, Pollard K R and Zhu J B 1990 Improving the efficiency

of emission tomography using variance reduction techniques *IEEE Trans. Nucl. Sci.* **NS-37** 749–53

[18] Zubal I G and Harrell C H 1991 Voxel based Monte Carlo calculations of nuclear medicine images and applied variance reduction techniques *Information Processing in Medical Imaging* ed A C F Colchester and D J Hawkes (Berlin: Springer) pp 23–33

[19] Yanch J C, Dobrzeniecki A B, Ramanathan C and Behrman R 1992 Physically realistic Monte Carlo simulation of source collimator and tomographic data acquisition for emission computed tomography *Phys. Med. Biol.* **37** 853–70

[20] Wang H, Jaszczak R J and Coleman R E 1992 Solid geometry-based object model for Monte Carlo simulated emission and transmission tomographic imaging systems *IEEE Trans. Med. Imaging* **MI-11** 361–72

[21] Smith M F 1993 Modelling photon transport in non-uniform media for SPECT with a vectorized Monte Carlo code *Phys. Med. Biol.* **38** 1459–74

[22] Raeside D E 1976 Monte Carlo principles and applications *Phys. Med. Biol.* **21** 181–97

[23] Andreo P 1991 Monte Carlo techniques in medical radiation physics *Phys. Med. Biol.* **36** 861–920

[24] Haynor D R, Harrison R L and Lewellen T K 1991 The use of importance sampling techniques to improve the efficiency of photon tracking in emission tomography simulations *Med. Phys.* **18** 990–1001

[25] Smith M F, Floyd C E Jr and Jaszczak R J 1993 A vectorized Monte Carlo code for modeling photon transport in SPECT *Med. Phys.* **20** 1121–7

[26] Martin W R and Brown F B 1987 Status of vectorized Monte Carlo for particle transport analysis *Int. J. Supercomput. Appl.* **1** 11–32

[27] Miura K 1987 EGS4V vectorization of the Monte Carlo cascade shower simulation code EGS4 *Comput. Phys. Commun.* **45** 127–36

[28] Hockney R W R J C 1981 *Parallel Computers* (Bristol: Hilger)

[29] Floyd C E Jr, Jaszczak R J and Coleman R E 1986 Image resampling on a cylindrical sector grid *IEEE Trans. Med. Imaging* **MI-5** 128–31

[30] Hubbell J H 1969 Photon cross sections attenuation coefficients and energy absorption coefficients from 10 keV to 100 GeV *National Bureau of Standards Report* NSRDS-NBS 29

[31] Anger H O 1964 Scintillation camera with multichannel collimators *J. Nucl. Med.* **5** 515–31

[32] Marsaglia G, Zaman A and Tsang W W 1990 Toward a universal random number generator *Stat. Prob. Lett.* **9** 35–9

[33] James F 1990 A review of pseudorandom number generators *Comput. Phys. Commun.* **60** 329–44

[34] Kahn H 1956 Applications of Monte Carlo *RAND Research Memorandum* RM-1237-AEC

[35] Smith M F, Floyd C E Jr, Jaszczak R J and Coleman R E 1992 Evaluation of projection pixel-dependent and pixel-independent scatter correction in SPECT *IEEE Trans. Nucl. Sci.* **NS-39** 1099–105

[36] Smith M F, Floyd C E Jr, Jaszczak R J and Coleman R E 1992 Three-dimensional photon detection kernels and their application to SPECT reconstruction *Phys. Med. Biol.* **37** 605–22

[37] Smith M F and Jaszczak R J 1994 Simultaneously constraining SPECT activity estimates with primary and secondary energy window projection data *IEEE Trans. Med. Imaging* **MI-13** 329–37

[38] Coffey J L, Cristy M and Warner G G 1981 Specific absorbed fractions for photon sources uniformly distributed in the heart chambers and heart wall

of a heterogeneous phantom (MIRD Pamphlet No 13) *J. Nucl. Med.* **22** 65–71

[39] Snyder W S, Fisher H L Jr, Ford M R and Warner G G 1969 Estimates of absorbed fractions for monoenergetic photon sources uniformly distributed in various organs of a heterogeneous phantom (MIRD Pamphlet No 5) *J. Nucl. Med.* **10** (Supplement 3) 5–52

[40] Smith M F and Jaszczak R J 1994 Generalized dual-energy-window scatter compensation in spatially varying media for SPECT *Phys. Med. Biol.* **39** 531–46

[41] Smith M F and Jaszczak R J 1994 Three-dimensional photon scatter in non-uniform media for a simulated ^{99m}Tc SPECT myocardial perfusion study *J. Nucl. Med.* **35** 17P (abstract)

[42] Heitler W 1954 *The Quantum Theory of Radiation* 3rd edn (Oxford: Clarendon)

CHAPTER 9

THE SIMSPECT SIMULATION SYSTEM

Marie-Jose Belanger,
Andrew B Dobrzeniecki and
Jacquelyn C Yanch

9.1 INTRODUCTION

SimSPECT was developed at MIT's Whitaker College Biomedical Imaging and Computation Laboratory (WCBICL). The SimSPECT system models the full tomographic imaging of photons released from physically realistic, nonuniform, asymmetric nuclear medicine source geometries [1]. It is based on MCNP (Monte Carlo for neutron–photon transport), a Monte Carlo code developed at Los Alamos Scientific Laboratory [2]. Disabling the neutron transport portion of MCNP and keeping only the photon transport, we then altered several input and output aspects of the code to allow full tomographic acquisition of any three-dimensional source geometry. Nuclear medicine collimators of various types from simple pinhole collimators to complicated geometries such as parallel and cone beam collimators are modelled realistically and completely [1, 3]. Collimator properties such as hole dimensions, number of holes and packing geometry are set by the user, as are parameters such as energy windows, pixel resolution and the number of tomographic views. Any three-dimensional (3D) source and scattering object can be specified with any degree of detail allowing for extremely realistic patient models in terms of both geometric dimensions and material composition.

Details of SimSPECT implementation, code verification, and run-time requirements are given below. Further information regarding the capabilities of this simulation package is given by way of two examples of current applications.

111

9.2 THE SIMSPECT CODE

The user-defined SimSPECT input file specifies the size, shape and energy spectrum of the radiation source, the composition and configuration of the source and scatter media using MCNP geometrical primitives, the collimator type and geometry and the number of projections to acquire. Once the input file has been processed, photon generation and transport are initiated.

The essential photon tracking and interaction algorithms contained in SimSPECT are from the embedded Monte Carlo transport code MCNP. MCNP contains photon cross-section tables for material with $Z = 1$–94 [4]. The data in the photon interaction tables allow MCNP to automatically account for coherent and incoherent scattering (with modification to the Thompson and Klein–Nishina differential cross section by appropriate form factors to take electron binding effects into account), photoelectric absorption with characteristic x-ray emission and pair production with local emission of annihilation quanta [2]. Thousands of hours of development and use have rendered MCNP a transport code which is extensively 'debugged' and validated [2, 5]. The latest version of SimSPECT incorporates MCNP 4A which includes the transport physics and tracking of electrons.

When using the Monte Carlo approach to simulating tomographic acquisition, forced detection via importance sampling offers an excellent way to improve the number of photons contributing to the final image per unit central processing unit (CPU) time. It is unclear, however, what effect this may have on the noise characteristics in the final image, or how accurately a realistic collimator can be modelled with this method. The SimSPECT approach is to maintain analogue photon physics within the acquisition of a single projection view but to make use of two techniques for increasing the speed of simulating multiple projections: (i) photon cloning and (ii) use of multiple CPUs. Figure 9.1 illustrates both techniques. One CPU becomes the photon *generator*, while second and third (or more) CPUs become the photon *consumers*. The photon *generator* initiates and tracks a photon within the source and scatter media. When the photon leaves this space and passes through a virtual sphere, its position, direction, energy and scatter order are saved. The photon is then cloned and redirected to all collimators with which that photon would have interacted. The photon cloning process is illustrated in figure 9.1. (The effect of this method of variance reduction on the noise characteristics of the final image has been evaluated as described below.) In SimSPECT, the interaction of the cloned photon with each of these collimators is carried out by one or more photon *consumers* using separate CPUs. Processes communicate via IPC (Interprocess Communication) sockets. This method achieves a high degree of coarse-grained parallelism, and can produce true tomographic simulations in reasonable times.

Photon transport in the detector crystal, light pipe and photomultiplier tubes (PMTs) is not simulated. Instead, the camera is modelled as a plane.

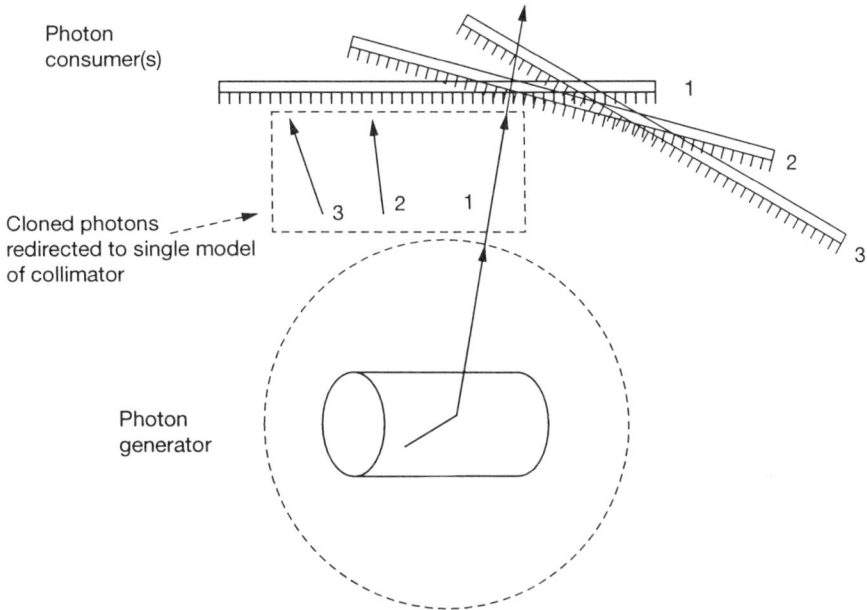

Figure 9.1 *An illustration of the generation of tomographic data in Sim-SPECT. Photons are generated and tracked in the object on the first CPU: the photon generator. When the photon passes through a virtual sphere, the photon's position, energy and scatter order are recorded. The photon is then cloned and redirected to all collimator positions with which that photon could have interacted. Photon tracking through the collimator is carried out by one or more separate CPUs termed the photon consumers.*

The 2D position of all photons successfully crossing this plane is recorded in a ListMode file as described in the next section.

9.3 THE LISTMODE OUTPUT

In SimSPECT, when a photon reaches the detection plane it is not, at this point, counted to a given pixel in the relevant projection image. Instead, ListMode files are generated (one for each projection view) in which details of each photon's history are sequentially stored [6]. The photon's absolute energy, its position as it crosses the detector plane, the number of scattering events in the phantom and the number of scattering events in the collimator are all recorded in the ListMode file. The energy of each photon in the ListMode file is sampled from the energy resolution of the camera which is assumed to have a full width at half maximum (FWHM) of 12%. This can be adjusted by the user. (For example this can be made very small to permit examination of the actual photon energy spectrum prior to modification by the NaI(Tl) crystal and PMTs.) The individual photons can subsequently be grouped in any number of ways using different experimental acquisition parameters including pixel resolution, energy window, scatter order and count level. This technique is similar to a method described by Haynor *et al* in which lists of photons leaving the phantom are obtained and stored prior to interaction with a particular collimator type [7]. The photons from a single list can then be made to interact with many different collimators and the effect of the collimator geometry on the resultant image can be examined. In SimSPECT, the collimator is part of the initial Monte Carlo simulation. Photons can later be grouped in different ways in order to examine the effect of energy window, image resolution, count level etc. The advantage of this method of photon storage is that it allows a single set of simulation data to be reused for many types of experiment involving various energy windows, image resolution etc.

In order to generate images from the ListMode files, a post-acquisition algorithm called SimVIEW is used. The SimVIEW system, also developed at MIT's WCBICL, is an image visualization and processing tool which converts ListMode data into pixel-based images through user-specified directives regarding final image resolution, intrinsic energy resolution and spatial resolution of the crystal, energy window, scatter orders or ranges and final count levels. While a single SimSPECT simulation can take many hours of CPU time (see below), generating images using SimVIEW takes only seconds. Matlab can also be used to generate the final image.

9.4 THE REQUIREMENTS

The SimSPECT system is programmed in a modular style with changes, removals and additions having been made to the core Monte Carlo transport module, MCNP. The SimSPECT system comprises 10 000 lines of C code and the embedded MCNP module comprises 90 000 lines of FORTRAN code. Operating system calls are made via Unix functions and the IPC communication in the inter-processor uses the Unix socket library. All graphics calls are made through the X-window system and the associated libraries (OSF/Motif, X Toolkit Intrinsics and X Extensions). The single executable SimSPECT file is approximately 2.5 Mb in size and requires 15–35 Mb of random-access memory (RAM) when running (depending on the complexity of the patient/phantom model and the collimator) [6]. SimSPECT can be easily stopped during a simulation in order to view photon accumulation in the image, and then restarted. The simulation can also be restarted without loss of data in the event of a network failure or a CPU crash.

Generating simulated SimSPECT output files is computationally expensive. For example, initiating 100×10^6 140 keV photons and generating 60 tomographic ListMode projection files requires 2.6 CPU d total on two processors. This simulation results in approximately 10 000 counts per projection image using a low-energy high-resolution (LEHR) collimator model and a $\pm 10\%$ energy window. The time is normalized to a machine with a reduced-instruction-set-computer- (RISC-) based CPU with a rating of 3.0 MFLOP. Images of localized hot sources will be acquired far faster than those of broadly distributed sources since adequate signal in each pixel representing the object will require fewer total photons.

It is difficult to compare the SimSPECT run times with those for other nuclear medicine simulation or modelling systems. SimSPECT is clearly less efficient than some other systems that have been developed. However, its reliance on complete simulation of transport physics, analogue transport methods, use of 3D asymmetric and heterogeneous media and physically accurate collimator models, has given SimSPECT the capability to produce simulation data that are not available via other packages. In addition, the use of ListMode files allows many 'experimental data sets' to be acquired with only one initial Monte Carlo run.

9.5 VALIDATION

SimSPECT relies on the physics programmed into the MCNP code for the modelling of photon transport. MCNP is a state-of-the-art particle transport code permitting extremely accurate simulation of photon interactions. Whalen *et al* [5] present extensive validation of the photon interaction physics under a variety of source and transport conditions. The SimSPECT system, with its user-defined model of the source, transport media and collimator, has also been extensively verified.

Verification has been carried out in a number of ways. First, the accuracy with which SimSPECT models the geometric responses of two collimator types (parallel hole and cone beam) was investigated. Next, the nature of the noise generated in the simulated data was examined and was compared to the noise in experimentally acquired data to determine whether the photon cloning method used in SimSPECT resulted in distortions of the noise characteristics. Finally, a number of simulations of various phantom geometries were generated and compared with those obtained experimentally. These evaluations are described in detail below.

9.5.1 Collimator response

The geometric response of two collimator types has been investigated. First, a low-energy general-purpose parallel hole collimator consisting of 18 000 hexagonally packed, 41 mm long holes was modelled in its entirety [1]. A simulated point source of ^{99m}Tc was 'imaged' at different distances from the face of the collimator. The FWHM of the resultant image of the point source was measured (an indication of spatial resolution) and plotted against distance from the collimator. For a parallel hole collimator, image resolution should be a linear function of collimator dimensions (hole diameter and length) and of source–collimator distance. A plot of the FWHM of point sources imaged via SimSPECT versus distance from the collimator generated a straight line, indicating that the geometric response of the collimator was accurately reflected in the simulation [1].

Modelling of a cone beam collimator was also carried out. Accuracy of collimator simulation was first evaluated by measuring the increase in sensitivity as a function of source–collimator distance, relative to the sensitivity of a parallel hole collimator [3]. For this experiment, two collimators extensively described in the literature [8] were modelled. Sensitivity was found to increase as the point source was moved further away from the collimator. This increase reached a factor of two at a distance of roughly 20 cm, in agreement with experimentally observed results.

Further verification of the accuracy of SimSPECT in modelling SPECT using a cone beam collimator was carried out by simulation of the 'slice-to-slice cross-talk' phantom. Reconstruction of projection data acquired using a cone beam collimator leads to artifacts which become more apparent with distance from the centre of the image. If simulated reconstructed cone beam data show the same artifacts as seen with real cone beam imaging this would further verify the accuracy of the simulation model. The phantom used consisted of seven parallel discs, 2 cm thick, of equal activity lying parallel to the axis of rotation. Projection data acquired at a collimator–phantom centre distance of 25 cm showed seven hot bars of roughly equal pixel intensity. The reconstructed image shows both spatial distortion away from the centre of the image and an underestimate of activity which becomes worse with distance away from the centre. This artifact is very similar to that observed experimentally and provides further validation of the accuracy of the cone beam collimator simulation with SimSPECT. Further details of this analysis can be found in [3].

9.5.2 Noise simulation

SimSPECT is largely an analogue Monte Carlo package in that photons are born in the source object and then tracked through the object and collimator. The vast majority of photons do not contribute to the final image as is the case in the real imaging situation. SimSPECT is almost entirely physically realistic in the generation of projection data and does not rely on forced detection or other methods of importance sampling, which may produce images with noise levels and noise characteristics no longer representative of

Table 9.1 *RMS values from a circular ROI within a uniform source distribution in a 128 × 128 planar projection. P_{RMS} is the predicted RMS value from Poisson statistics and avg is the average count per pixel. For both SimSPECT and experimental data, a circular ROI containing 2128 pixels was used to evaluate the RMS.*

	SimSPECT			Experimental	
avg	RMS	P_{RMS}	avg	RMS	P_{RMS}
6.74	0.377 ± 0.011	0.385	6.77	0.392 ± 0.011	0.384
13.52	0.269 ± 0.008	0.272	13.59	0.269 ± 0.008	0.272
20.23	0.219 ± 0.007	0.222	20.42	0.225 ± 0.007	0.221
26.92	0.189 ± 0.005	0.193	27.22	0.194 ± 0.0006	0.192

realistic imaging. However, SimSPECT does deviate from physically realistic photon transport when simulating tomography. In order to generate efficient tomographic acquisitions, the simulation clones and redirects any photon successfully escaping the phantom to all possible collimators with which the photon could potentially have interacted. To investigate the effect of this method of variance reduction on the noise characteristics of the final image, a study was undertaken to compare the noise generated in images created by SimSPECT with the noise levels and characteristics of experimentally acquired data [9, 10].

The root mean square of the noise (RMS) and the noise power spectrum (NPS) were two statistical quantities used to characterize the noise. The RMS of the noise measures local noise fluctuations while the NPS assesses the texture of the noise in spatial frequency. The noise was examined by comparing both synthetic and experimental images of large cylindrical homogeneous phantoms. Details of experimental data acquisition are provided in [10].

The RMS of the noise was estimated by taking the square root of the variance of a chosen region of interest (ROI) and dividing by the mean. Planar projections from uniform hot phantoms of constant depth contain only Poisson noise when examined far from the edges of the phantom. Thus, the expected RMS in such a ROI should be equal to $1/\sqrt{\bar{N}}$ where $\bar{N}$ is the projection count level. The RMS of the noise of both simulated and experimental planar images was evaluated and compared with the expected value. Results are given in table 9.1. The RMS found in SimSPECT planar projections compared well with values predicted by Poisson statistics and the values of RMS found in experimental planar projections.

Unlike projections of a phantom of constant depth, projections from a homogeneous cylindrical ('or curved') phantom have non-stationary Poisson noise due to the curvature of the phantom. As a result, the noise level varies from the centre to the edges of the projected phantom. In a reconstructed slice from a series of such projections, the RMS of the noise decreases from the centre to the edges. This variation is further increased by the effect of

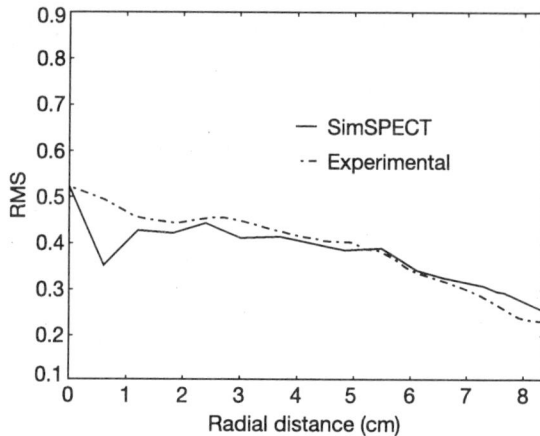

Figure 9.2 *Radial RMS profiles of SimSPECT and experimental tomo-graphic reconstruction images (64 × 64). Profiles were averaged over 12 or 20 adjacent slices in the case of simulated or experimental data, respectively [10].*

attenuation as was previously observed by Gillen [11]. The average RMS of the noise in the reconstruction can be predicted using the relationship derived by Huesman [12]. This relationship assumes that the projections have a stationary Poisson noise equal to the average count level in the projection. Consequently, for a cylindrical phantom, it only provides an estimate of the RMS of the noise in the reconstruction and the effect of attenuation is not considered:

$$\mathrm{RMS} = k \, (v^{3/4}/\sqrt{N_{\mathrm{tot}}}) \tag{9.1}$$

where N_{tot} is the total number of counts in the slice and v is the number of pixels containing information in the reconstructed slice. The constant k, calculated to be 1.2 in tomographic backprojection using a Ramp filter, varies with the type of reconstruction filter and the cut-off frequency used.

The RMS profiles for both synthetic and experimental reconstruction are shown in figure 9.2. The profiles were obtained by using a group of pixels from adjacent slices located at the same specific radius and calculating the RMS within this pixel group. This procedure compensated for the variation of count level due to attenuation effects. The mean RMS of the noise from the RMS profile was then determined and used in equation (9.1) to obtain a value for the constant k. The estimated values of RMS and k are shown in table 9.2.

The NPS was then used to examine the texture of the noise in both simulated and experimental images [10]. The first comparison was carried

Table 9.2 *Mean RMS and k values from profiles through a uniform reconstructed cylindrical source distribution using filtered backprojection with a Ramp reconstruction filter [10].*

	SimSpect		Experimental	
Filter	RMS	k_{RMS}	RMS	k_{RMS}
Ramp	0.39 ± 0.01	1.16 ± 0.04	0.40 ± 0.01	1.20 ± 0.04

out using projection data. The NPS was estimated within a large ROI far from the edges of the phantom, using the square modulus of the 2D discrete Fourier transform, a method originally developed by Welch [13]. The mean pixel count was subtracted to generate a noise-only image and then an apodizing window was applied to reduce truncation errors. The ROI mean was then subtracted again in order to remove the zero frequency of the noise. The estimated NPSs for both synthetic and experimental projections are given in figure 9.3. The expected white spectrum is illustrated by the solid flat line corresponding to the mean count level. As shown, the NPS from the SimSPECT acquisition is within expected spectral variance, as is the NPS obtained from experimental projection data [10].

The NPS from reconstructed slices were calculated within a circular ROI centred at the origin of the reconstructed object, keeping well away from the edges. To remove the effects of attenuation, the mean count level in different annuli was calculated using adjacent slices and then subtracted from each pixel in the annular region. This was repeated for all regions to obtain a noise-only ROI. The resulting ROI was then apodized and the

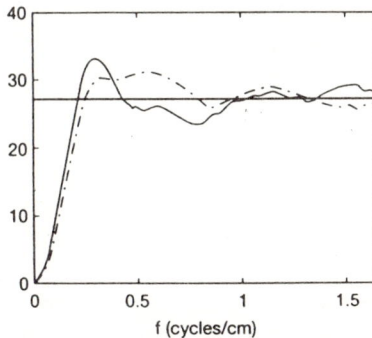

Figure 9.3 *NPSs of experimental and SimSPECT generated planar images. The flat solid line represents the average count level of both Sim-SPECT and experimental planar projections and therefore the expected NPS. The solid curve represents the NPS from SimSPECT; the broken curve represents the NPS from experimental projections. The expected statistical fluctuation of the NPS is 0.37 times the count level (flat solid line) [10].*

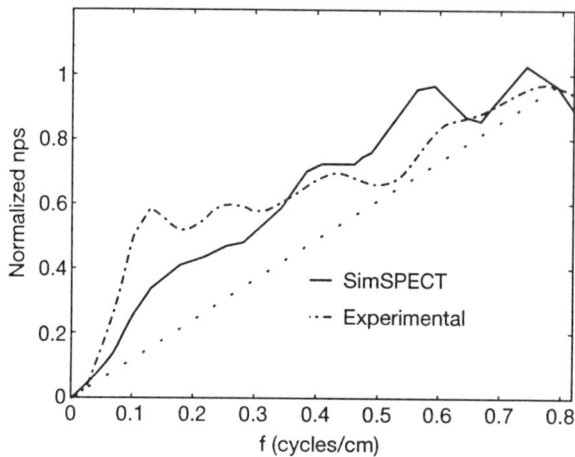

Figure 9.4 *NPSs of tomographic reconstructions of SimSPECT and experimental projections of a uniform hot cylinder. For both acquisitions, a filtered backprojection was carried out using a Ramp filter with a cut-off frequency of 0.82 cycles cm⁻¹. The dotted line represents the theoretical NPS obtained if the projections contained only white noise [10].*

global mean was subtracted to remove the zero frequency of the noise. The NPSs obtained from both synthetic and experimental data were then normalized by dividing the magnitude of the NPSs by the projection mean count level. This step removed the dependence of the NPSs on the projection count level.

The NPSs from SimSPECT and experimental data are shown in figure 9.4. Both spectra are in general agreement within the expected spectrum variance; both also deviate from the theoretical NPS for the filtered backprojection of stationary noise, which has the shape of a Ramp filter. This deviation is expected since the noise in the tomographic projections of interest is not truly stationary. The NPSs from the experimental data contain a low-frequency peak at 0.13 cycles cm⁻¹, which is thought to be a centring artifact due to the difficulty of accurately locating the centre of the ROI prior to the NPS estimate [10].

This noise analysis demonstrated that the mean noise *level* of the simulated images is preserved both in planar acquisition and in tomographic acquisition using the directional cloning technique. The NPSs from synthetic planar images were found to be within spectrum variance of experimental noise spectra and within spectrum variance of the expected constant spectrum of stationary noise.

Although the estimates of spectral variance for both experimental and simulated data were large, it can be concluded that the directional cloning

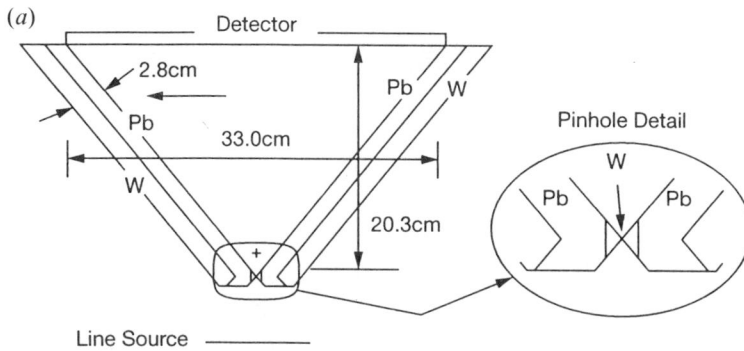

(a)

(b)

Pinhole 0 - 1.1 MeV   Shielded pinhole 0 - 1.1 Mev

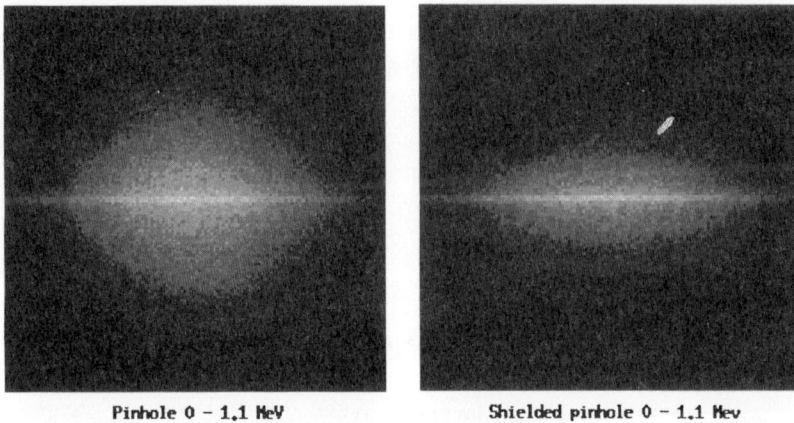

Figure 9.5   *(a) A model of a tungsten shielded collimator. (b) A compari-son between simulated images obtained from a standard 1 mm pinhole collimator and a tungsten-shielded pinhole collimator using an open energy window: 0–1.1 MeV. (c) A comparison between images obtained from a standard pinhole and a tungsten-shielded pinhole collimator using a 20% energy window centred at 204 keV. (d) A perpendicular profile through each line source imaged in (c).*

of photons did not distort the NPS to a magnitude greater than the variance of the NPS estimates.

9.5.3 Comparison of experimental and simulated data

Initial assessment of the ability of SimSPECT to accurately simulate tomo-graphic acquisition of a typical nuclear medicine phantom was carried out by comparing real and simulated SPECT data of a water-filled cylindrical phantom containing four hot spheres of varying diameters [1]. All experi-mental parameters (collimator type, source–collimator distance, number of

(c)

Pinhole 0.163 - 0.245 MeV   Shielded pinhole 0.163 - 0.245 Mev

(d)

Max = 12.000, @ [64,63]   Max = 11.000, @ [64,63]

Pinhole 0.163 - 0.245   Shielded pinhole 0.163 - 0.245

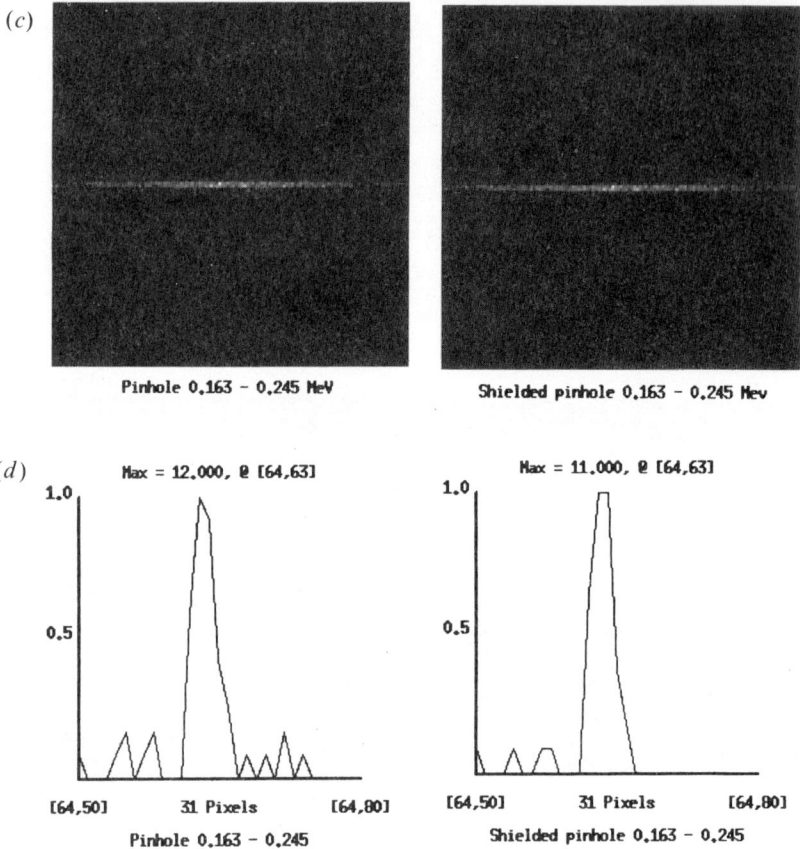

Figure 9.5   *Continued.*

projections etc) were modelled in SimSPECT. Both experimental and simulated projections were reconstructed via filtered backprojection using the same filter.

Transaxial slices through the reconstructed object were generated; simulated and experimental images were qualitatively compared by drawing profiles through images of the hot spheres. The similarity of the images (see [1]) and of line profiles through even the smallest sphere (1.3 cm diameter) indicates that the SimSPECT simulation system is accurately modelling the realistic SPECT imaging of asymmetric sources.

9.6 CURRENT APPLICATIONS

SimSPECT is well suited to quickly examining different collimator designs. For example, SimSPECT has recently been used to develop collimation strategies for imaging with ^{95m}Tc. This isotope has a half-life of 61 d and emits a 204 keV photon. Unfortunately, it also emits a number of higher-energy photons ranging from 253 to 1039 keV. These photons result in substantial septal penetration of standard lead collimators. Figure 9.5 compares the performance of a standard 1 mm lead pinhole with the same pinhole surrounded by a 14 mm tungsten shield. The source is a 0.1 mm line located 6 cm from the pinhole collimator containing 50% ^{99m}Tc and 50% ^{95m}Tc. [14]. With SimSPECT, all materials and both isotopes are incorporated accurately into the simulation. Figure 5(*b*)–(*d*) illustrates the ability of the tungsten shield to block the high-energy photons from ^{95m}Tc.

The use of positron emitters in SPECT is also being modelled via SimSPECT in order to better understand the potential improvements that different collimator designs can have on 511 keV gamma camera imaging. With SimSPECT it is a simple matter to alter collimator material or collimator dimensions to investigate the tradeoffs between reduced septal penetration and reduced sensitivity. SimSPECT is also currently being used to improve the detection of atherosclerotic lesions in human arteries in nuclear medicine imaging. This investigation was initiated by recent advances in the development of ^{99m}Tc–based radiopharmaceuticals that can target atherosclerotic lesions. This work uses SimSPECT to investigate the effect of lesion size, artery size and depth, background organ uptake, collimator geometry and energy window on the 'successful' detection of arterial lesions in single-photon nuclear medicine images.

REFERENCES

[1] Yanch J C, Dobrzeniecki A B, Ramanathan C and Berhman R 1992 Physically realistic Monte Carlo simulations of source, collimator and tomographic data acquisition for emission computed tomography *Phys. Med. Biol.* **37** 853–70

[2] Briesmeister J F 1986 A general Monte Carlo code for neutron and photon transport *Los Alamos Scientific Laboratory Technical Report* LA-7396-M.

[3] Yanch J C and Dobrzeniecki A B 1993 Monte Carlo simulation in SPECT: complete 3D modeling of source, collimator and tomographic data acquisition *IEEE Trans. Nucl. Sci.* **NS-40** 198–203

[4] Hubbell J H, Veigele W J, Briggs E A, Brown R T, Cromer D T and Howerton R J 1975 Atomic form factors, incoherent scattering functions and photon scattering cross sections *J. Phys. Chem. Ref. Data* **4** 471

[5] Whalen D J, Hollowell D E and Hendricks J S 1991 MCNP: photon benchmark problems *Los Alamos National Laboratory Technical Report* LA-12196

[6] Dobrzeniecki A B and Yanch J C 1993 Computational and visualization technique for Monte Carlo based SPECT *MIT Technical Report* MIT-WCBICL-93-01

[7] Haynor D R, Harrison D L and Lewellen T K 1991 The use of importance sampling techniques to improve the efficiency of photon tracking in emission computed tomography *Med. Phys.* **18** 990–1001

[8] Jaszczak R J, Greer K L, Coleman R E 1988 SPECT using a specially designed cone-beam collimator *J. Nucl. Med.* **29** 1398–1405

[9]   Bélanger M-J, Dobreniecki A B and Yanch J C 1993 Noise characteristics of a SPECT simulation system *Proc. IEEE Medical Imaging (San Francisco, CA)* ed L Klaisner (Piscataway, NJ: IEEE)

[10] Bélanger M-J, Dobrzeneicki A B, Yanch J C 1994 Noise characteristics of images generated by a SPECT simulation system *MIT Technical Report* MIT-WCBICL-94-01

[11] Gillen J 1992 A simple method for the measurement of local statistical noise levels in SPECT *Phys. Med. Biol.* **37** 1573–79

[12] Huesman R H 1977 The effect of a finite number of projection angles and finite lateral sampling of projections on the propagation of statistical errors in transverse section reconstruction *Phys. Med. Biol.* **23** 511–21

[13] Welch P D 1967 The use of fast Fourier transform for the estimation of power spectra: a method based on time averaging over short, modified periodigrams *IEEE Trans. Audio-Elec.* **AU-15** 70–3

[14] Lazewatsky J and Dobrzeniecki A B 1995 Gamma imaging with ^{95m}Tc and ^{99m}Tc, a design study *J. Nucl. Med.* **36** 180P

CHAPTER 10

MONTE CARLO SIMULATION OF PHOTON TRANSPORT IN GAMMA CAMERA COLLIMATORS

Daniel J de Vries and Stephen C Moore

10.1 INTRODUCTION

While the spatial resolution of the collimator is a limiting factor in determining the resolution of a gamma camera system, other factors related to the collimator design can also affect the quality of an image. When imaging low- or medium-energy radiations in the presence of higher-energy emissions, the high-energy photons can lose sufficient energy to be detected in the window set for acquisition of one of the lower energies. This 'contamination' of an image by a higher-energy photon can result from scattering in the collimator, or from penetration through the collimator followed by scattering in the detector components. These sources of degradation in image quality can arise when imaging a radionuclide with multiple emissions, such as ^{111}In, ^{123}I and ^{67}Ga, or when acquiring simultaneous, dual-isotope images from pairs of radionuclides.

To obtain a collimator that is appropriately designed for a particular radionuclide and gamma camera, a complete measure of the point spread function (PSF) of the collimator is necessary. Three PSF components may be defined to represent the several possible event topologies for photons passing through the collimator. The geometric component consists of photons which were collimated properly by passing through the holes. The penetration component consists of photons which have passed through, either partially or completely, one or more septa without being attenuated (i.e. scattered or absorbed). The collimator scatter component consists of photons that have scattered at least once in septal material prior to exiting the collimator.

Equations which approximately describe the geometric properties of parallel hole collimators were derived by Anger [1]. These equations give values for spatial resolution, efficiency and the probability of single-septum penetration, given the thickness or length of the collimator, the hole size and width or thickness of the septa. The shape of the geometric component of the PSF can be determined by analytical calculations [2, 3], and the penetration component can be determined by numerical ray tracing techniques [4–6]. However, a mathematical description of the collimator scatter contribution has eluded researchers.

Given that Monte Carlo simulation techniques have been used successfully to examine photon transport in radiology and nuclear medicine [7, 8], we investigated the application of these techniques to the study of collimator design. Some Monte Carlo programs have approximated the effect of the collimator by calculating the probability of a photon being detected within a solid angle of acceptance of a collimator hole. However, our collimator design studies required a simulation that would produce all components of the PSF, including collimator scatter, by accounting for the interactions in the collimator. Furthermore, simulated images had to be produced within a period of time that would make a design optimization process feasible.

Therefore, we developed a Monte Carlo simulation of photon transport that accounted for penetration of and scatter in the collimator [9]. Such a simulation could potentially be inefficient and time consuming, due to the low probability (less than 10^{-3} at 140 keV) of collimated photons reaching the detector with sufficient energy to contribute to the image. To address this problem, our code was designed for use with an array processor, which propagated up to 1024 photons in parallel. 'Arrays' of photons were processed through the source and phantom, the collimator and the scintillation detector. Then, with the use of variance reduction techniques, it was possible to obtain accurate estimates of the total radial PSF in less than a day.

Validation was performed [9] by imaging mono-energetic 'point' sources of ^{99m}Tc, ^{51}Cr and ^{85}Sr (140, 320 and 514 keV, respectively) on a General Electric StarCam with low-energy, general-purpose (LEGP) and medium-energy (ME) collimators. Comparisons of measured and simulated PSFs demonstrated the validity of the model and the significance of collimator scatter in the degradation of image quality.

10.2 METHODS

Originally, our simulation code was implemented in FORTRAN/77 on a 32-bit, Unix workstation from Sun Microsystems, which was equipped with an array processor (AP) from SKY Computers. After determining that the quality of the numbers was acceptable for this application, the random number generator, *rand*, provided with the Sun FORTRAN library, routines was used. The program consisted of several modules, which simulated the source and phantom, collimator and detector, through which photons were propagated to produce projection images. Although this version required an array processor, the collimator subroutine has recently been adapted for use with the SIMIND [10] Monte Carlo code.

10.2.1 Photon interactions in matter

The photon interactions that were simulated include the photoelectric effect, Compton scattering and coherent scattering. The importance of including coherent scattering was indicated by the cross sections for lead, a material typically used for collimators and camera shielding. At about 150 keV, the probability of occurrence of coherent scattering is comparable to that of Compton scattering.

For simulating interactions in various materials, fitted cross sections were generated from data published by McMaster *et al* [11]. For each material, the mass attenuation coefficients and the relative probabilities of each type of interaction were stored in lookup tables, which were loaded into memory from files at the beginning of a simulation. Values were indexed by energy in increments of 1 keV for energies from 1 to 520 keV. To determine whether a photon would interact within an object, the inverse transform method [12] was used (equation (1.5) in Chapter 1).

If an interaction was to take place within the given material, the discrete inversion technique [13] was then used to determine which interaction to simulate. Using the cross sections for the material, the interval [0, 1] was divided into subintervals, each having a fractional range determined by the probability that a particular interaction would occur. A uniformly distributed random number on the interval [0, 1] was then generated and the subinterval which included that number determined the type of interaction.

The storage of attenuation coefficients and the probabilities of the several interactions in lookup tables improved efficiency by reducing both the number of computations required and random numbers used. A similar approach was used for the scatter angles.

From the unpolarized differential cross sections for Compton and coherent scatter, polar scatter angles were generated and stored in lookup tables for each material. (As explained in [14], polarized cross sections were not required.) Using the Klein–Nishina formula and the Thomson scattering formula, Compton and coherent scatter angles were calculated, respectively, and were modified by atomic form factors from Hubbell *et al* [15]. (The computation of the differential cross sections for compound or composite materials assumed that molecular interference is negligible.) Due to the need for the sines and cosines of the scatter angles in computations, cosines were stored in the lookup tables, in place of the angle, to reduce the number of computations required during a simulation. The Compton and coherent scatter lookup tables were indexed by energy in increments of 10 keV. At each energy in the table, cosines of 60 'equiprobable scatter angles' [16] on the interval $[-1, 1]$ were stored.

10.2.2 Variance reduction techniques

Direction biasing was used for simulating the emission of a photon from the location of a decay in the radioactive source. Although other Monte Carlo simulations have achieved increased efficiency by limiting the solid angle into which a photon is emitted, such that photons would be likely to pass through the collimator holes, for our applications it was necessary to irradiate most of the face of the collimator in order to examine penetration and scatter.

Consider a photon emitted from a point source at the origin of the coordinate system, as defined in figure 10.1. At emission the photon was assigned its initial direction, given by the direction cosines of the unit momentum vector. For isotropic emission, an invariant differential element of the solid angle from a point is given by $d\Omega = d(\cos \theta) \, d\varphi$, where θ is the polar angle and φ is the azimuthal angle. In figure 10.1, the unit momentum vectors of the photons can be described by the direction cosines: $v_x = \sin \theta \cos \varphi$, $v_y = \sin \theta \sin \varphi$, $v_z = \cos \theta$.

When a PSF was produced from a simulated point source, the circular symmetry of the source was used to obtain a smooth image profile by radial averaging of the image intensity. However, the direct simulation of isotropic decay resulted in a large pixel variance near the origin of the radial PSF. This was due to the fact that fewer pixels, and therefore fewer photons, were used in the radial average at the peak in comparison to pixels at each larger radius from the peak. Therefore, direction biasing, a variance reduction technique, was used to increase the sampling of small polar angles at emission.

Although the standard method for simulating isotropic decay is to select values for θ and $\cos \theta$ at random from uniform distributions, we selected uniformly distributed values for θ instead of values for $\cos \theta$. To compensate for the resulting bias in direction, each photon began its history with an assigned weight, between zero and one, depending on the probability of being emitted in the chosen direction. By giving the photon a weight of $\sin \theta$, this procedure was formally equivalent to choosing $\cos \theta$ uniformly and, therefore, produced a uniform distribution of photon weights. By selecting values for θ rather than $\cos \theta$, however, the estimates near the origin of the radial PSF were obtained with improved precision.

When propagating photons through an attenuating phantom, it would be inefficient to allow them to be absorbed in photoelectric interactions. This problem was addressed by the variance reduction technique called survival biasing [17]. Given the known probability of occurrence of the photoelectric effect, this interaction can be simulated *implicitly* by forcing the interaction

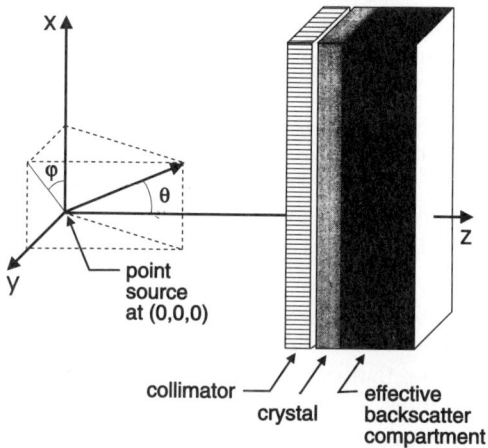

Figure 10.1 *Geometry of the components of the photon transport simulation, showing the spherical decay angles θ and φ. The point source can be replaced by other sources, which include various attenuating media. (From de Vries D J et al 1990 Development and validation of a Monte Carlo simulation of photon transport in an Anger camera. IEEE Trans. Med. Imaging* **MI-9** *431 © 1990 IEEE.)*

to be a scatter event, and then multiplying the weight of the photon by the probability that the event was *not* photoelectric absorption. Propagation of the photon was then continued.

Similarly, the explicit simulation of photoelectric absorption in the collimator would result in the propagation of many gamma photons that would never reach the detector. Furthermore, for many applications of the simulation, it is unlikely that the characteristic x-rays, Auger electrons and potential bremsstrahlung produced by photoelectric interactions in the collimator would have energies high enough to be detected within an energy window of interest, even if these radiations survived the collimator to reach the detector. Therefore, an interaction in the collimator was also forced to be a scatter event, and the weight of the photon was again multiplied by the probability that the interaction was not the photoelectric effect.

The use of survival biasing required that some photons were discarded if they did not escape the phantom or collimator after a specified number of interactions. With each interaction the energy and weight of a photon decreased, and eventually became insignificant, so that continued propagation would be unproductive. However, simply discarding a photon on the basis of an energy or weight cut-off could bias the results of a simulation [17]. Therefore, the effect of the number of interactions simulated before applying a cut-off, also referred to as the order of scatter in this context, was examined by comparing simulated radial PSFs of collimator scatter as the order of scatter was incrementally increased. The minimum order of scatter required was selected as the one after which additional interactions produced no significant change in comparison to statistical variations.

Collimator scatter PSFs were produced from photons that exited the back of the collimator after scattering in the septa and prior to entering the simulated detector. When the PSF from photons which scattered once was compared with the PSF from photons that scattered up to twice, a significant change was seen for a 320 keV source, as shown in figure 10.2. When the maximum number of interactions simulated in the collimator was increased, any differences in the PSFs were small compared to statistical variations. With the photon energy of the source decreased to 140 keV, there was no observable change in the radial PSF when two or more scatters were simulated. This demonstrated that the number of collimator interactions required for a simulation depended on the energy of the isotope. As a precaution, we used an order of scatter that was one greater than the observed minimum number required. For example, up to two collimator interactions were simulated for a source energy of 140 keV, and up to four interactions were simulated for 320 keV.

In contrast to the inefficiency of losing photons to absorption in the phantom and collimator, absorption in the detector is desirable. Therefore, the photoelectric effect was *explicitly* simulated, in addition to scatter, in the detector. It was assumed that primary electrons, characteristic x-rays and Auger electrons produced by photoelectric interactions deposited all of their energy at the interaction site. (This approximation is valid insofar as the ranges of these radiations are small compared to the intrinsic spatial resolution of the gamma camera.)

As the energy of the photon incident upon the detector increased, the probability that the photon would penetrate through the detector without interacting and contributing to the image increased. Therefore, a variance reduction technique was used to force detection. In view of the fact that the probability of interaction along the path of a photon through the detector is calculable, the location of an interaction was chosen by generating an exponentially distributed random distance to interaction along the initial path through the detector. Then the photon weight was multiplied by $[1 - \exp(-\mu d)]$, which is the probability that an interaction would occur along the path of length d.

10.2.3 Simulated sources and phantoms

For collimator design studies, the simulation of simple geometric sources was required. Point sources were used for the validation procedure and for producing the PSF for various collimator designs. In addition to a point in air, a point in an attenuating phantom was necessary. For this purpose, and for estimating the collimator efficiency with an extended source, cylindrical

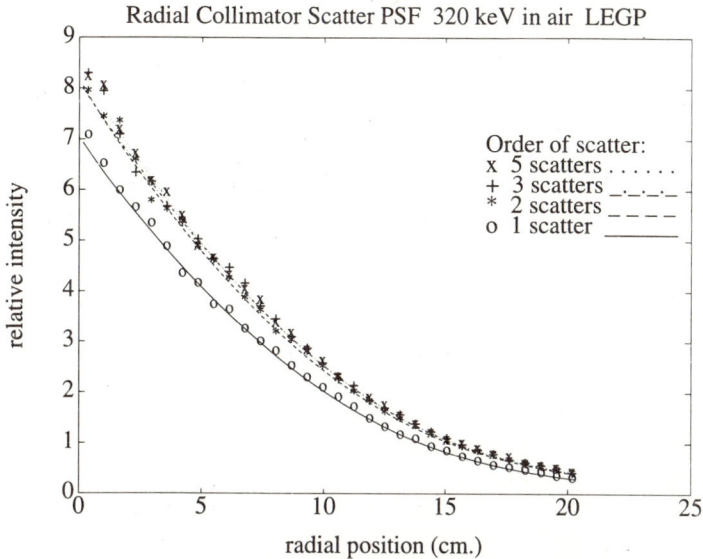

Figure 10.2 *A comparison of radial collimator scatter PSFs for various orders of scatter (i.e., one scatter in septa, up to two scatters in septa etc) at 320 keV with an LEGP collimator. PSFs were produced simply from collimated photons without explicitly simulating detection. (From de Vries D J et al 1990 Development and validation of a Monte Carlo simulation of photon transport in an Anger camera. IEEE Trans. Med. Imaging MI-9 432. © 1990 IEEE.)*

phantoms were simulated, with water approximating the attenuating effect of soft biological tissues. The contents of the cylinder could be specified to contain activity for the simulations that required an extended source.

At emission, each photon was assigned its initial spatial coordinates, energy, direction and the weight which accounted for the bias in the initial direction. For a source within a phantom, the intersection of the path of a photon with the boundary of the cylinder was calculated to determine the shortest distance to the point of exit. This distance was compared to an exponentially distributed random distance to a point of interaction. If the distance to exit was greater than the distance to an interaction, a scattering interaction was simulated, and the weight of the photon was modified according to the survival biasing scheme. A counter was used for each photon to record how many times it scattered in the phantom.

Unless a point source is positioned far from the collimator, the intensity in the image will vary depending on the position of the source with respect to the pattern of the collimator structure, i.e. the septa and the holes. To eliminate the effect of the position of the source with respect to the collimator, a spatially averaged PSF was produced by incrementally translating the point source over one spatial period of the collimator structure during each run of the simulation [3]. For each photon emitted, x and y offsets were added to the coordinates of the radioactive decay. The offsets were subtracted when the photons exited the collimator.

10.2.4 Simulated collimators

The simulation approximated the effect of parallel hole collimators that have been cast (in contrast to those made of laminated foil). The design of the collimator could be altered by changing the septal material, or by changing the geometric parameters, which defined the septal thickness, hole area and hole depth.

For ease of implementation, the only shapes allowed for holes were square or rectangular. (The way in which square holes can be used to approximate other hole geometries is discussed in the section on validation.) The rectangular volume of the collimator was defined by orthogonal planes, which represented the horizontal and vertical septa. The spatial coordinates of septal planes were stored in arrays at the beginning of a simulation.

When a photon entered the collimator, the direction cosines were used to calculate path lengths to the intersections with the planes corresponding to the front or back (x plane), left or right (y plane) and top or bottom (z plane) surfaces defining the volume of the collimator (figure 10.1). Of these intersections, the one which gave the minimum path length to exit from the collimator was used to determine the total distance the photon would travel through the collimator if no interactions occurred in the septa. If the path of the photon intersected any septa between the entrance and exit points, it was necessary to determine how much of the path was contained in septal material.

Therefore, all septal planes intersected by the photon had to be identified. Intersections with septal planes were quickly calculated with an algorithm similar to the one described by Siddon [18] for fast calculation of the radiological path for CT voxel arrays. The range of planes intersected by a photon trajectory was determined by calculating the minimum and maximum array indices corresponding to the first and last planes intersected, for both horizontal and vertical septa.

A flag was used to indicate whether a photon would contribute to the geometric, penetration or scatter component of the collimator PSF. If the path of the photon did not intersect any septa, the path through the collimator was contained entirely within a hole, and the flag was set to indicate that the photon belonged to the geometric component. On the other hand, if a photon travelled entirely within a septum, or if its path intersected one or more septa, the distance travelled in septal material was calculated and compared to an exponentially distributed random distance to an interaction site.

In the latter case, path lengths from the point of entrance into the collimator to each plane intersected were calculated and stored in one of two arrays: one for intersections with horizontal planes and one for intersections with vertical planes. Pointers were used to step through the arrays, and flags were used to indicate whether the segment of the path under consideration was in a hole or in a septum. At each intersection, if the region traversed up to that point was contained in a septum, the distance travelled in septal material was accumulated. Each time this distance was increased, it was compared to the randomly generated distance to an interaction site, until the photon either interacted or exited the collimator.

If the photon passed through part or all of one or more septa, but exited the collimator without an interaction, the flag was set to indicate that the photon had penetrated through the collimator. If a photon interacted in the septal material, a scatter interaction was forced to occur, according to the survival biasing technique previously described, and the flag was set to indicate that the photon had scattered in the collimator. If it was determined that a photon exited from the front or side surfaces of the collimator, its history was terminated.

The circular symmetry of the source/phantom geometry was exploited to improve the efficiency further by simulating a second camera directly opposed to (180° from) the first. The spatial coordinates and direction cosines of any photon which hit the second collimator were rotated to the corresponding position on the face of the first collimator (through which all photons were then propagated). This approach minimized the number of photons which missed a collimator, while allowing for a large solid angle of emission from the source to sufficiently irradiate the face of a collimator. Moreover, with a source located off centre (closer to one collimator than the other), it was possible to acquire two different projections simultaneously by recording which of the two cameras had been hit by a photon.

10.2.5 Simulated detector

A gamma camera was simulated with a detector that approximated a scintillation camera with an NaI crystal. A rectangular volume was defined by the intersection of six orthogonal planes. The path length for an incident photon to pass through and exit from the detector was determined by finding the minimum distance to an intersection with one of the six planes. An exponentially distributed random distance to a point of interaction was generated and compared to the distance to the exit point.

If an incident gamma photon was absorbed by the photoelectric effect, its energy was deposited in the crystal. If a photon interacted in the crystal by Compton scattering, the energy imparted to the recoil electron was deposited in the crystal. Following the first (forced) interaction along the initial path through the detector, the photon proceeded according to 'natural' processes (i.e. without further biasing). If energy was deposited at multiple sites (e.g. Compton scatter followed by photoelectric absorption), the centroid of the positions of energy deposition was calculated and the summed energy was considered to be deposited at this point.

Depending on the specified image matrix size, pixel coordinates were calculated from the spatial coordinates of the point of energy deposition. The intrinsic spatial resolution of the camera was simulated by 'blurring' the x and y coordinates of the point of energy deposition position, using a Gaussian-distributed random variable to add offsets to the coordinates. A resolution of 4 mm full width at half maximum (FWHM) was simulated. Similarly, the energy resolution of the camera was approximated with Gaussian blurring, based on the measured resolution of the gamma camera used for our validation experiments. This resulted in a resolution of 11% FWHM for energies of 140 keV or greater.

Depending on which subroutine was linked to the simulation code, a choice of two different formats was provided for the simulation results. With the format provided by one routine, two-dimensional arrays were used to store each component of an image (i.e. geometric, penetration and collimator scatter). Pixels of each array contained sums of photon weights, as a result of the variance reduction techniques used to represent the probability of various events throughout the history of a photon. When a photon was detected with an energy that fell within specified window thresholds, the weight of the photon was added to a pixel. In addition, given the radial symmetry of the sources and phantoms, one-dimensional radial arrays (histograms) of weights, counts and squares of weights were produced for each component, with bins indexed using the radial position (i.e. $(x^2 + y^2)^{1/2}$) of a detected photon. This allowed direct determination of the profile of the radial PSF, and of the statistical uncertainty of the value at each point of the profile (see appendix 10.2).

The other format for simulation results was 'list-mode' output, which allowed greater freedom for post-processing of data. With this format, information about each photon that deposited energy within a specified range was written to a file. The information saved for each photon included

- a byte for the pixel coordinates
- a byte for the index for the radial arrays used for PSFs
- a byte for the weight of the photon
- a byte for the energy
- a bit indicating whether the energy had been scaled prior to storage in byte format
- a bit indicating whether scattering occurred in the phantom
- a bit indicating which of two collimators the photon had entered
- two bits indicating to which component (geometric, penetration or collimator scatter) the photon contributed

- a bit indicating whether the photon had deposited less than peak energy and had Compton scattering in its history
- a bit indicating whether the photon had backscattered *into* the crystal
- a bit indicating from which of two possible photopeaks the photon had originated.

Initially, we simulated a detector consisting solely of the NaI crystal. However, preliminary comparisons of PSF profiles from simulated and experimental measurements of a 320 keV (^{51}Cr) point source, acquired in a 20% energy window centred at 159 keV (for ^{123}I), indicated that the simulation had substantially underestimated the experimentally measured number of detected events (figure 10.3(a)).

Examination of the energy spectrum revealed that the energy window included a region that contained a peak caused by two classes of backscattered photons. For 320 keV photons, backscattering in the crystal could deposit up to 178 keV, while photons detected after backscattering from camera components behind the crystal could deposit up to 142 keV. The second class of photons, which backscattered *into* the crystal, had not been accounted for by the simulation. This class included photons that backscattered from the light pipe, photomultiplier tubes, mu-metal magnetic shielding or other structures in the real camera and then deposited all or part of their remaining energy in the NaI crystal.

At medium to high gamma energies where Compton scattering is the predominant interaction, materials found in the components behind the crystal (such as iron, aluminium and Pyrex® glass) have approximately the same Compton scatter mass attenuation coefficients. Therefore, the linear attenuation coefficients for each material depend primarily on the physical density.

Although explicit simulation of each component was considered impractical, the similarity of the attenuation coefficients motivated the search for a single layer of material behind the simulated crystal (the 'backscatter compartment' in figure 10.1), which would approximate the appropriate backscatter. Therefore, it was necessary to determine an effective density and thickness for this layer, based on the distribution of mass and composition of components in the camera we were trying to simulate. Appendix 10.1 explains how values were estimated for these parameters from measurements of camera components. Using cross-section data for Pyrex® glass, the effective backscatter compartment was simulated with a slab of material having a thickness of 6.6 cm and a density of 66% of that of Pyrex®.

10.2.6 User interface

The initialization for each run of the simulation was handled primarily with a program which presented menus to the user for the specification of

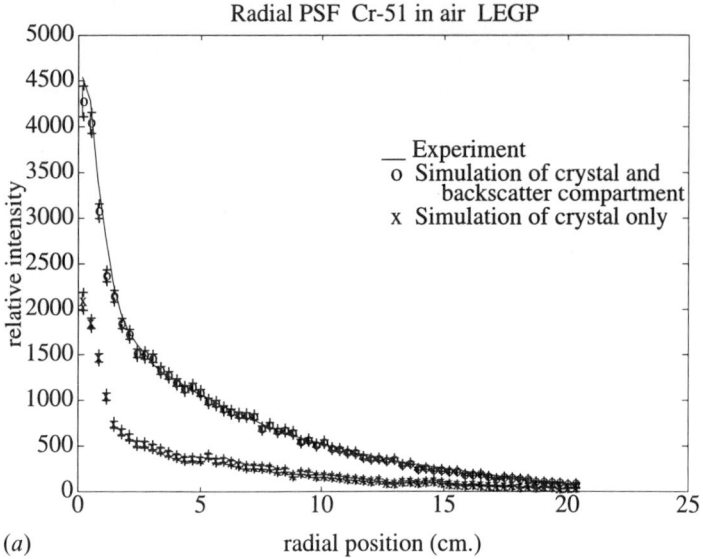

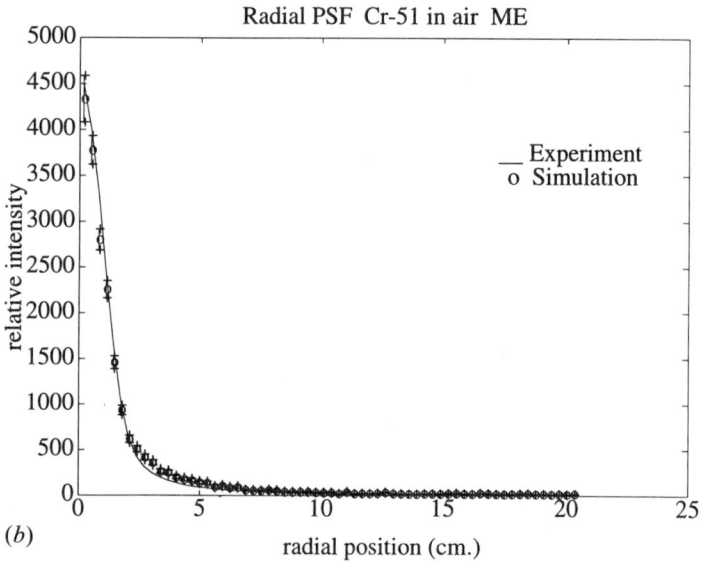

Figure 10.3 *A comparison of experimental and simulated PSFs for the LEGP and ME collimators for 320 keV (^{51}Cr using a 20% ^{123}I energy window). Simulated PSFs were scaled by the ratio of the number of photons emitted by the experimental source to the number emitted by the simulated source. Error bars indicate the statistical uncertainty (± 1 SD) for simulated values. (a) LEGP collimator (note the significance of backscatter in the gamma camera.) (b) ME collimator. (From de Vries D J et al 1990 Development and validation of a Monte Carlo simulation of photon transport in an Anger camera. IEEE Trans. Med. Imaging **MI-9** 434, 436. © 1990 IEEE.)*

Table 10.1 *Collimator parameters. Data in this table are from de Vries D J et al 1990 Development and validation of a Monte Carlo simulation of photon transport in an Anger camera. IEEE Trans. Med. Imaging **MI-9** 435. © 1990 IEEE.*

	LEGP (cm)	ME (cm)
Septal thickness (s)	0.03	0.14
Hole size (d)	0.25	0.34
Hole depth (a)	4.1	4.2
Source-to-collimator distance (b)	12	12
Field of view	41	41

Collimators: simulated, Pb with square holds; experimental, cast Pb with hexagonal holes.

parameters for source, phantom, collimator and detector. This information was stored in an ASCII file which was read at the start of a simulation. Some information, such as how many decays to simulate, was entered at prompts at the start of a run. Various supporting programs were written to process the output of a simulation, producing results such as energy spectra, images and statistics.

10.2.7 Validation of the simulation program

Using point sources of several mono-energetic radionuclides, the simulation program was validated by comparing simulated and experimentally measured radial PSF profiles. For experimental measurements, a GE Star-Cam was used to acquire projection images with two different cast, hexagonal hole collimators: an LEGP collimator and an ME collimator. By pipetting a small volume of a radionuclide into a micro-centrifuge tube suspended from a stand at 12 cm from the collimator, point sources of ^{99m}Tc (140 keV), ^{51}Cr (320 keV) and ^{85}Sr (514 keV) were approximated.

Projection images were acquired in a 20% energy window centred on 140 keV for ^{99m}Tc and in a 20% energy window centred on 159 keV (simulating a window used for ^{123}I) for the two higher-energy radionuclides. The images from the 320 and 514 keV photons were used to examine the effect of the high-energy gamma rays from ^{123}I (and possibly from ^{124}I contamination) on the 159 keV image. Table 10.1 shows the collimator parameters used for the experiments. The crystal thickness was 0.935 cm, simulated image size was 128×128 and the pixel size was about 0.32 cm^2.

The activity of the ^{99m}Tc source was measured with a dose calibrator, with an accuracy estimated at $\pm10\%$ accuracy due to the geometry of the approximated point source. The activities of the other sources were measured by New England Nuclear Medical Products (Dupont) with an accuracy of $\pm4\%$. These activities were used to calculate the number of gamma ray photons emitted during the image acquisition.

For the comparison of simulated and experimental PSFs, a scale factor for the simulated results was calculated from the ratio of the number of photons emitted during the experiment to the number emitted in a simulation run. Simulated PSFs were acquired by looping through the photon propagation process typically 5000–10 000 times, with an array of 1024 photons emitted per loop. Due to the high counting rate during the experiment, a dead-time correction was applied to the [85]Sr image.

Although the penetration component of an image exhibits a pattern dependent upon the shape and pattern of the collimator holes, the shapes of the average geometric transfer functions for square and hexagonal hole collimators are almost identical [3]. Square hole and hexagonal hole collimators with the same open area and lead content (mass of lead per unit area) would be expected to have almost identical *radially averaged* PSFs.

10.3 RESULTS AND DISCUSSION

After including the effective backscatter compartment, the radial PSF of the simulated 320 keV photons, acquired in a 20% photopeak window centred at 159 keV, showed good agreement with the measured radial PSF for both collimators (figure 10.3). The statistical uncertainty of the simulated data points is indicated by the error bars showing ±1 standard deviation.

The ability to examine separately the components of an image based on the events in the collimator is demonstrated in figure 10.4. Of the photons detected in the ^{123}I energy window from a 320 keV point source, those which scattered at least once in the LEGP collimator comprised 27% of the total PSF, while photons which penetrated one or more septa comprised 70% of the total PSF. In comparison, the 320 keV photons which scattered at least once in the ME collimator before detection in the ^{123}I energy window comprised 18% of the total PSF, while photons which penetrated one or more septa comprised 33%. These results and those for the other radionuclides are summarized in table 10.2. The table also includes the simulation run times for the hardware-dependent code developed for use with the AP, based on obtaining sufficient statistics for smooth simulated PSF component profiles. Depending on the photopeak energy and the collimator efficiency, simulation run times of about 48 h or less produced useful PSFs, consisting of the sum of all components. (The time required increased as the photon energy and collimator lead content were increased, or as the collimator efficiency was decreased.)

For additional information and discussion about the results of these experiments, see [9]. For examples of applications see Chapter 13 and [19].

Table 10.2 *PSF component contributions, statistical uncertainties and run times. Data in this table are from de Vries D J et al 1990 Development and validation of a Monte Carlo simulation of photon transport in an Anger camera. IEEE Trans. Med. Imaging* **MI-9** *437.* © *1990 IEEE.*

Energy (keV)	% geometric	% penetration	% coll. scatter	Run time (h)
LEGP collimator				
140	94.5 ± 0.11	3.6 ± 0.11	1.9 ± 0.04	9.5
320	3.5 ± 0.05	69.5 ± 0.33	27.0 ± 0.31	16.0
514	0.5 ± 0.01	66.2 ± 0.18	33.3 ± 0.18	20.5
ME collimator				
320	49.6 ± 0.86	32.8 ± 1.01	17.6 ± 0.59	14.5
514	3.5 ± 0.06	48.7 ± 0.41	47.8 ± 0.40	39.5

(Note: 320 keV and 514 keV images were acquired in a 20% ^{123}I window (159 keV).)

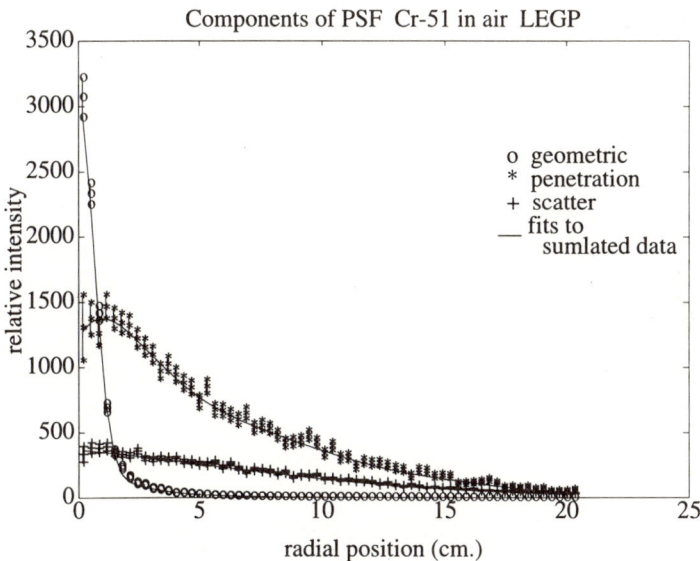

Figure 10.4 *Geometric, penetration and scatter components for simulated PSFs for 320 keV photons using a 20% ^{123}I energy window and the LEGP collimator. Error bars indicate the statistical uncertainty (± 1 SD). For better visualization, solid lines show curves fit to the data. The geometric component was fitted with a Gaussian plus an exponential, and the penetration and collimator scatter components were fitted with polynomials. (From de Vries D J et al 1990 Development and validation of a Monte Carlo simulation of photon transport in an Anger camera. IEEE Trans. Med. Imaging* **MI-9** *436.* © *1990 IEEE.)*

10.4 FUTURE WORK

The collimator module that has recently been adapted for use with the SIMIND code (described in Chapter 11) is going through a process of testing and validation. This standard FORTRAN subroutine can be linked in place of the SIMIND collimator. Through values passed as parameters, the routine receives the required information about the collimator and the photon entering it. It returns new information (e.g. energy, position and direction) about the collimated photon, and returns a code indicating whether the photon belongs to the geometric, penetration or collimator scatter class.

Future additions to the collimator code may include the production of characteristic x-rays by photo-electric absorption in the collimator, in order to study the effects on dual-radionuclide imaging with isotopes such as ^{201}Tl and ^{99m}Tc. Also, for higher-energy sources the addition of pair-production interactions may be of interest. Given that all interactions are processed in a separate subroutine of the collimator code and cross sections are stored in files, such additions may require relatively few modifications.

10.5 CONCLUSION

For some applications, the simulation of photon interactions in gamma camera collimators can be important. When imaging with a photopeak of a radionuclide which also emits high-energy photons (e.g. ^{123}I, ^{111}In, ^{67}Ga), the degradation of the image by photons which scatter in the collimator can be equal in importance to the degradation caused by septal penetration. In addition to these effects, it was demonstrated that under certain conditions, high-energy photons can make a significant contribution to an image due to backscattering from detector components located behind the scintillation crystal. The program described in this chapter provides a means for studying these effects.

APPENDIX 10.1 CONSTRAINTS FOR DETERMINATION OF EFFECTIVE VALUES FOR THE BACKSCATTER COMPARTMENT

To determine effective values for the density and thickness of the single-layer backscatter compartment used to approximate the real camera, a three-compartment model of the components behind the NaI crystal was first developed, based on the estimated distribution of mass in the GE StarCam used for the validation experiments. The measured length of a photomultiplier tube (PMT) was ~10 cm and, with mu-metal shielding, its total mass was ~200 g. Therefore, the first section of the model represented the 1.25 cm

Pyrex® light pipe, the optical coupling material and the Pyrex® entrance window of the PMT. The thickness of this section was estimated at 1.5 cm over a hexagonal tube face area of 28.7 cm². The photocathode end of the PMT and mu-metal shielding comprised the second section of the model. Its thickness was 3.5 cm and its mass was estimated at 110 g. The remainder of the PMT was represented by the third section.

Assume for this model that there are three slabs of material with thicknesses x_1, x_2 and x_3, and linear attenuation coefficients μ_1, μ_2 and μ_3. For a photon path which enters the first slab normal to the face, the total probability of an interaction anywhere within the three regions is given by

$$K = 1 - e_1 e_2 e_3 \tag{10.1}$$

where $e_1 = \exp(-\mu_1 x_1)$, $e_2 = \exp(-\mu_2 x_2)$ and $e_3 = \exp(-\mu_3 x_3)$.

The probability density function (pdf) for an interaction at a distance x is given by

$$
\begin{aligned}
p(x) &= [\mu_1 \exp(-\mu_1 x)]/K &&\text{for region 1} \\
p(x) &= [e_1 \mu_2 \exp(-\mu_2 x)]/K &&\text{for region 2} \\
p(x) &= [e_1 e_2 \mu_3 \exp(-\mu_3 x)]/K &&\text{for region 3.}
\end{aligned}
\tag{10.2}
$$

By integrating this pdf with respect to x over the thickness of the three compartments, it is straightforward to show that the mean interaction distance is given by

$$E(x) = (f_1 + e_1 f_2 + e_1 e_2 f_3)/K \tag{10.3}$$

where

$$
\begin{aligned}
f_1 &= (1 - e_1)/\mu_1 - x_1 e_1 \\
f_2 &= x_1(1 - e_2) - x_2 e_2 + (1 - e_2)/\mu_2 \\
f_3 &= (x_1 + x_2)(1 - e_3) - x_3 e_3 + (1 - e_3)/\mu_3.
\end{aligned}
$$

Now consider an 'equivalent' model with one compartment consisting of a single slab of material of thickness x_{eff} and linear attenuation coefficient μ_{eff}. The values of x_{eff} and μ_{eff} can be determined by two constraints. The first requires that the mean interaction distance in the single (effective) compartment, $E(x_{\text{eff}})$, be equal to $E(x)$, as defined above; and the second requires that the probability of an interaction, $1 - \exp(-\mu_{\text{eff}} x_{\text{eff}})$, be equal to K, as defined above. Based on these constraints and the values given in the first paragraph, a thickness of 6.6 cm and a density of 66% of that of Pyrex® were used for the effective backscatter compartment.

APPENDIX 10.2 ESTIMATE OF STATISTICAL UNCERTAINTY FOR SIMULATED POINT SPREAD FUNCTIONS

For estimates of the statistical uncertainty of the value at each bin of the radial PSF, let W be a random variable representing the weight of a detected photon, and N be a random variable representing the number of photons contributing to a bin. The value in a bin, denoted by the random variable Y, is calculated by the summation

$$Y = \sum_{i=1}^{N} W_i. \tag{10.4}$$

The variance of Y is then defined as

$$\sigma_Y^2 = E(N)\sigma_w^2 + \sigma_N^2 E^2(W). \tag{10.5}$$

The notation E is the expected value of a random variable. If N is large, the mean and variance of the random variable N are given by

$$E(N) = N \tag{10.6}$$

and

$$\sigma_N^2 = N. \tag{10.7}$$

The mean of the variable W is

$$E(W) = Y/N \tag{10.8}$$

and the variance is given by

$$\sigma_{W^2} = \frac{1}{(N-1)} \sum_{i=1}^{N} (W_i - E(W))^2 = \frac{1}{(N-1)} \sum_{i=1}^{N} [W_i^2 - Y^2/N]. \tag{10.9}$$

The standard deviation of Y is obtained by substituting equations (10.5)–(10.8) into (10.4) and taking the square root

$$\sigma_Y = \sqrt{\frac{N}{(N-1)} \left[\left[\sum_{i=1}^{N} (W_i^2) - \frac{Y^2}{N} \right] + \frac{Y^2}{N} \right]}. \tag{10.10}$$

For $N \gg 1$, equation (10.10) can be simplified to

$$\sigma_Y = \sqrt{\sum_{i=1}^{N} W_i^2}. \tag{10.11}$$

If each of the photons has equal weight, then equation (10.11) is simplified to

$$\sigma_Y = \sqrt{NE(W)^2} = Y/\sqrt{N}. \tag{10.12}$$

This shows, as expected, that the fractional error for a bin value, σ_Y/Y, is inversely proportional to the square root of the number of photons contributing to the bin. (The equations in appendix 10.2 are taken from an article by de Vries *et al* [9].)

REFERENCES

[1]   Anger H O 1964 Scintillation camera with multichannel collimators *J. Nucl. Med.* **5** 515–31

[2]   Barrett H H and Swindell W 1981 *Radiological Imaging. The Theory of Image Formation Detection and Processing* 2 vols (New York: Academic)

[3]   Metz C E, Atkins F B and Beck R N 1980 The geometric transfer function component for scintillation camera collimators with straight parallel holes *Phys. Med. Biol.* **25** 1059–70

[4]   Muehllehner G and Luig H 1973 Septal penetration in scintillation camera collimators *Phys. Med. Biol.* **18** 855–62

[5]   Beck R N and Redtung L D 1985 Collimator design using ray-tracing techniques *IEEE Trans. Nucl. Sci.* **NS-32** 865–9

[6]   Newiger H and Jordan K 1985 Optimization of collimators for imaging positron emitters by a gamma camera *Eur. J. Nucl. Med.* **11** 231–4

[7]   Chan H and Doi K 1983 The validity of Monte Carlo simulation in studies of scattered radiation in diagnostic radiology *Phys. Med. Biol.* **28** 109–29

[8]   Raeside D E 1976 Monte Carlo principles and applications *Phys. Med. Biol.* **21** 181–97

[9]   de Vries D J, Moore S C, Zimmerman R E, Mueller S P, Friedland B and Lanza R C 1990 Development and validation of a Monte Carlo simulation of photon transport *IEEE Trans. Med. Imaging* **MI-9** 430–8

[10] Ljungberg M and Strand S-E 1989 A Monte Carlo program for the simulation of scintillation camera characteristics *Comput. Methods Programs Biomed.* **29** 257–72

[11] McMaster H W, DelGrande N K, Mallett H, Hubbell J H and National Bureau of Standards 1969 Compilation of x-ray cross sections (Livermore, CA: University of California—Livermore)

[12] Rubinstein R Y 1981 *Simulation and the Monte Carlo method* (New York: Wiley)

[13] Lewis T G and Smith B J 1979 *Computer Principles of Modeling and Simulation* (Boston, MA: Houghton)

[14] Rosenthal M S and Henry L J 1990 Scattering in uniform media *Phys. Med. Biol.* **35** 265–74

[15] Hubbell J H, Veigele W J, Briggs E A, Brown R T, Cromer D T and Howerton R J 1975 Atomic form factors, incoherent scattering functions and photon scattering cross sections *J. Phys. Chem. Ref. Data* **4** (3) 471–538

[16] Kalendar W A 1979 Determination of the intensity of scattered radiation and the performance of grids in diagnostic radiology by Monte Carlo methods *PhD Thesis; Wisconsin Medical Physics Report* WMP-102

[17] Williamson J F 1988 Monte Carlo simulation of photon transport phenomena: sampling techniques *Monte Carlo Simulation in the Radiological Sciences* ed R L Morin (Boca Raton, FL: Chemical Rubber Company)

[18]   Siddon R L 1985 Fast calculation of the exact radiological path for a three-dimensional CT array *Med. Phys.* **12** 252–5
[19]   Gagnon D, Luperriere L, Pouliot N, de Vries D J and S C Moore 1992 Monte Carlo analysis of camera-induced spectral contamination for different primary energies *Phys. Med. Biol.* **37** 1725–39

CHAPTER 11

THE SIMIND MONTE CARLO PROGRAM

Michael Ljungberg

11.1 INTRODUCTION

The Monte Carlo simulation code, SIMIND [1], describes a clinical SPECT camera and can easily be modified for almost any type of calculation or measurement encountered in SPECT imaging. The entire code has recently been changed to Fortran-90 and includes versions that are fully operational on VAX-VMS systems, various Unix systems such as DEC and Sun, and on MS-DOS (Lahey LF90 Compiler). Most of the main code structure is similar for all of the operating systems, but in cases where the operating system becomes unique, additional information on the code as it pertains to the specific system is provided. The code is strict Fortran-standard with only two system-dependent routines, namely a *getarg* function that reads the command line arguments from the operating system and a spawn command routine that calls the operating system and performs certain commands, such as 'dir'. These two routines are, however, often included by the different compiler vendors in their system libraries. Apart from these routines, SIM-IND also uses the ANSI escape sequences for cursor positioning and screen handling.

The SIMIND system consist of two main programs, named CHANGE and SIMIND. The CHANGE program provides a menu-driven way of defining the system that will be simulated and of writing the data to external files. The actual Monte Carlo simulation is performed by the program SIMIND, that reads the input files, created by CHANGE, and writes results to the screen or to different data files. In this way, several input files can be prepared and loaded into a command file for submission to a batch queue, a convenient way of working since Monte Carlo simulations by default are time consuming. Figure 11.1 shows a flowchart that describes the different file structures for SIMIND and CHANGE.

Included in SIMIND there is a logical flag system for the user to turn on and off different features, such as SPECT simulation, phantom simulation,

145

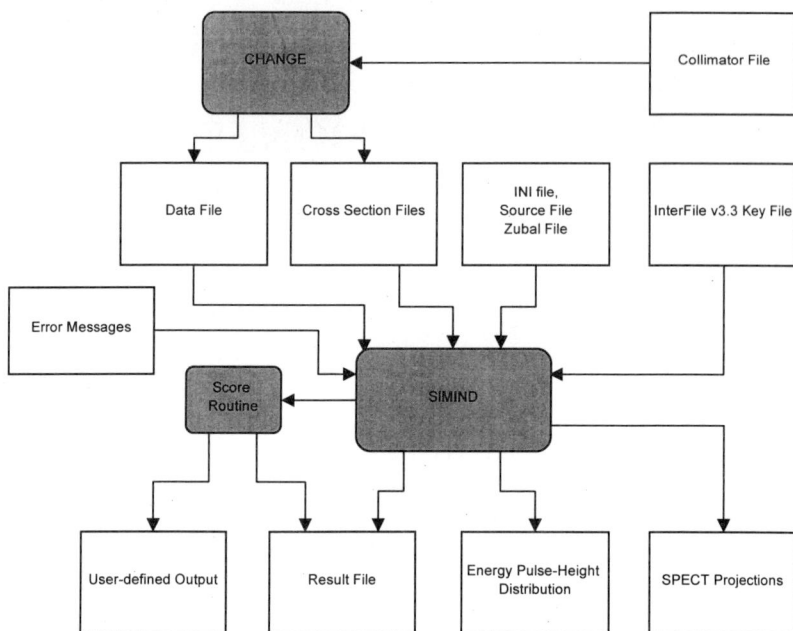

Figure 11.1 *A flowchart describing the file structure for the SIMIND and CHANGE programs.*

energy resolution etc, without the need to redefine input variables. This can also be performed from the command line.

The new FORTRAN-90 language allows dynamic allocation of memory, which is very convenient since arrays can be defined at execution and thus matrix sizes and energy pulse-height distribution array length can be defined in CHANGE and read in as ordinary variables. Instead of COMMON blocks the new concept of MODULES is used. Dynamical allocated arrays thus significantly reduce the need for recompilation. Many default values, earlier defined as PARAMETER statements are now in a special initialization file that the user can easily modify.

A standard file format is sometimes needed to be able to import simulated data to commercial imaging processing systems easily. In the nuclear medicine community, concern has been raised over the development of an appropriate mechanism for the transfer of nuclear medicine image data files between computer systems from different vendors. This has resulted in a specification (Interfile V3.3) for an intermediate file format with a list of key-value pairs for the header data associated with nuclear medicine image data files [2]. Many vendors have adopted this format as an aid to exchange data. Included in SIMIND is such a possibility of writing images using the Interfile V3.3 file format.

11.2 BASIC ALGORITHM

The Monte Carlo program SIMIND is based on the use of uniformly distributed random numbers for modelling the random processes of the different photon interactions. In short, the code works as follows: photons emitted from simulated decay in the phantom are followed step by step towards the scintillation camera. Since the history of acquired events is not lost as is true in practical imaging, important parameters that are not accessible by measurements (for instance, the numbers of scatter interactions in the phantom for a particular photon history or scatter order, scatter angles, imparted energy etc) can always be deduced from simulated data.

Several variance reduction techniques have been employed, such as forced interaction and detection, importance sampling and limited-direction emission. This is then compensated by a proper calculated photon history weight (see Chapter 2 for more details about variance reduction methods).

Cross-section tables for the photoelectric interactions, coherent and incoherent scattering and pair production have been generated by the XCOM program, written by Berger and Hubbell [3]. Form factors and scattering functions, used for the coherent scattering sampling, and the correction for bounded electrons are taken from Hubbell *et al* [4, 5]. Included in the cross-section table is the discontinuity at the K shell, an effect important for material with relatively high atomic number.

The simulation of the scintillation camera collimator is based on geometrical calculations to check whether the photon will pass through a collimator hole. The shape of the collimator holes is taken into consideration. Included in CHANGE is a large data base that covers most of the commercial collimators available for both SPECT and planar imaging. No explicit simulations of photon interactions and scattering in the collimator are included. An approximation of the septal penetration of photons through the edges of the hole walls is carried out by calculating an effective collimator thickness from its physical thickness by the formula $d_{eff} = d_{phys} - 2/\mu(h\nu)$. The possible collimator types are the parallel hole collimator, slant hole collimator and fan beam collimator. Work has been done to include an analytical collimator, developed by Frey and Li, University of North Carolina, Chapel Hill. The collimator model is based on an analytic formulation of the geometric response for parallel [6] and converging beam [7] collimators. A modification of the collimator, developed by DeVries and described in Chapter 10, has also been made in order for it to be linked to SIMIND.

If the photon passes through the collimator, it is followed in the NaI(Tl) scintillation crystal until it is absorbed or escapes the crystal. Characteristic x-rays, emitted from photoelectric absorption, can be simulated. Simulation of pair production is included when the photon energy is above the energy threshold $2mc^2$. The imparted energy is scored at each interaction site and the X and Y coordinates for the image are calculated from the centroid of these energy depositions.

A layer of different material can be defined around the scintillator volume to simulate, for example, an aluminium protecting cover. Here, photon interactions are simulated but no scoring of imparted energy is made.

In most situations, the backscattering from the light guide and photomultipliers is not considered important. However, when simulating radionuclides that emit multiple high-energy photons, some of these photons can be backscattered into the crystal and contribute to an energy window in the lower-energy region. To include the effect from these events in a simulation, a volume can be defined with an effective thickness that corresponds to the equipment behind the NaI(Tl) crystal. If a photon penetrates the crystal, it is then followed in this volume until escape or backscattering back into the NaI(Tl) crystal.

SPECT simulation can be performed for both circular and noncircular orbits. Projections are stored as floating point matrices, since variance reduction techniques are used. Simulations of 360 and 180° rotation modes can be selected with an arbitrary start angle.

Statistical variations in the energy signal are simulated by convolving the imparted energy in the crystal from every photon history 'on line' with an energy-dependent Gaussian function (with a full width at half maximum (FWHM) that depends on $1/\sqrt{E}$). The imparted energy and the spatial location of each interaction site are calculated. If the simulated energy signal is within a pre-defined energy window the apparent position of the event is calculated from the centroid for the imparted energy in the scintillation crystal and is stored as corresponding projection matrices for further evaluation.

11.3 INTERACTIONS IN THE PHANTOM

In figure 11.2 the different steps when simulating the photon interactions in the phantom are shown. A photon history starts by sampling a decay position within the source volume. A random number then determines whether the emitted photon is a primary photon that escapes from the phantom without further interaction or whether the photon will be scattered inside the phantom. For the primary case, the photon is 'forced' to penetrate the phantom without interaction. The probability for this occurrence is assigned to the photon history weight (PHW) (the probability for the photon to travel the simulated path) and the photon is followed in a direction towards the detector system.

For the case of a scattered photon, a sample is taken of how many times the photon will be scattered before escaping from the phantom. An isotropic direction is sampled and the distance to the phantom surface along the direction is determined. The probability of interaction within this distance is calculated and the photon is then 'forced' to interact within this distance

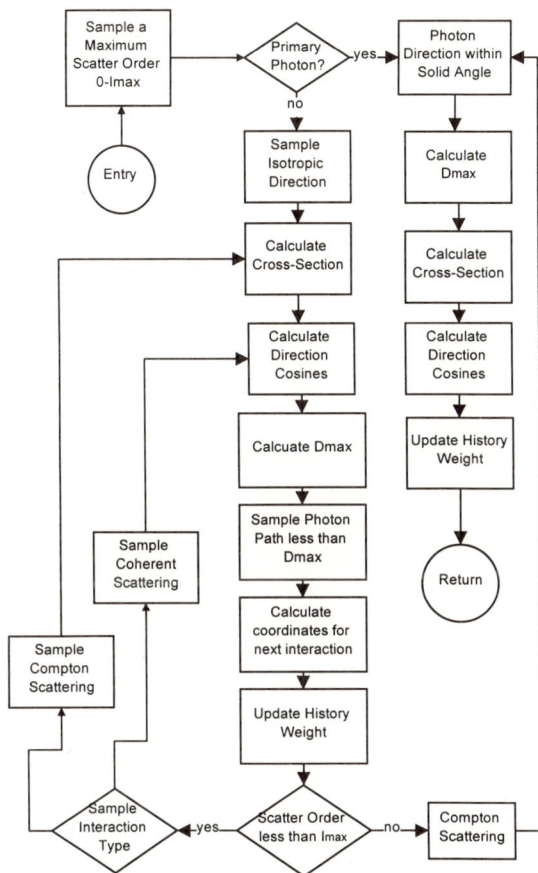

Figure 11.2 *A flowchart describing the steps simulating photon interaction in the phantom.*

by sampling a truncated photon path. The PHW is adjusted for the truncated path length. The type of interaction is then selected and the photon is followed by the technique described above, until the appropriate number of scattering events has been simulated. No photoelectric interaction is allowed. In the last interaction, the photon is 'forced' to be Compton scattered in a direction towards the detector and to escape without any further interaction.

The calculation of the path length from an interaction point to the boundary is performed in two different ways, depending on the phantom type being used. For the simple intrinsic phantoms, the path length is calculated analytically using equations obtained for the ellipsoids, cylinders and rectangular phantoms. When simulating clinically realistic nonhomogeneous objects, the computer phantom is defined by integer matrices where each matrix describes the density distribution in a particular transversal section of the object. The reason SIMIND works with density distribution and not attenuation distribution is that calculating the attenuation as a function of photon energy is necessary. If the maps were scaled to attenuation values then one would have to have a set of maps stored on discs for each photon energy. Now, the attenuation coefficient is the product of the density values from the maps, and the total and partial mass attenuation coefficients read into a table allocated by the program. In the recent version of SIMIND, two different phantom materials can be selected (mainly used to differentiate between bone tissue and muscle tissue). The selection of which material will be used depends on the magnitude of the particular density value. If this value is greater than some predefined threshold, the second table will be used.

To obtain a particular photon path in the voxel phantom, the photon is followed using small steplengths. An attenuation coefficient is calculated by accumulating the product of the mass attenuation coefficient and the density values that are closest to the path. This is done until the phantom surface is reached. From this total attenuation coefficient, a truncated attenuation coefficient is then sampled. The photon is once again followed from the starting point along the path in the current direction. The next interaction coordinates will then be at the location where the new accumulated attenuation coefficient exceeds the sampled attenuation coefficient.

11.4 PROGRAMMING SIMIND

The native SIMIND program can simulate different types of detector parameter and create images, SPECT projections and energy pulse-height distributions. To take full advantage of the Monte Carlo code, however, some programming is necessary. Generally, this can be a major task if the user is not familiar with the particular source code. Furthermore, problems will occur keeping track of different versions if the author of the main code makes modifications. An option to link a user-written subroutine has therefore been included. This has the feature of separating codes, written by a user, from the main distributions of SIMIND. By communicating with the main code using subroutines and global common blocks rather than adding source codes directly into the main program, the user can access all important variables without requiring an extensive knowledge of how the actual SIMIND code is written. The advantage is thus the ability to write special routines for different simulations and simply link the compiled versions of these to the main program when needed. The user can write three basic

Table 11.1   *The different stages in the photon history where the scoring routine is called.*

iopt	Stage in the history
0	After reading data from the SMC file.
1	Before starting the main loop of the program.
2	After generation of the decay point X_0, Y_0, Z_0 and the direction. If phantom is selected then after escape from the phantom. If SPECT is selected then after the phantom/source has been rotated.
3	After n photon interactions in the crystal.
4	After the photon history has been terminated.
5	After a complete simulation of a SPECT projection.
6	After simulating the total number of photon histories.
7	During printout, so the user can add results to the result file.
8	After an interaction in the phantom. The call is made after calculation of new cross-section data and direction cosines but before calculating the coordinates for the new interaction point.

types of subroutine: the *score* routine, the *source* routine and the *isotope* routine.

11.4.1 The score routine

A call is made to the scoring routine at several distinctive stages during the simulation of a photon history which enables the user to decode an integer flag and thus determining where the photon is 'located' at the time when the call was made. An integer variable is assigned a value depending on the stage of the photon history. The call to the subroutine from SIMIND is of the form *call score(iopt,lun)* where *lun* is a logical unit number connected to the results file and used for output and *iopt* is an integer flag that tells the user at which stage in the photon history simulation the call to the scoring routine was made. The user can then decode the iopt variable and make further calculation depending on the value of the variable. The values of iopt and the corresponding stages for the call in the code are listed in table 11.1.

Consider the following routine where the imparted energy at each inter-action point, BIDRAG, is accumulated and the average energy for each photon history is calculated using the variable PHOTON, which is the number of histories per projection. An example of a command script for a simulation where an511.smc is the input file is shown overleaf. A simulation study is made for (i) a case with the source at various depths (varied by data given by switch /**18:**) and a hot- and cold-sphere phantom simulation with and without phantom interaction. The switch/**SC:** determines the maximum number of scatter orders in the phantom.

```
Subroutine score(iopt,lun)

Include 'simind.fcm'          ! Connect to simind through common
block
Real sum,average

If(iopt.eq.0) then            ! Just after reading smc values
     Sum = 0                  ! Initialize the counter
Elseif(iopt.eq.3) then        ! After each interaction
     Sum = sum + bidrag       ! Integrate energy
Elseif(iopt.eq.6) then        ! All histories terminated
     Average = sum / photon   ! Calculate average
Elseif(iopt.eq.7) then        ! Print result in *.res file
     Write(lun,10),average    ! using the lun variable
     Format(' Average was found to be :',f12.3)
End if
End
```

```
% simind an511/sc:6/18:9.50/iwb:2 an511a1
% simind an511/sc:6/18:6.00/iwb:2 an511a2
% simind an511/sc:6/18:2.50/iwb:2 an511a3
% simind an511/sc:6/18:0.00/iwb:2 an511a4
% simind an511/sc:6/18:-2.5/iwb:2 an511a5
% simind an511/sc:6/18:-6.0/iwb:2 an511a6
% simind an511/sc:6/18:-9.5/iwb:2 an511a7/sp:1

% simind jaz511/sc:6/iwb.3  jaz511ap/hot/tr:11
% simind jaz511/sc:6/iwb.3  jaz511aa/hot/fa:11

% simind jaz511/sc:6/iwb.3  jaz511bp/cold/tr:11
% simind jaz511/sc:6/iwb.3  jaz511ba/cold/fa:11
```

11.4.2 The source routine

The aim of this routine is simply to give SIMIND Cartesian coordinates x_0, y_0, z_0 that describe where the decay occurred. The user can then write source routines that are more complex than the geometries included in SIMIND (i.e. cylinders and spheres). Provided with the SIMIND code are source routines simulating the Jaszczak cylindrical phantom and a routine for generating decays according to a source map defined by integer maps. The latter are used to simulate more complex source distributions.

More complex sources can be simulated by using a source map. This map is a two-byte integer matrix where each pixel value shows the number of decays to be simulated and the relative pixel location indicates the location for those decays. A user can then by an external program define any type of source geometry. An example of such a simulation is shown in figure 11.3 where a map, created by a CT study of the elliptical De Luxe rod insertion (Data Spectrum), was made and used to sample decays.

Ideal   Total   Primary

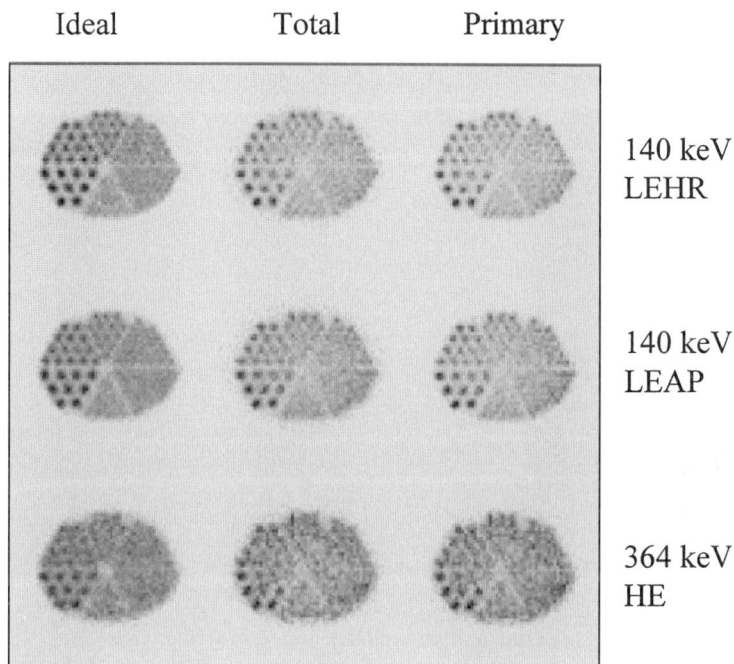

140 keV
LEHR

140 keV
LEAP

364 keV
HE

Figure 11.3 *Simulation of a rod phantom for three different collimators. The left row shows the 'air' case where no scattering in the phantom is simulated. The middle row shows images obtained from both primary and scattered events and the right row shows images obtained from scattered photons only.*

Also included in SIMIND is access to two anthropomorphic voxel-based computer phantoms, developed by Zubal and Harrell [8] and described in more detail in Chapter 3. The phantoms consist of a set of byte images where each pixel value is a code that defines a particular organ. These codes can then be used to assign a value, corresponding to, for example, density or activity. The treatment of these images has been made simple in SIMIND by using a separate ASCII file that defines the relative activity concentration for each organ. Based on this information, the code calculates the number of decays for each organ and normalizes this to the total number of photon histories intended to be simulated. Figure 11.4 shows a whole-body simulation of a ^{99m}Tc bone scintigraphy with simulated decay in the bones only using the Zubal phantom. The left image shows the photo-peak (126–154 keV). The middle image shows the scatter part of the photo-peak and the right image was obtained in a Compton window between 92 keV and 126 keV. This CW window is often used as a scatter window in scatter subtraction correction methods.

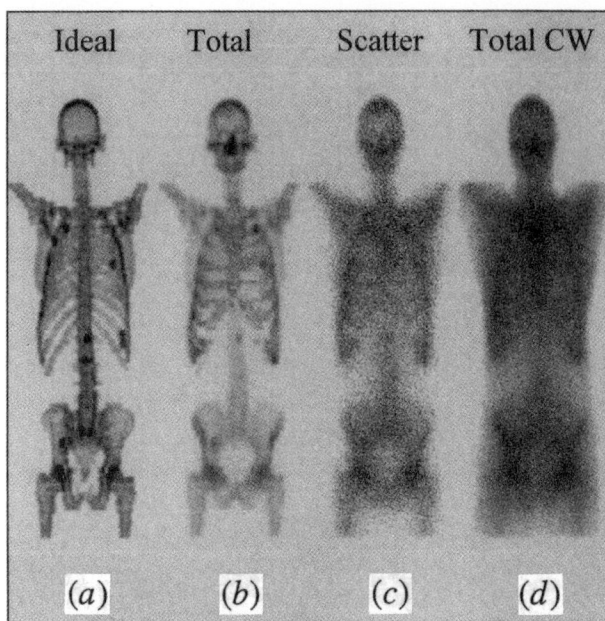

Figure 11.4 *Simulation of a simulated whole-body bone scan using the Zubal phantom and multiple metastases. Images are shown for the 'air' case (a), total (b), scatter in the photo-peak (c) and total events in a lower Compton window (d). Energy window settings for the photo-peak were 154–126 keV and for the Compton window 126–92 keV.*

11.4.3 The isotope routine

This routine is used to give SIMIND a value of the photon energy. Such a routine can include decay schemes of various complexity. A logical flag in SIMIND controls when to stop the calls to the isotope routine. The difference between this technique and the alternative of running several independent simulations of different photon energies and summing the results is the addition of summation of pulse pileup events due to simultaneous multiple-photon emission.

11.5 THE USER INTERFACE

A SIMIND command at the operating system prompt consists of the following parts: the program name, an input file, an output file and control switch (optional). Consider the following example of a command: `$ simind input output/01:140/fa:1`. The input file in the command line above, `input.smc`, defines a file created by CHANGE that contains the data for the particular simulation. If not explicitly given, the output base file name, `output`, will be equal to the input file name but, since SIMIND creates multiple result files, these are separated by different extensions. The user can control the input parameters, originally created by the program CHANGE, at the command level. Each value, given from an index in CHANGE, can be overridden by a matching a corresponding control switch at run time. The reason for this option is that only one data file (for example, `input.smc`) is necessary for multiple simulations. The main advantage of such user interfaces is the easy setup of multiple simulations. In the example above, index 01 in CHANGE (which is the photon energy) is changed to 140 and flag 1 (which controls the data presented to the screen) is set to false. A larger example of such a command file is shown in the second text box on page 152.

11.6 TRANSMISSION IMAGING

Included in SIMIND is a possibility to simulate transmission studies (TCT) with simple-photon emission computed tomography (SPECT) with either a parallel hole collimator or a fan beam collimator [9]. The difference between SPECT and TCT is mainly the fact that the source now is outside the phantom in a location opposite the camera head.

Figure 11.5 shows schematically the geometry for the transmission simulation. The phantom is defined by integer matrices and is thus placed between

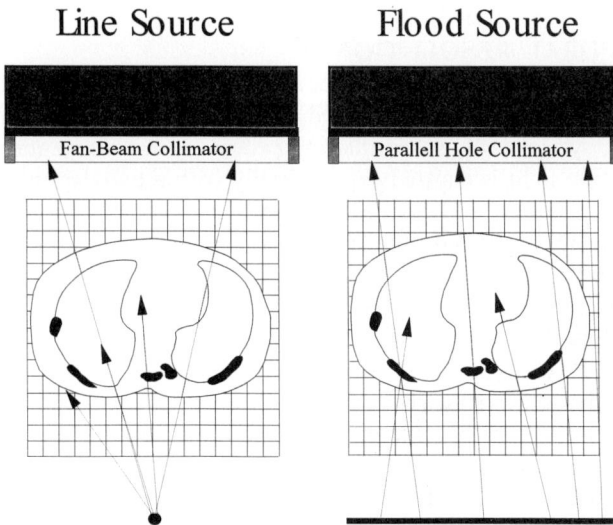

Figure 11.5 *A schematic figure showing the setup for the fan-beam collimator transmission simulation and the parallel hole collimator transmission simulation. (Reprinted with permission from Ljungberg M et al 1994 IEEE Trans. Nucl. Sci. **NS-41** 1577–84. ©1994 IEEE.)*

the source and the camera. An emitted transmission photon is traced into the phantom matrices and it is checked whether or not it will strike the phantom. If it will, the entrance matrix cell is calculated and the simulation is continued as described in the previous section. If not, the photon is regarded as being a primary photon and is continued towards the camera without interaction in the phantom. It is also necessary to define here the scatter order. The transmission source can be a line or a flood source at different distances away from the phantom. The emission angle for the transmission photon can be selected to be forced into a cone or a fan (which matches a situation where collimation is performed with a slit only in the axial direction).

Simultaneous transmission/emission SPECT is possible to simulate. The ratio between the activity in the transmission source and the emission source is given in CHANGE. In the scoring routine, the user can decode the origin of a detected photon (emission or transmission), which allows studies of cross-talk effects between transmission and emission energy windows. The output of the transmission simulation is transmission projections or/and blank projections (without the object *in situ*). The latter can also be correlated to the transmission projection. The advantage with this option is that artefacts in the reconstructed images due to large deviations in the logarithm calculation can be avoided.

11.7 TEMPORAL RESOLUTION

A scoring routine has been developed to simulate the temporal resolution of a SPECT system [10]. A scintillation process is not momentary. Instead the scintillation light is emitted exponentially with a two-exponential decay constant of 0.23 μs for 60% of the light and 1.5 μs for the remainder. Since an event and its coordinate are calculated from the centroid of the light emitted from the different interaction sites there will be pulse pileups when imaging high activities. This is because the contribution of light from earlier not completely decayed events. The possibility to simulate this has been included in SIMIND by a special scoring routine.

The contribution from earlier events cannot be calculated directly due to the implementation of the different variance reduction techniques. Instead, the simulations are divided into two main steps. In the *first part*, a preselected number of photon histories are simulated and the imparted energy, then photon weight, scattering order and spatial coordinates for the event are stored in an array. In the *second part*, a distribution function is calculated from the photon weights stored in the array. Properly Poisson distributed events are then sampled from that distribution by a selecting from the relative probability of each event using a uniform random number.

The time between each event is sampled from the equation $\Delta t = \ln(R)/$ cps where cps is the expected count rate in the whole energy spectrum. The energy signal is calculated from the equation $H_i = E_i + H_{i-1} \exp(-\Delta t/ \tau)$ where E_i is the signal for the energy imparted from the last interaction, H_i is the total energy signal and τ is the decay constant. The spatial coordinates are also affected by pulse pileup since these are calculated from the centroid of the scintillation light. To obtain results that are comparable with experimental data, a subroutine defining the specific camera features, such as the type of pileup rejection, the dead-time characteristics and the type of pulse-tail extrapolation, can be linked into the code.

11.8 EXAMPLES OF RESEARCH WORK UNDERTAKEN WITH SIMIND

In the early work on SIMIND [11], we investigated different scatter correction techniques by evaluating the scatter distribution only. We compare the scatter distribution from the CW correction method [12] and the convolution-subtraction method [13] with true scatter. In a development of the convolution-subtraction method to include depth-dependent scatter functions, we used SIMIND to calculate scatter line spread functions (SLSFs) as functions of different locations in both a cylindrical water phantom [14] and for a nonhomogeneous case [15]. The MC simulation allowed us here to obtain scatter distribution and accurate scatter/total ratios for the different locations which then made us model the scatter distribution in SPECT projection by using a reconstructed image as an estimate of the source distribution. The technique of separating scatter from primaries was also used in a later paper [16] where four different scatter correction methods were investigated using the Hoffman bitmap source. Other researches have also been using SIMIND for scatter correction evaluation of [99m]Tc [17, 18], [201]Tl [19, 20] and [123]I [21].

The ability to image activity distributions emitting photons at high fluence rates is of great interest in both diagnostic and therapeutic nuclear medicine. Count losses, energy pulse pileup and mispositioning in scintillation camera imaging are related phenomena when registrations with high photon fluencies (see Chapter 6). These effects have theoretically been evaluated at our laboratory in an earlier study with a simplified MC code [22] and recently by an implementation of the SIMIND code [10]. Here, studies of how well multiple-window scatter correction methods perform under a high-count-rate situation were made. The relative fraction of events originating from photons, of different scattering orders, was evaluated as a function of count rates. The fraction of multiple-scattering events was found to increase as the count rate becomes higher and the number of first-order scatter events is

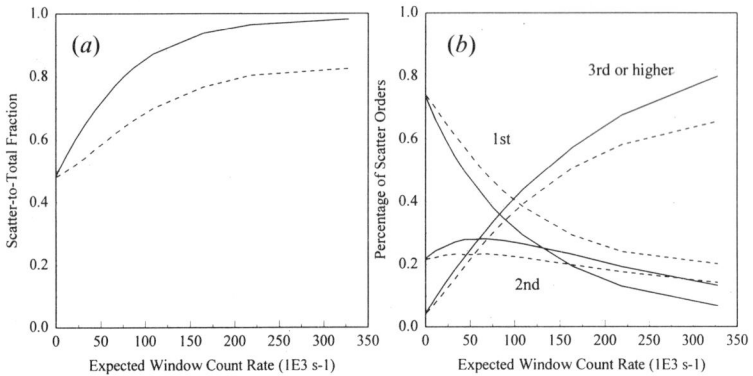

Figure 11.6 *Simulations of the scatter-to-total fraction in the photopeak window without pileup rejection (solid) and with pileup rejection (dashed) are shown in (a). (b) The scatter events, separated into components of first, second and third or higher orders of scattering. (Reprinted with permission from Ljungberg M et al 1995 Conf. Record IEEE Medical Imaging Conf. pp 1682–6. ©1995 IEEE.)*

subsequently decreased because of the shift of the energy pulse-height distribution towards higher energies, due to pulse pileup. Figure 11.6 shows some results from such a simulation.

The code has been used extensively by the group in North Carolina at Chapel Hill as a tool for developing scatter response functions (see for example [23–25]) and for evaluation of iterative reconstruction methods [26]. The work with the scatter functions is further described in a subsection in Chapter 15 in this book. In a close collaboration with the group of Professor King at the University of Massachusetts Medical Centre, Worcester, MA, SIMIND has been used to develop and evaluate the dual-peak window (DPW) scatter correction method [27–29]. SIMIND has also been used to evaluate the possibility of segmenting Compton scatter images for outlining object contour and major structures, such as the lungs [30, 31]. This approach is described further in Chapter 14. The code has also been used to study the contribution of scatter in cardiac images from photons emitted from the liver [32]. Here it could be shown that 'hot' livers could introduce artefacts in polar maps, which could be judged as functional defects. Monte Carlo works with SIMIND have also studied the question of whether the DPW scatter correction of SPECT images can improve the detection of cold lesions in the liver [33].

Li *et al* [20, 34] have studied the effect on the downscatter of ^{99m}Tc transmission photons into an energy window used to image ^{201}Tl photons using a triple-headed camera fan beam SPECT system where one head was used for the transmission data. Monte Carlo simulations were performed with

SIMIND for a similar system using a 3D MCAT anthropomorphic computer phantom and events from primary and scattered ^{99m}Tc photons were separated to analyse the fraction of downscatter quantitatively. A figure of merit between the detection efficiency and scatter fractions in the presence of downscatter was then evaluated.

Relatively few works have been done to evaluate transmission SPECT by Monte Carlo simulations. Wang *et al* [35, 36] have developed Monte Carlo methods for simulating transmission SPECT. One special issue here was to develop a composite object model of a complex nonhomogeneous object, such as the human body, in order to present an alternative to existing simple geometries or voxel-based geometries. The aim is here to improve computation speed and memory size and therefore present a method for simulation of clinically significant objects.

Welch *et al* [37] have studied the downscatter problem for a [153]Gd transmission using SIMIND. The aim of the work was to investigate the usefulness of dual-energy transmission imaging using the Picker PRISM 300 triple-detector gamma camera system fitted with three 65 cm focal length fan beam collimators. The transmission source was a static Gd-153 line source and two transmission windows were used, one centred over 44 keV and the other over 99 keV. Part of the study consisted of a Monte Carlo simulation of the system using SIMIND. The aim of the simulation was to investigate the contamination of the data in the 44 keV transmission window by 99 keV photons from the Gd source. The simulation showed that, while there was some cross contamination (downscatter), the amount was low and the distribution could be efficiently estimated by using the data from the two 'back' detectors (those detectors not being used for transmission measurements). Figure 11.7 show a graph of this downscatter distribution.

Recently, we have studied the imaging properties of [18]FDG with scintillation cameras and SPECT [38]. Because of the thin NaI(Tl) crystal, there is a chance that photons can be scattered away from the crystal and only partly deliver energy. This results in a correctly positioned event that appears in the Compton region of the energy pulse-height distribution. The aim of the work was then to evaluate the fraction of 'primaries' in the Compton region and to investigate whether these events could add to the photopeak data to increase the sensitivity. Work in progress is the development of a scatter correction technique for the data acquired in this Compton window [39].

The group at the University of Utah has in collaboration with the University of North Carolina, Chapel Hill, used SIMIND to allow for transmission simulation and in particular evaluated the problem of downscatter into transmission windows [34, 37, 40]. In a recent publication, SIMIND has been used to evaluate the noise properties and collimator selection for rCBF studies [41]. Work has also been ongoing to evaluate properties of a concave collimator for SPECT [42] by using SIMIND to establish optimal parameters for such a collimator regarding spatial resolution and sensitivity.

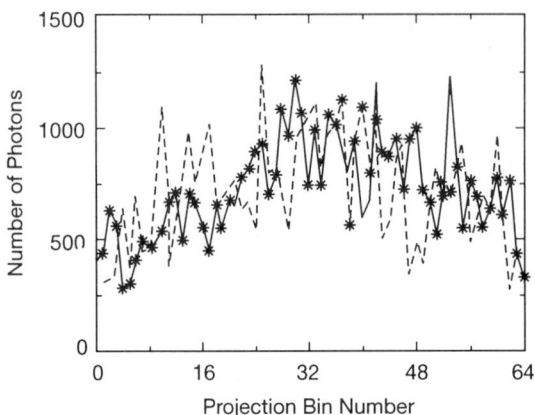

Figure 11.7   *Profiles through the middle of the phantom of the true down-scatter data and the estimated downscatter data, obtained from two detectors and the 44 keV energy window. (Reprinted with permission from Welch et al 1995 IEEE Trans. Nucl. Sci. **NS-42** pp 2331–8. ©1995 IEEE.)*

Studies with SIMIND have also been performed to evaluate perfusion lung scan regional anatomy [43].

A recent work at our laboratory has focused on the possibility of using Compton-scattered data for registration of ^{131}I-SPECT and CT images in the abdominal region [44]. Here, SIMIND has been used to evaluate the optimal energy window location for the Compton window by investigating the scatter orders of the photons that create events in different parts of the energy spectrum.

REFERENCES

[1]   Ljungberg M and Strand S-E 1989 A Monte Carlo program simulating scintillation camera imaging *Comput. Methods Programs Biomed.* **29** 257–72

[2]   Todd-Pokropek A, Cradduck T D and Deconinck F 1992 A file format for the exchange of nuclear medicine image data: a specification of Interfile version 3.3 *Nucl. Med. Commun.* **13** 673–99

[3]   Berger M J and Hubbell J R 1987 *XCOM: Photon Cross-sections on a Personal Computer* NBSIR 87-3597 (Washington, DC: National Bureau of Standards)

[4]   Hubbell J H, Veigele J W, Briggs E A, Brown R T, Cramer D T and Howerton R J 1975 Atomic form factors, incoherent scattering functions and photon scattering cross sections *J. Phys. Chem. Ref. Data* **4** 471–616

[5]   Hubbell J H and Overbo I 1979 Relativistic atomic form factors and photon coherent scattering cross section *J. Phys. Chem. Ref. Data* **8** 69–105

[6] Metz C E, Atkins H L and Beck R N 1980 The geometric transfer function component for scintillation camera collimators with straight parallel holes *Phys. Med. Biol.* **25** 1059–70.

[7] Tsui B M W and Gullberg G T 1990 The geometric transfer function for cone and fan beam collimators *Phys. Med. Biol.* **35** 81–93

[8] Zubal I G and Harrell C R 1993 Voxel based Monte Carlo calculations of nuclear medicine images and applied variance reduction techniques *Image Vision Comput.* **10** 342–8

[9] Ljungberg M, Strand S-E, Rajeevan N and King M A 1994 Monte Carlo simulation of transmission studies using a planar source with a parallel collimator and a line source with a fan-beam collimator *IEEE Trans. Nucl. Sci.* **NS-41** 1577–84

[10] Ljungberg M, Strand S-E, Rotzen H, Englund J E and Tagesson M 1995 Monte Carlo simulation of high count rate scintillation camera imaging *Conf. Records IEEE Medical Imaging Conf. (Norfolk, VI, 1994)* (Piscataway, NJ: IEEE) pp 1682–6

[11] Ljungberg M, Msaki P and Strand S-E 1990 Comparison of dual-window and convolution scatter correction techniques using the Monte Carlo method *Phys. Med. Biol.* **35** 1099–110

[12] Jaszczak R J, Greer K L, Floyd C E, Harris C C and Coleman R E 1984 Improved SPECT quantification using compensation for scattered photons *J. Nucl. Med.* **25** 893–900

[13] Axelsson B, Msaki P and Israelsson A 1984 Subtraction of Compton-scattered photons in single-photon emission computed tomography *J. Nucl. Med.* **25** 490–4

[14] Ljungberg M and Strand S-E 1990 Scatter and attenuation correction in SPECT using density maps and Monte Carlo simulated scatter functions *J. Nucl. Med.* **31** 1559–67

[15] Ljungberg M and Strand S-E 1991 Attenuation and scatter correction in SPECT for Sources in a nonhomogeneous object: A Monte Carlo study *J. Nucl. Med.* **32** 1278–84

[16] Ljungberg M, King M A, Hademenos G J and Strand S-E 1994 Comparison of four scatter correction methods using Monte Carlo simulated source distributions *J. Nucl. Med.* **35** 143–51

[17] Luo J Q, Koral K F, Ljungberg M, Floyd C E Jr and Jaszczak R J 1995 A Monte Carlo investigation of dual-energy-window scatter correction for volume-of-interest quantification in ^{99m}Tc SPECT *Phys. Med. Biol.* **40** 181–99

[18] Naude H, van Aswegen A, Herbst C P, Lötter M G and Pretorius P H 1996 A Monte Carlo evaluation of the channel ratio scatter correction method *Phys. Med. Biol.* **41** 1059–69

[19] Li J Y, Tsui B M W, Frey E C and Gullberg G T 1996 Deconvolution scatter compensation for ^{201}Tl fan-beam cardiac SPECT *Conf. Records IEEE Medical Imaging Conf. (San Francisco, CA, 1995)* (Piscataway, NJ: IEEE) pp 1175–9

[20] Li J Y, Tsui B M W, Welch A, Frey E C and Gullberg G T 1995 Energy window optimization in simultaneous ^{201}Tl TCT and SPECT data acquisition *IEEE Trans. Nucl. Sci.* **NS-42** 1207–13

[21] Luo, J Q, Koral K F, Ljungberg M, Fessler J A, Koeppe R A and Kuhl D E 1994 Monte Carlo study of circumferential variation in cortex activity in ^{123}I SPECT *J. Nucl. Med.* **35** 82P (abstract)

[22] Strand S-E and Lamm I-L 1980 Theoretical studies of image artifacts and counting losses for different photon fluence rates and pulse-height distributions in single-crystal NaI(Tl) scintillation cameras *J. Nucl. Med.* **21** 264–75

[23]   Frey E C and Tsui B M W 1993 A practical method for incorporating scatter in a projector–backprojector for accurate scatter compensation in SPECT *IEEE Trans. Nucl. Sci.* **NS-40** 1107–16

[24]   Frey E C, Ju Z-W and Tsui B M W 1993 A fast projector–backprojector pair modeling the asymmetric, spatially varying scatter response function for scatter compensation in SPECT imaging *IEEE Trans. Nucl. Sci.* **NS-40** 1192–7

[25]   Frey E C and Tsui B M W 1993 Modeling the scatter response function in inhomogeneous scattering media *Conf. Records IEEE Medical Imaging Conf. (San Francisco, CA, 1993)* (Piscataway, NJ: IEEE) pp 1184–8

[26]   Frey E C, Tsui B M W and Ljungberg M 1993 A comparison of scatter compensation methods in SPECT subtraction-based techniques versus iterative with accurate modeling of the scatter response *Conf. Record IEEE Medical Imaging Conf. (San Francisco, CA, 1993)* (Piscataway, NJ: IEEE) pp 1035–7

[27]   King M A, Hademenos G J and Glick S J 1992 A dual photopeak window method for scatter correction *J. Nucl. Med.* **33** 605–13

[28]   Hademenos G J, Ljungberg M, King M A and Glick S J 1993 A Monte Carlo investigation of the dual photopeak window scatter correction method *IEEE Trans. Nucl. Sci.* **NS-40** 179–85

[29]   Hademenos G J, King M A, Ljungberg M, Zubal I G and Harrell C R 1993 A scatter correction method for ^{201}Tl images: a Monte Carlo investigation *IEEE Trans. Nucl. Sci.* **NS-40** 1179–86

[30]   Pan T-S, Ljungberg M, King M A and DeVries D J 1996 Attenuation correction of cardiac perfusion images in SPECT using Compton scatter data: a Monte Carlo investigation *IEEE Trans. Med. Imaging* **MI-15** 13–24

[31]   Pan T-S, King M A, DeVries D J and Ljungberg M 1993 Segmentation of the body and lungs from Compton scatter and photopeak window images in SPECT: a Monte Carlo investigation *Conf. Records IEEE Medical Imaging Conf. (San Francisco, CA, 1993)* (Piscataway, NJ: IEEE) pp 1657–61

[32]   King M A, Xia W, DeVries D J, Pan T S, Villegas B J, Dahlberg S, Tsui B M W, Ljungberg M and Morgan H T 1996 A Monte Carlo investigation of artifacts caused by liver uptake in single-photon emission computed tomography perfusion imaging with ^{99m}Tc labeled agents *J. Nucl. Cardiol.* **3** 18–29

[33]   DeVries D J, King M A, Soares E J, Tsui B M W and Metz C E 1997 Evaluation of the effect of scatter correction on lesion detection in hepatic SPECT imaging *IEEE Trans. Nucl. Sci.* **44** 1733–40

[34]   Li J Y, Tsui B M W, Welch A and Gullberg G T 1995 Energy window optimization in simultaneous SPECT/TCT data acquisition using Monte Carlo simulations *Conf. Records IEEE Medical Imaging Conf.* pp 1587–91

[35]   Wang H L, Jaszczak R J and Coleman R E 1993 A new composite model of objects for Monte Carlo simulation of radiological imaging *Phys. Med. Biol.* **38** 1235–62

[36]   Wang H J, Jaszczak R J and Coleman R E 1992 Solid geometry-based object model for Monte Carlo simulated emission and transmission tomography imaging systems *IEEE Trans. Med. Imaging* **MI-11** 361–72

[37]   Welch A, Gullberg G T, Christian 'P E, Li J Y and Tsui B M W 1995 An investigation of dual energy transmission measurements in simultaneous transmission emission imaging *IEEE Trans. Nucl. Sci.* **NS-42** 2331–8

[38]   Ljungberg M, Ohlsson T, Sandell A and Strand S-E 1996 Scintillation camera imaging of positron-emitting radionuclides in the Compton region *Conf.*

Records IEEE Medical Imaging Conf. (San Francisco, CA, 1995) (Piscataway, NJ: IEEE) vol 2 pp 977–81

[39] Ljungberg M, Danfelter M, Strand S-E, King M A and Brill B A 1997 Scatter correction in scintillation camera imaging of positron-emitting radionuclides *Conf. Records IEEE Medical Imaging Conf. (Anaheim, CA, 1995)* (Piscataway, NJ: IEEE) vol 3, pp 1532–6

[40] Welch A, Gullberg G T, Christian P E, Datz F L and Morgan H T 1995 A transmission-map-based scatter correction technique for SPECT in inhomogeneous media *Med. Phys.* **22** 1627–35

[41] Ärlig Å, Jacobsson L, Larsson A, Ljungberg M and Wikkelsö C 1997 Collimator selection for rCBF-studies and triple-headed SPECT evaluated using noise-resolution plots *Nucl. Med. Commun.* **18** 655–61

[42] Kimiaei, S, Ljungberg M and Larsson B S 1997 Evaluation of optimally designed concave–planar collimators in single photon emission tomography *Eur. J. Nucl. Med.* **24** 1398–1404

[43] Hutson K, Pate J, Roberts K, Bond H, Gould H R and Smith G T 1994 Perfusion lung scan regional anatomy defined by Monte Carlo simulation *J. Nucl. Med.* **35** 239P (abstract)

[44] Sjögreen K, Ljungberg M, Erlandsson K, Floreby L and Strand S 1997 Registration of abdominal CT and SPECT images using Compton scatter data *Proc. Information Processing in Medical Imaging* at press

CHAPTER 12

MONTE CARLO IN SPECT SCATTER CORRECTION

Kenneth F Koral

12.1 THE BASIC PROBLEM

The subject of this chapter is the use of Monte Carlo simulation to solve a problem in single-photon emission computed tomography (SPECT). The problem arises from those counts in a projection which are mispositioned because of gamma ray Compton scattering within a patient or object. If these counts are reconstructed as standard, they lead to loss of contrast and resolution and incorrect quantification in the final image set. The incorrect quantification is illustrated with ^{99m}Tc in phantoms: the activity in a focal region is typically higher by 18% when there is no 'tissue background' and by 23% when the 'tissue background' activity concentration is one-fifth of the focal region activity concentration. The activity error increases linearly with further background activity concentration increases [1]. The extent of the contrast deficit is illustrated by ^{201}Tl in ten patients: the percentage of the volume of the left-ventricular wall that is characterized as below normal in blood flow increases from 12.4 to 20.2% after a correction for scattering [2]. The existence of scattered gamma rays, therefore, calls for a 'correction' (alias a 'compensation'). The technique of Monte Carlo simulation of gamma ray scattering has been applied both to assess the problem and to solve it.

12.2 SOLUTION BY ENERGY DISCRIMINATION?

One might assume that energy discrimination could separate out scattered gamma rays from those which are unscattered and eliminate the scattered ones. The highest energy for a scattered gamma ray occurs when the scattering angle approaches zero degrees; at exactly zero degrees, the energy of the scattered gamma ray is equal to that of the unscattered photon. The differential cross section for gamma ray scattering is nonzero at zero degrees.

165

However, on the plus side, the exit solid angle for the scattered gamma ray approaches zero at zero degrees. The product, therefore, also approaches zero. It would seem that fortuitously the observed energy of scattered photons would be less than that of the unscattered gamma ray and they could be eliminated by energy discrimination. However, all Anger cameras have finite energy resolution. (With a sodium iodide crystal, the resolution is around 10% at 140 keV.) This finite energy resolution smears the true energy of a gamma ray over a range of detected energies, some lower but some higher. Therefore, the detected energy of a given scattered gamma ray can be even greater than the true energy of the unscattered photon. Moreover, due to the same finite resolution, the detected energy of a given unscattered gamma ray can be lower than its true energy, leading to further mixing. In summary, energy discrimination by itself cannot solve the problem.

12.3 USE OF THE MONTE CARLO METHOD

A comprehensive review of scatter correction methods, irrespective of their relationship to Monte Carlo simulation, has been published recently by Buvat *et al* [3]. Below, the use of Monte Carlo in individual cases of correction will be detailed. In general, the forte of Monte Carlo simulation is that it allows exact separation of detected gamma rays which have undergone scattering and those which have not. The ways in which that information is used are quite varied as will be seen under the headings and subheadings that follow.

12.4 SCATTER CORRECTION WITHOUT SUBTRACTION

There are three methods that 'correct' for scattered gamma rays without subtraction. Two of them assign the scattered counts to their true origin during tomographic reconstruction. These two methods are related, have been applied to ^{99m}Tc, and require knowledge obtained by Monte Carlo simulation. The third uses Wiener filtering of the projection data. It has been applied to ^{201}Tl and does not need Monte Carlo simulation information.

The first method of the three [4] was originally called 'inverse Monte Carlo'. For a given geometry, Monte Carlo simulation establishes the probabilities for a gamma ray to be scattered and then detected with a given energy in a given projection pixel, relative to the probabilities for similar detection without scatter. These probabilities are what need to be known to do the 'inversion' that is the scatter-included reconstruction problem [4]. How they are to be known for a particular patient is one of the practical problems of this otherwise elegant technique. The other is the memory storage needed to contain all the probability values when one includes these relatively unlikely but possible events.

The second method is an extension of the scatter-included reconstruction problem that employs two abutted energy windows [5]. Solution by the method of generalized matrix inversion was applied to Monte Carlo simulated data from a cylindrically symmetric phantom. For the one region of interest evaluated, there was no improvement over the older method (which was already quite accurate) and substantial improvement over the dual-energy-window method [6] only for the lower of two projection count densities [5].

The third method uses Wiener filtering to sharpen the resolution of the projection data [7], thus, hopefully, relocating scattered counts nearer to the location of the projection of their origin [8]. (The Wiener filter is designed not to increase noise. Such noise increase is a drawback of subtraction techniques unless compensation steps are taken. Such compensation, however, is provided, for example, by regularization methods [9].) The Wiener filter thus reduces the effects of scatter and improves relative quantification [7]. To obtain absolute quantification in the corrected image relative to a calibration, scatter needs 'to be present in roughly similar quantities in the object to be quantified and the reference' [10]. Calibration of a patient image by the image of a known-activity point source in air would not be appropriate because there would be scatter for the patient but not for the point source [10].

12.5 SCATTER CORRECTION BY SUBTRACTION

Correction by subtracting away those counts which are from scattered gamma rays has been applied to many isotopes.

12.6 ^{99m}Tc

12.6.1 Correction techniques

Many of the Compton scatter correction techniques applied to ^{99m}Tc have been tested with Monte Carlo. Some of them need it for a crucial setup step. This fact illustrates the power of the Monte Carlo technique.

The original correction scheme applied to SPECT, the dual-energy-window method, has a key parameter, the k value [6]. This parameter has a basic definition as the ratio of the number of scattered counts within the main photo-peak window divided by the number of scattered counts within a lower-energy-scatter (alias monitor) window. In operation, this parameter multiplies an image reconstructed from the scatter window projections before it is subtracted from an image reconstructed from the photo-peak window projections. Choosing the value for this parameter is somewhat subjective. Monte Carlo (as well as experiment) has been used to make and examine the choice. Simulating a 5 cm long line source within a circular cylinder and summing counts in a very thick (12.8 cm) transverse slice all the way to the edge of the projection, Floyd *et al* obtained a value of $k = 0.59$ when the line was on the cylinder axis and $k = 0.51$ when it was 8 cm off axis [11]. It was true, however, that in the reconstructed image 0.5 was far too small for proper correction at the line source location. Later, Ljungberg *et al* used Monte Carlo to show that a single k value was not accurate for correction of most of the pixels in a projection profile through a hot point source in a cold cylinder and was indeed too small at exactly the source location [12]. Still later, Luo *et al* showed that the energy spectrum from a hot sphere in a cold cylinder is much less intense throughout the scatter window than the spectrum from a cold sphere in a hot cylinder [1] when the two spectra are normalized to the same number of counts in the photo-peak window region. This is as shown in figure 12.1. The spectrum is for a circular region of interest in the projection opposite the sphere as is of interest when the goal is quantifying the activity in the sphere. Since the two spectra are different in the region of the scatter window while their sum is the same in the region of the photo-peak window, the k value is obviously quite different for these two source distributions. As a general consequence of the difference, the k value to quantify a hot sphere that is immersed in a warm-cylinder 'background' varies strongly with the activity level of the background compared to the activity level of the sphere [1]. To adapt the dual-energy-window method to the task of focal quantification in cases where there is background activity, Luo and Koral have introduced an iterative, background-adaptive version. Using Monte Carlo tests, they report rapid convergence and good preliminary results [13].

Earlier Koral *et al* simulated energy spectra within small spatial regions in a projection using Monte Carlo [14]. (The underlying hope was that newer tomographs would provide such energy spectra at each pixel, at least as an option, perhaps with list-mode acquisition.) Their test object was a sphere within a cylinder. The spectra were easily generated using Monte Carlo because the simulation provided the detected energy. The number and energy limits of the separate energy bins that made up the spectrum were chosen in advance. The authors then fitted these spectra with a model that assumed the scattered spectrum could be represented by a polynomial. The fitting essentially separated the counts into a scattered total and an unscattered total. The authors used a definition of scatter fraction, SF, as the number of scattered events within a window divided by the number of unscattered counts within that same window. This value, with a standard 20% photo-peak window, was calculated from the 'spectral fitting' result and compared to the true value from the simulation to test the effectiveness

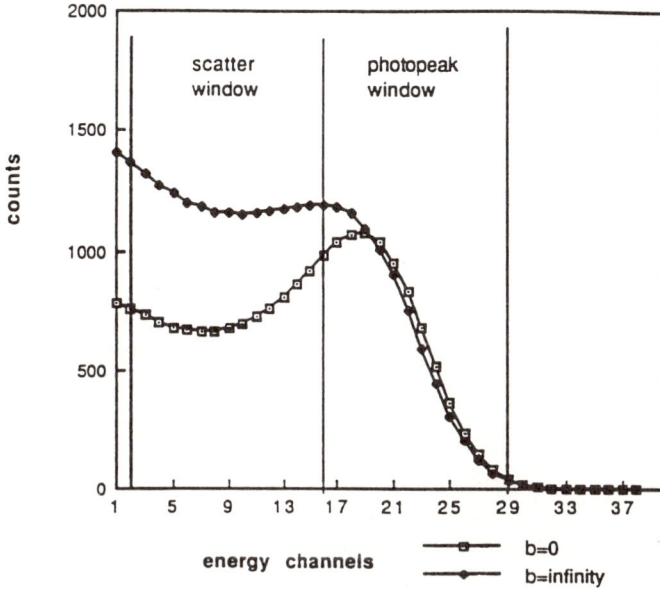

Figure 12.1 *Results from ^{99m}Tc Monte Carlo simulation. The energy spectra of scattered gamma rays are compared for two projection images distinguished by being from different source distributions. The b value in the legend is equal to the ratio of the activity concentration in a large cylinder background divided by the activity concentration within a sphere placed in the cylinder. The first spectrum is from a hot sphere immersed in a cold-cylinder background. It has a b value of zero. The second spectrum is from a cold sphere in a hot-cylinder background. It has a b value of infinity. Only gamma rays within a circular region of interest opposite the sphere and of the same diameter have been included in the spectra. Only these counts would be of interest for quantifying the activity of the sphere. The spectra have been normalized so that the numbers of scattered counts within the 20% photo-peak window are equal. With that normalization, it is seen that the spectra within the scatter window are very different. Importantly, it is also obvious that the total numbers of counts within the scatter window are very different. The k value, defined as the ratio of the number of scattered counts within the the main photo-peak window divided by the number of scattered counts within the lower-energy scatter window, then, is obviously quite different for these two source distributions. (Permission to reproduce this figure from [1] was granted by the authors and by the Institute of Physics Publishing Limited.)*

of the method [14]. They found that the average difference in the SFs, averaged over all evaluated points, was 16.2%. Later, a regularized, deconvolution-fitting approach was tested in the same way [15]. The deconvolution-fitting algorithm did not assume a polynomial shape for the scatter component of the energy spectrum but did require knowledge of the camera energy response function (i.e. the detected 'energy' spread function for any true gamma ray energy). Neither approach was constrained to explicitly prevent negative values in the scatter spectrum solution, although such values in the solution were truncated to zero. Guidance about the regularization parameter, lambda, required for the deconvolution-fitting algorithm was obtained through extensive Monte Carlo testing [15].

Ogawa *et al* also generated tomographic data sets for simple geometric phantoms [16] by Monte Carlo simulation. Their three-window method employs the usual photo-peak window and adds two windows, one on either side, of known, narrow width. They reduced the size of the local spatial regions for which the scatter was to be independently estimated to single pixels. A straight line connecting the count levels in the two narrow windows estimated the scatter in the main window. They emphasized the practicality of their correction method: computation is simple and many modern cameras do or will allow for three energy windows. In their testing, they used images reconstructed from the unscattered photons as their 'gold standard'. For their 'cold-spot' phantom, the visual diameter of the spot is right with correction whereas it was too small without. A profile across the image shows slight overcorrection at the bottom of the cold spot [16].

With the method of Ljungberg and Strand [17], Monte Carlo simulation is an essential part of the correction as it is used to generate required 'scatter line spread functions'. These functions, computed at given rotation angles with the point source at a given spatial location, are one dimensional, giving the number of scattered counts versus distance. The distance axis can be thought of as in a plane that (i) contains the point source and (ii) is perpendicular to the axis of rotation. The object itself is three dimensional; the projection image has the normal two dimensions. The function becomes one dimensional by integration along the camera face in the direction parallel to the axis of rotation. (The exact rationale for the desirable dimension reduction by integration is not clear. Moreover, it would seem that an object whose activity changes rapidly in the integration direction might not be well corrected.) Knowledge of the functions for the object is required. The method is thus like inverse Monte Carlo. In real cases, the knowledge may be only approximate. In the research on the correction method to date, no explicit tests of the error generated by inconsistent functions have been carried out. With the functions, an uncorrected image is reconstructed from the original projection data. This image is used to provide the weights for the convolution. For each pixel, the scatter line spread function whose point source location was closest to the pixel is employed. The result from the multiple convolutions is an estimate of the scatter component. These estimates are subtracted from the original projections and then a corrected image reconstructed. These manipulations can be rationalized as an iterative procedure which, in the present implementation, is stopped after only one iteration. A special attenuation correction is combined with the scatter correction. In [17], Monte Carlo results from a heart phantom are compared to similar results from a phantom containing 'gaseous ^{99m}Tc', that is, ^{99m}Tc that neither scatters nor attenuates. The energy window had a width of 25%. The 80 mm diameter ventricle is on the axis of a much larger cylinder or is displaced off axis by as much as 100 mm (for the largest, 300 mm diameter, cylinder). Reconstructed values for the heart wall are typically improved by correction from 30% of expected to 95%. Note that this change reflects both attenuation and scatter correction, however. For the 250 mm diameter, intermediate-sized cylinder, the reconstructed value averaged over five displacements is exactly 100% of expected (the range appears to be no more than 98–102%) [17].

The dual-photo-peak-energy-window method was tested on Monte Carlo-simulated data for (i) a point source in a nonuniformly attenuating medium and (ii) a circular-cylinder phantom with a ring of spherical sources [18]. The dual-photo-peak method (previously called the split-photo-peak method) uses the ratio of counts in two nonoverlapping energy windows within the photo-peak region as input to a regression relation. This relation is assumed to have the form

$$SF(i) = AR(i)^B + C \qquad (12.1)$$

where A, B and C are constants to be determined, $SF(i)$ is the scatter fraction within the full photo-peak window, evaluated at pixel i, and $R(i)$ is the ratio of the counts in the lower window over the counts in the upper window for the same pixel. (See the section on comparison of correction methods and figure 12.4 for a comment on this regression equation.) A pixel-by-pixel estimate of the scatter in the summed windows is obtained and subtracted to yield an estimate of the primary [18]. The same Monte Carlo that generated the test data was used to simulate point sources in a uniform elliptical phantom. Data from these point sources were used to evaluate the coefficients in the regression equation. This is an example of following what I call the 'consistency rule'. This rule tells one to combine experimental data only with other experimental data or to combine simulation data only with other simulation data. In the above case, to have used a regression equation from experimental data would violate needed consistency. To evaluate the results from the phantoms, the difference between the object and image is computed pixel by pixel and summed. Using a 10% upper window and a 10% lower window with a symmetric split of the photopeak, the difference measure is better for the dual-photopeak method compared to the dual-energy-window method for the point sources in the nonuniform attenuator and for a ring of hot spheres. It is not better for a ring of cold spheres. A different proportioning of the split (10% upper, 15% lower, upward shifted) does not alter the general trend of the results [18].

Naude et al [19] simulated results from disc sources of four different diameters placed at various depths in water. Their scatter correction method is similar to the split-photo-peak technique but replaces the regression equation with certain ratios that are assumed to have fixed values. The method

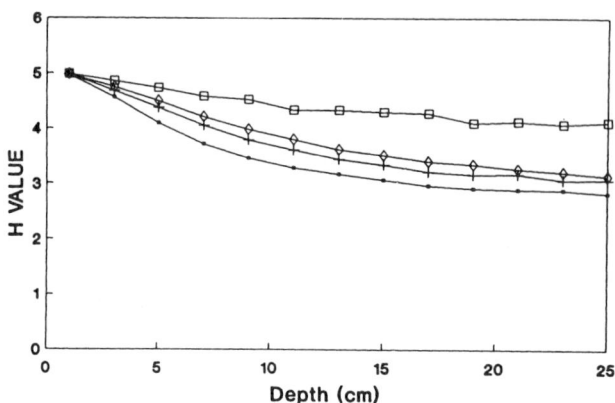

Figure 12.2 *Results from a ^{99m}Tc simulation: variation of one of the parameters in the channel ratio method with the diameter of the disc source and its depth in a water bath. The parameter, H, is assumed constant in the method but is shown to vary with diameter and depth. The symbols for different disc diameters are as follows: dots for 5.2 cm, the + signs for 7.4 cm, the triangles for 9.0 cm and the squares for 16.4 cm. The variation with depth decreases as the disc diameter increases from 5.2 to 16.4 cm. (Permission to reproduce this figure from [19] was granted by the authors and by the Institute of Physics Publishing Limited.)*

is called the channel ratio method [20]. Their simulations tested the variance of the H parameter, the ratio of the number of scattered counts in the lower half of the photo-peak to the number of scattered counts in the upper half of the photopeak [19]. The results, shown in figure 12.2, reveal that H decreases as the depth increases but approaches an asymptotic value as the depth exceeds 15 cm. The change is smaller when the diameter of the disc source is larger. A type of average H value was also calculated for all the cases shown in figure 12.2. Then the authors investigated the effect on quantification from using the average H value of 4.117, instead of the true H value. With an 'air reference', use of the 4.117 value caused an error of only 4.4%, averaged over the individual cases. Thus, at least for these cases, the consequence of the H value not being constant as assumed is fairly mild.

Gagnon *et al* [21] used a Monte Carlo program that (i) employed perfect spatial and energy resolution, (ii) modelled results from a point source in water and (iii) included backscattering from the Pyrex behind the sodium iodide crystal in the Anger camera. They required a 'good' event to have the correct location, within the uncertainty of a 2 mm by 2 mm pixel. Then, they characterized how many events were 'good' even though they had the 'wrong' energy, for example. They argue that 'primary' and 'scattered' are too rough an indication of the information content for a particular event [22].

12.6.2 Reconstructed activity from scatter and comparison of correction methods

Monte Carlo has been used to give the pattern of reconstructed activity using only the scattered gammas and to compare several of the correction methods that use subtraction [23]. The simulated phantoms include (i) a 20.8 cm diameter cylinder containing a circle of six spheres of diameter ranging from 8 mm to 3.0 cm. These spheres were either radioactive in a nonradioactive surround or the reverse, and (ii) a Hoffman three-dimensional brain phantom simulating normal blood flow (grey matter to white matter to ventricles being in the proportion of four to one to zero). 'Each simulated SPECT projection was pre-filtered' before reconstruction by 'filtered backprojection using a modified ramp filter' [23]. The four correction methods were the dual energy window with $k=0.5$, the dual photo-peak window, the three window (especially in a form that has only two windows and so is like the dual energy window with $k=0.5$) and the Ljungberg–Strand method that employs the scatter line spread functions.

For the hot spheres, the true scatter pattern is a low-resolution replication of the true activity distribution. Two of the four correction schemes duplicate this pattern; the dual-energy-window method adds activity around the edge of the cylinder and the scatter-line-spread-function method yields a cold central circle (see the top row in figure 12.3). For the cold spheres, the true scatter pattern does not match the true activity, perhaps because of a defect in the attenuation correction. The four correction methods all differ from this pattern in slight but different ways (see the bottom row in figure 12.3). For the brain phantom the true scatter pattern again is a low-resolution replication of the true activity distribution, probably because the pattern is more like a hot-sphere pattern than a cold-sphere pattern. For this phantom, all correction schemes are similar except that the scatter line spread function method suffers from artifactual strength along the sides of the brain.

According to the authors, the comparison results 'indicate that the differences in performance between different types of scatter correction technique are minimal for ^{99m}Tc brain perfusion imaging. Thus, a user may select a correction method that is easy to implement on a particular system' [23]. They also conclude, however, that it is 'important to perform a test of scatter correction methods for source distributions which closely match the clinical application to which they will be applied' [23].

A second Monte Carlo study has recently compared 'two subtraction based compensation techniques with image reconstruction using iterative reconstruction including an accurate scatter model' [24]. The subtraction methods were the dual-energy-window method and a method based on convolution of an image with a scatter response function [25]. The scatter

SCATTER IMAGES - HOT SPHERES

CW   DPW   TW   SLSF   TRUE

SCATTER IMAGES - COLD SPHERES

CW   DPW   TW   SLSF   TRUE

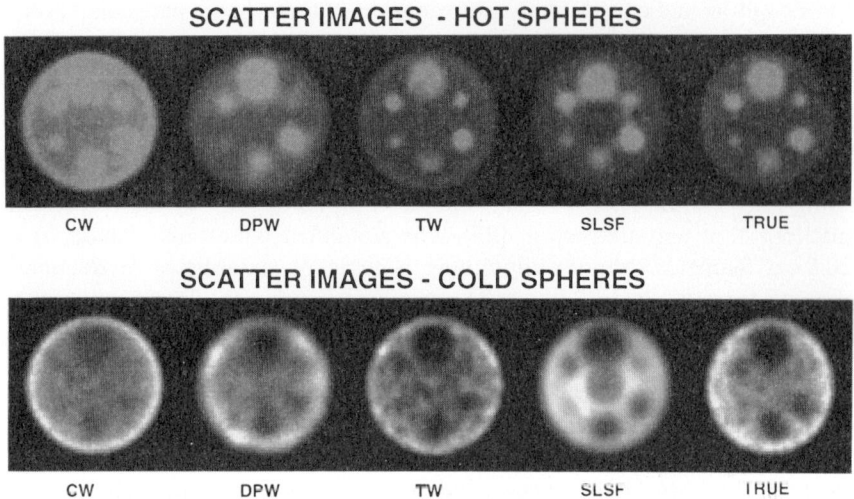

Figure 12.3 *Single-slice images achieved by filtered backprojection reconstruction of only and all scattered gamma rays. Two ^{99m}Tc source distributions have been simulated by Monte Carlo. The result from the true projection of scattered gamma rays is shown on the right for both activity distributions. The results from four scatter estimation techniques are also shown. Here, CW refers to the dual-energy-window method, DPW to the dual-photo-peak-energy-window method, TW to the three-window method and SLSF to the scatter line spread function method. (Permission to reproduce this figure from [23] was granted by the authors and by the Society of Nuclear Medicine.)*

response function is defined as the spatial variation of scattered gamma rays about an origin corresponding to a given strength of unscattered gamma rays at that origin. The convolution method assumes that the scatter response function is spatially invariant. Actually, this function is now known to be 'highly dependent on the object shape and source position' according to Frey *et al* [24]. The use of an 'accurate scatter model' referred to above is an attempt to deal with such variability by a technique called 'slab derived scatter estimation' [24] in an approach similar to inverse Monte Carlo. Among the results from the study are that (i) even judging by the entire image rather than a limited region of interest, the optimum k in the dual-energy-window method varies for the case of hot rods compared to that of cold rods and (ii) the extra work of the more complicated and more accurate scatter model produces better results by most of the figures of merit employed.

A third Monte Carlo study has recently compared nine correction techniques that are based on two or more windows [26]. The test data are generated for a single, top-view projection of a cylindrical ^{99m}Tc–water solution phantom containing four small cylinders of activity concentration 2, 4,

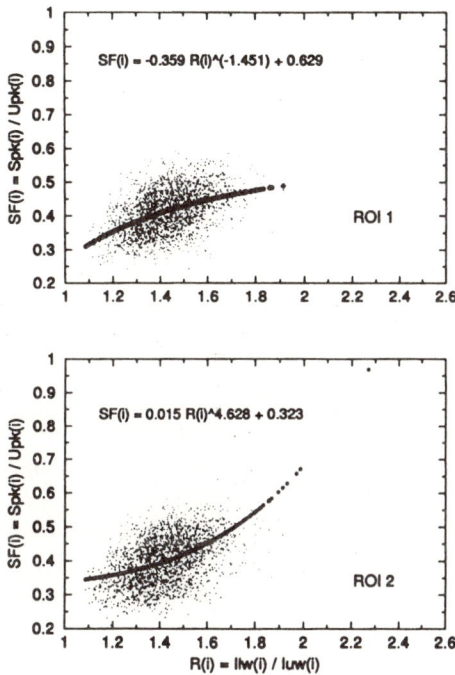

Figure 12.4 *Simulation result with the dual-photo-peak window method for a ^{99m}Tc source. The circular projection is evaluated with a central ROI (ROI 1) and a larger ROI extending to the edge of the circle (ROI 2). The scatter fraction, SF(i), is plotted against the ratio, R(i), of the number of counts falling into the lower-energy subwindow divided by the number of counts falling into the higher-energy subwindow. Each dot corresponds to a pixel, in the projection image. Depending on which ROI is chosen, one obtains a very differently shaped power law relation. (Permission to reproduce this figure from [26] was granted by the authors and by the Society of Nuclear Medicine.)*

6 and 8 relative to the background concentration of the cylinder. Results for the dual-photo-peak-energy-window method are shown in figure 12.4. It is seen that the regression equation that is characteristic of the method depends on the ROI which is evaluated. That is, ROI 1(1943 pixels) covered the central part of the circular projection while ROI 2 (2257 pixels) included the edges of the circle. The functional form of the curve for the two ROIs is quite different: one curve is cupped upward while the other is cupped downward. However, the actual variation of the two functions in the region containing most of the data points does not appear to be drastically large.

In addition to the dual-photo-peak-energy-window method, three of the nine methods under comparison have been described above. They are the dual-energy-window, the triple-energy-window and the channel ratio methods. One of the nine methods involves no correction at all and is presented merely to characterize the correction problem. Two others are closely related to methods I have already presented. These two are the method of Logan and McFarland [27] and the technique of Bourguignon *et al* [28]. The last two, FADS or the 'constrained factor analysis of dynamic structures' method [29] and a variation on this technique by Buvat *et al* [30], are of a type not mentioned previously. They require more than two windows and involve a fit to the data as do the methods in [14] and [15], which were not tested. All the methods are compared with respect to relative quantification, absolute quantification and signal-to-noise ratio. Comments are made on the strengths and shortcomings of each method. A pixel-by-pixel analysis of the accuracy of the estimate of the number of unscattered photons gives values of around 10% when the counting statistics are good and worse (15–45% for the dual-energy-window method, for example) and variable depending upon method when the counting statistics are poor. Here good counting statistics means the number of unscattered counts per pixel is greater than 130 and poor counting statistics means the number is between 130 down to 30. The newer factor analysis approach [30] tends to outperform the other methods. However, the authors acknowledge the need for the 'more sophisticated acquisition mode (30 energy windows)' [26] that is not yet available on the majority of commercial cameras.

12.6.3 Accuracy of Monte Carlo

As for the faithfulness of the Monte Carlo simulation, Rosenthal and Henry [31] and the present author with his collaborators [14] have both noticed that the spectrum from an experimental point source has a larger-magnitude low-energy tail than is obtained with some Monte Carlo programs. One possibility to be considered is that some or all of these counts originate from gamma rays that are scattered in what is supposed to be the scatter-free source. In our work at least, however, the mass of that source is so small that this possibility seems unlikely. Rosenthal and Henry see a potential source for these counts as 'scatter from the photomultiplier tubes back into the NaI(Tl) crystal, [31], whereas it is our opinion that the effect is caused by characteristics of the camera electronics rather than by true lower-energy gamma rays [14]. In any case, it is important to follow the 'consistency rule'. As applied here, it requires that a simulation of a complex phantom should be matched with a simulation of a scatter-free source.

12.7 ^{201}Tl

Ljungberg and Strand repeated the tests of their scatter attenuation correction method on their heart phantom filled with ' ^{201}Tl', a pseudo-isotope with a single gamma emission at exactly 75 keV [17]. A 30% energy window

Table 12.1 *Gamma ray yields for*
^{131}I *(energies greater than or equal*
to 364.4 keV)

Energy (keV)	Number per nuclear disintegration
364.4	8380×10^{-4}
404.8	6×10^{-4}
502.9	29×10^{-4}
636.7	657×10^{-4}
643.0	14×10^{-4}
722.8	174×10^{-4}

was used. Resulting reconstructed values, again averaged over five displacements of the ventricle from the symmetry axis of the enclosing cylinder, are slightly too small for the smallest and largest cylinder sizes, being 88 instead of 100%. For the intermediate size, the average value is closer to right at 94% [17].

12.8 ^{111}In

Ljungberg and Strand also repeated their heart-phantom tests on ^{111}In, concentrating only on the emission at 247 keV and neglecting the lower-energy emission at 172 keV which should also be a viable experimental procedure [17]. A 25% energy window was employed. The resulting values have almost the same good accuracy as the ^{99m}Tc results quoted above.

12.9 ^{131}I

No Compton scatter correction techniques have used simulated data from ^{131}I for development or testing. This fact is due to the lack of a proven, general purpose ^{131}I simulation package although Bice *et al* have simulated results from ^{131}I for a square-hole collimator with a fairly slow program [32]. One of the complicating factors is the presence of low-yield, high-energy gamma rays as shown in table 12.1 [33]. These have a combined strength of 880 compared to 8380 for the primary 364 keV gamma ray. These gamma rays penetrate the lead septa of conventional high-energy collimators fairly easily and so produce an even higher number of counts than would be expected from their relative yield. Bice *et al* state that 'of the primary 364 keV photons that make the photo-peak window, about 49% of them either scattered from or penetrated collimator septa' if the counts from the entire camera face are totalled [32]. Another complicating factor with a standard-thickness crystal ($\frac{1}{4}$ or $\frac{3}{8}$ in thick) is partial deposition in the crystal. Using experimental measurements, Pollard *et al* assessed the source of gamma rays in a 268–320 keV ^{131}I scatter window [34]. For their report, they combine scatter from the collimator and from other crystal-surrounding materials with nonscatter that undergoes partial energy deposition. They define this combination as 'camera induced scatter'. They estimate that 60% of the counts are from 'camera induced scatter' of primary photons, 20% are from 'camera induced scatter' of higher-energy photons and only 20% are from scatter within the object [34]. Therefore, the ideal Monte Carlo program needs to include a good collimator penetration computation and a good camera detection computation. It also should be fast, general purpose and proven, if possible.

Yanch and Dobrzeniecki say that with their program SimSPECT 'the collimator is physically modelled in its entirety' [35]. However, convincing proof that it correctly accounts for penetration from the high-energy gamma rays of ^{131}I has not as yet been presented. The modular program of Harrison *et al* has its generation and scatter component available but the detection with collimation module has not as yet been developed and tested with [131]I [36]. Two groups, de Vries *et al* [37] and Gagnon *et al* [21, 22] have programs which are said to simulate gamma rays up to 512 keV. However, 8% of the gamma emissions of ^{131}I exceed even this energy. The conclusion is that the ideal [131]I Monte Carlo simulation program is not yet available.

12.10 ALTERNATIVES TO MONTE CARLO

In checking for the amount of scatter either before or after a correction has been applied, two experimental alternatives to Monte Carlo are available. For a point source, one simply measures first with the point source in the scattering medium and then again at the exact same location in air (that is, the scattering medium has been removed). For more complicated objects, however, there is no way to avoid self-scatter and self-attenuation, assuming the isotope is not available in a gaseous form. The second experimental approach is to use a high-purity germanium detector. This detector has much higher energy resolution than NaI(Tl) and so scattered and unscattered photons are almost completely separated by their detected energies. The complications of measuring scatter fractions with this detector are detailed in [38]. Since the method does have its own complications, Monte Carlo is still very handy for the development and testing of scatter correction schemes.

12.11 ADDITIONS TO MONTE CARLO

Although it might be possible for Monte Carlo to simulate them, there are complexities in the detection problem which are probably too camera specific for Monte Carlo to be worthwhile. The example that comes to mind is local shift of energy spectra with rotation angle [39]. If the effect of such shifts are to be taken into account, it would probably be best to add on a post-simulation, non-Monte Carlo module which would require and utilize input from previous experimental measurements on the specific camera.

12.12 CONCLUSIONS FROM MONTE CARLO

In summary, it has been found that several scatter correction methods are nearly equivalent with respect to scatter correction in ^{99m}Tc brain perfusion imaging. In other comparisons, the more complicated scatter correction method has yielded the best result when compared to simpler methods. Lastly, in some cases, assumptions that underlie the correction methods have been found to be violated. Examples include the dual-energy-window method and the channel ratio method. However, the importance of the violation needs also to be evaluated. In at least one case, the negative consequences of the violation have been judged to be fairly mild.

REFERENCES

[1] Luo J-Q, Koral K F, Ljungberg M, Floyd C E and Jaszczak R J 1995 A Monte Carlo investigation of dual-energy-window scatter correction for volume-of-interest quantification in ^{99m}Tc SPECT *Phys. Med. Biol.* **40** 181–99

[2] Floyd J L, Mann R B and Shaw A 1991 Changes in quantitative SPECT ^{201}Tl results associated with the use of energy-weighted acquisition *J. Nucl. Med.* **32** 805–7

[3] Buvat I, Benali H, Todd-Pokropek A and Di Paola R 1994 Scatter correction in scintigraphy: the state of the art *Eur. J. Nucl. Med.* **21** 675–94

[4] Bowsher J E and Floyd C E Jr 1991 Treatment of Compton scattering in maximum-likelihood expectation-maximization reconstructions of SPECT images *J. Nucl. Med.* **32** 1285–91

[5] Smith M F, Jaszczak R J and Coleman R E 1992 Simultaneously constraining SPECT activity estimates with primary and secondary energy window projection data *IEEE Medical Imaging Conf. Record (Orlando, FL, Oct. 25–31, 1992)* ed G T Alley (Piscataway, NJ: IEEE) pp 1175–7

[6] Jaszczak R J, Greer K L, Floyd C E, Harris C C and Coleman R E 1984 Improved SPECT quantification using compensation for scattered photons *J. Nucl. Med.* **25** 983–90

[7] Links J M, Jeremy R W, Frank T and Becker L C 1990 Wiener filtering improves quantification of myocardial blood flow with Tl SPECT *J. Nucl. Med.* **31** 1230–6

[8] Links J M 1995 Scattered photons as 'good counts gone bad' Are they reformable or should they be permanently removed from society? Editorial *J. Nucl. Med.* **36** 130–1

[9]   Wang X, Koral K F, Buchbinder S F, Petrick N, Clinthorne N H and Rogers
      W L 1992 Deconvolution-fitting Compton-scatter correction with both
      energy and spatial regularization *J. Nucl. Med.* **33** 925 (abstract)
[10] Links L M 1995 personal communication
[11] Floyd C E, Jaszczak R J, Harris C C, Greer K L and Coleman R E 1985
      Monte Carlo evaluation of Compton scatter subtraction in single photon
      emission computed tomography *Med. Phys.* **12** 776–8
[12] Ljungberg M, Msaki P and Strand S-E 1990 Comparison of dual-window and
      convolution scatter correction techniques using the Monte Carlo method
      Phys. Med. Biol. **35** 1099–110
[13] Luo J-Q and Koral K F 1994 Background-adaptive dual-energy-window cor-
      rection for Compton scattering in ^{99m}Tc SPECT *Nucl. Instrum. Methods
      Phys. Res.* A **353** 340–3
[14] Koral K F, Wang X, Zasadny K R, Clinthorne N H, Rogers W L, Floyd
      C E Jr and Jaszczak R J 1991 Testing of local gamma ray scatter fractions
      determined by spectral fitting *Phys. Med. Biol.* **36** 177–90
[15] Wang X and Koral K F 1992 A regularized deconvolution-fitting method for
      Compton-scatter correction in SPECT *IEEE Trans. Med. Imaging* **MI-11**
      351–60
[16] Ogawa K, Harata Y, Ichihara T, Kubo A and Hashimoto S 1991 A practical
      method for position-dependent Compton-scatter correction in single pho-
      ton emission CT *IEEE Trans. Med. Imaging* **MI-10** 408–12
[17] Ljungberg M and Strand S-E 1990 Scatter and attenuation correction in
      SPECT using density maps and Monte Carlo simulated scatter functions
      J. Nucl. Med. **31** 1560–7
[18] Hademenos G J, Ljungberg M, King M A and Glick S J 1991 A Monte Carlo
      investigation of the dual photo-peak window scatter correction method
      IEEE Medical Imaging Conf. Record (Santa Fe, NM, Nov. 2–9, 1991) ed
      G T Baldwin (Piscataway, NJ: IEEE) pp 1814–21
[19] Naude H, van Aswegen A, Herbst C P, Lotter M G, Pretorius P H 1996 A
      Monte Carlo evaluation of the channel ratio scatter correction method
      Phys. Med. Biol. **41** 1059–66
[20] Pretorius P H, van Rensburg A J, van Aswegen A, Lotter M G, Serfontein
      D E and Herbst C P 1993 The channel ratio method of scatter correction
      for radionuclide image quantitation *J. Nucl. Med.* **34** 330–5
[21] Gagnon D, Laperriere L, Pouliot N, deVries D J and Moore S C 1992 Monte
      Carlo analysis of camera-induced spectral contamination for different
      primary energies *Phys. Med. Biol.* **37** 1725–39
[22] Gagnon D, Pouliot N and Laperriere L 1991 Statistical and physical content
      of low-energy photons in holospectral imaging *IEEE Trans. Med. Imaging*
      MI-10 284–9
[23] Ljungberg M, King M A, Hademenos G J and Strand S-E 1994 Comparison
      of four scatter correction methods using Monte Carlo simulated source
      distributions *J. Nucl. Med.* **35** 143–51
[24] Frey E C, Tsui B M W and Ljungberg M 1992 A comparison of scatter
      compensation methods in SPECT subtraction-based techniques versus iter-
      ative reconstruction with accurate modeling of the scatter response *IEEE
      Medical Imaging Conf. Record (Orlando, FL, Oct. 25–31)* ed G T Alley
      (Piscataway, NJ: IEEE) pp 1035–7
[25] Axelsson B, Msaki P and Israelsson A 1984 Subtraction of Compton-scattered
      photons in single-photon emission computerized tomography *J. Nucl. Med.*
      25 290–4

[26] Buvat I, Rodriguez-Villafuerte M, Todd-Pokropek A, Benali H and Di Paola R 1995 Comparative assessment of nine scatter correction methods based on spectral analysis using Monte Carlo simulations *J. Nucl. Med.* **36** 1476–88

[27] Logan K W and McFarland W D 1992 Single photon scatter compensation by photo-peak energy distribution analysis *IEEE Trans. Med. Imaging* **MI-11** 161–4

[28] Bourguignon M H, Wartski M, Amokrane N, Berrah H, Riddell C, Valette H, De Dreuille O, Bendriem B, Delforge J and Syrota A 1993 Le spectre du rayonnement diffuse dans la fenêtre du photopic analyse et proposition d'une methode de correction *Medecine Nucleaire* **17** 53–8

[29] Mas J, Hannequin P, Ben Younes R, Bellaton B and Bidet R 1990 Scatter correction in planar imaging by constrained factor analysis of dynamic structures (FADS) *Phys. Med. Biol.* **35** 1451–65

[30] Buvat I, Benali H, Frouin F, Bazin J P and Di Paola R 1993 Target apex seeking in factor analysis of medical image sequences *Phys. Med. Biol.* **38** 123–38

[31] Rosenthal M S and Henry L J 1990 Scattering in uniform media *Phys. Med. Biol.* **35** 265–74

[32] Bice A N, Durack L D, Pollard K R and Eary J F 1991 Assessment of ^{131}I scattering and septal penetration in clinical gamma camera high energy parallel hole collimators *J. Nucl. Med.* **32** 1058–9 (abstract)

[33] Williams L E 1987 Nuclear structures and decay *Nuclear Medicine Physics* vol 1, ed L E Williams (Boca Raton, FL: Chemical Rubber Company) pp 1–42

[34] Pollard K R, Bice A N, Durack L D, Eary J F and Lewellen T K 1992 Camera-induced Compton scatter and collimator penetration in ^{131}I imaging *J. Nucl. Med.* **33** 889 (abstract)

[35] Yanch J C and Dobrzeniecki A B 1991 Monte Carlo simulation in SPECT complete 3D modeling of source collimator and tomographic data acquisition *IEEE Medical Imaging Conf. Record (Santa Fe, NM, Nov. 2–9, 1991)* ed G T Baldwin (Piscataway, NJ: IEEE) pp 1809–13

[36] Harrison R L, Haynor D R, Gillispie S B, Vannoy S D, Kaplan M S and Lewellen T K 1993 A public-domain simulation system for emission tomography: photon tracking through heterogeneous attenuation using importance sampling *J. Nucl. Med.* **34** 60P (abstract)

[37] deVries D J, Moore S C, Zimmerman R E, Mueller S P, Friedland B and Lanza R C 1990 Development and validation of a Monte Carlo simulation of photon transport in an Anger camera *IEEE Trans. Med. Imaging* **MI-9** 430–8

[38] Zasadny K R, Koral K F, Floyd C E Jr and Jaszczak R J 1990 Measurement of Compton scattering in phantoms by germanium detector *IEEE Trans. Nucl. Sci.* **NS-37** 642–6

[39] Koral K F, Luo J-Q, Ahmad W, Buchbinder S and Ficaro E P 1995 Changes in local energy spectra with SPECT rotation for two Anger cameras *IEEE Trans. Nucl. Sci.* **NS–42** 1114–9

CHAPTER 13

DESIGN OF A COLLIMATOR FOR IMAGING ^{111}In

Stephen C Moore, Daniel J deVries,
Bill C Penney, Stefan P Mueller,
and Marie Foley Kijewski

13.1 INTRODUCTION

We have used the Monte Carlo program [1] described in Chapter 10 to optimize the design of a medium-energy (ME) collimator for imaging ^{111}In. Renewed interest in improved imaging of this isotope has been stimulated by tumour imaging agents such as ^{111}In octreotide, as well as by various monoclonal antibodies used primarily for research. Because ^{111}In emits gamma photons of two different energies, the collimator design problem is more complex than it is for monoenergetic emitters such as ^{99m}Tc. For ^{111}In imaging, the usual problems of gamma ray absorption and scatter in the patient are compounded by the effects of photon interactions in the collimator and detector, and by contamination of the image in the 172 keV window by scattered 247 keV photons which have lost energy.

Monte Carlo simulation was an essential component of this work for several reasons. Most importantly, it is impossible to calculate analytically the contribution to an image of photons which scatter in the collimator and detector, even though patient scatter might be reasonably approximated by some model calculations. Although evaluation of images obtained experimentally with different collimators might allow prediction of a suitable collimator design for imaging ^{111}In, the number of real collimators available for testing with a given camera system is small, and most of these are likely to be far from optimal. Monte Carlo simulation permitted us to analyse a large number of possible designs without the expense of actually constructing many different collimators for testing.

The collimator design was optimized for detecting spherical lesions of unknown activity and size, within a given range of sizes, centred in a spherical background 'organ' of unknown activity which, in turn, was embedded in a larger cylindrical attenuator. In the remainder of this chapter, we shall describe the methods used to accomplish this optimization, along with results illustrating each step of the procedure. These steps consisted of

  (i) determination of collimator, detector and 'patient' spectral contributions from the 172 and 247 keV decay photons for collimators containing different amounts of lead,
 (ii) simulation of ^{111}In point spread functions (PSFs) in water for many different collimator designs, with and without contributions from patient scatter,
(iii) parametrization of collimator PSFs and efficiencies as smoothly varying functions of spatial resolution and collimator lead content,
 (iv) Cramer–Rao bound calculations of signal-to-noise ratios (SNRs) for the detection task described above, using the parametrized functions from step (iii), and
  (v) identification of the combination of collimator parameters providing the maximum detection SNR.

13.2 STUDY OF SPECTRAL COMPONENTS

The Monte Carlo program was set up to detect photons in two 20% energy windows, one centred at 172 keV and the other at 247 keV. As described in Chapter 10 and [1], we simulated collimators with square holes. For the purpose of comparing simulation results with those expected for hexagonal-hole collimators, we note that square-hole and hexagonal-hole collimators with identical values of spatial resolution (full width at half maximum (FWHM) of the radially averaged PSF), count sensitivity and lead content may be obtained by using collimators of the same thickness ($a_{hex} = a_{square}$), but slightly different hole size ($d_{hex} = 1.0746d_{square}$) and septal thickness ($s_{hex} = 1.0746s_{square}$). List-mode data storage enabled us to keep track of important aspects of each event's history during the propagation of photons through the source, collimator and detector. We simulated a GE StarCam^{TM} gamma camera; the detector code approximated photon backscatter from the materials behind the NaI(Tl) crystal by simulating a single material layer whose thickness and density were selected to provide the same mean interaction depth and total scatter fraction as we calculated for a multicompartment material model. The energy resolution was 11% (FWHM) of the total energy deposited per event. (We had previously measured this constant percentage energy resolution for several different radionuclides with photopeak energies greater than 140 keV.)

   We first simulated a point source in air, as well as a point source in a 22 cm diameter water cylinder. In both cases, the point was 17.4 cm from the front face of a simulated NuTech^{TM} ME collimator containing approximately

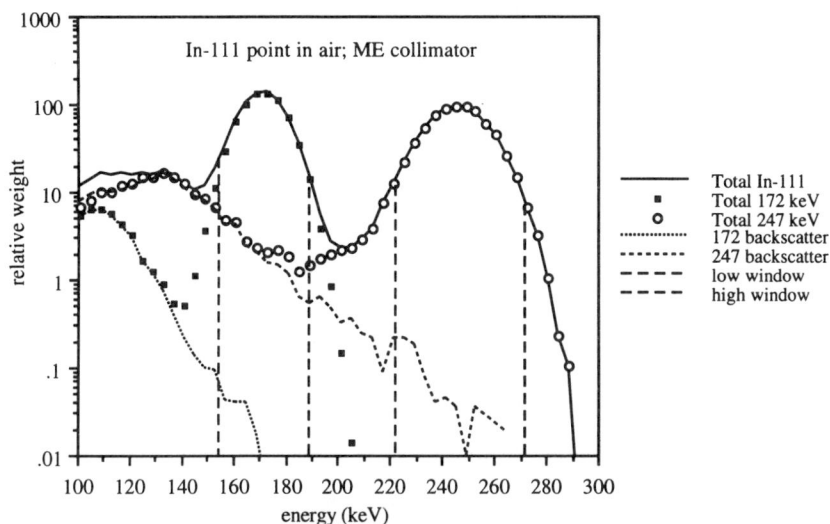

Figure 13.1 *Energy spectra simulated for a point source of ^{111}In in air, showing separately the contributions from 172 and 247 keV photons. The backscattered spectra are for photons which scatter off material behind the NaI(Tl) crystal before detection.*

14 g cm^{-2} of lead. Spectra were obtained separately for the 247 keV and the 171 keV photons emitted from ^{111}In in air (figure 13.1) and water (figure 13.2).

From the air spectra of figure 13.1, it is evident that backscattered 247 keV photons contribute at most a few per cent of the total counts under the 172 keV photopeak. However, from figure 13.2 it may be seen that 247 keV photons which are Compton scattered before detection (mostly in the water phantom) can contribute significantly to the 172 keV photopeak window. This source of scattered photons, in fact, exceeds the contribution from 172 keV scattered photons.

To obtain a better approximation to body imaging conditions, all additional simulations reported in this chapter were performed using a 30 cm diameter cylindrical water attenuator, the centre of which was located 16 cm from the collimator face. Spectra from a centred point source of ^{111}In were simulated for collimators with the same spatial resolution (1.0 cm FWHM), but with several different values of lead content. The three geometric parameters (*a*, *d* and *s*) used for each collimator simulation were calculated by maximizing the collimator efficiency for the desired values of lead content and spatial resolution. For each energy window, the fraction of detected photons which penetrated or scattered in the collimator was

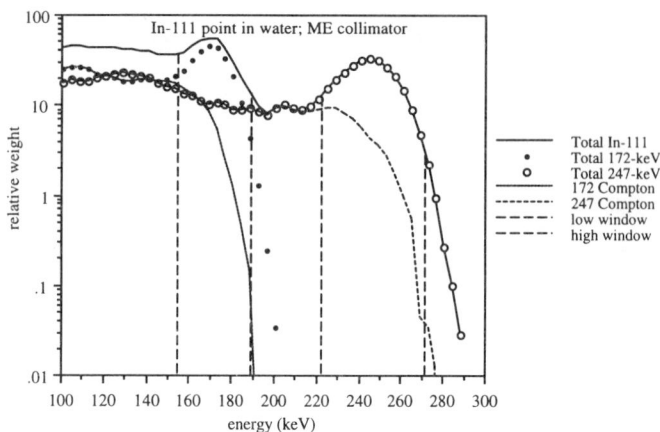

Figure 13.2 *Energy spectra simulated for a point source of ^{111}In in water, showing separately the contributions from 172 and 247 keV photons. The Compton spectra are for photons which scatter in the 22 cm diameter water phantom, the collimator or the detector.*

determined. These fractions are tabulated separately for the two energy windows (table 13.1) to provide an indication of the extent to which the collimator penetration and scatter components are affected by the lead content. (It should be noted that the total percentage of penetrating photons listed here is not equivalent to the 'single-septal penetration probability', another common indicator of the degree of collimator penetration.)

13.3 COLLIMATOR PSFS AND EFFICIENCIES: FUNCTIONAL PARAMETRIZATIONS

We simulated several different collimators, providing system resolution values in the range 0.75–2.25 cm FWHM. For most of these resolution values, we also simulated collimators of differing lead content in the range

Table 13.1. *^{111}In collimator penetration and scatter fractions for a point source centred in a 30 cm diameter water cylinder.*

Collimator lead content (g cm^{-2})	172 keV peak window		247 keV peak window	
	% penetration	% scatter	% penetration	% scatter
10	14.6	7.1	58.4	22.5
14	7.3	3.4	34.6	15.5
18	6.9	2.8	21.2	8.8
22	6.2	2.5	14.0	5.6
26	5.3	2.5	11.3	4.4

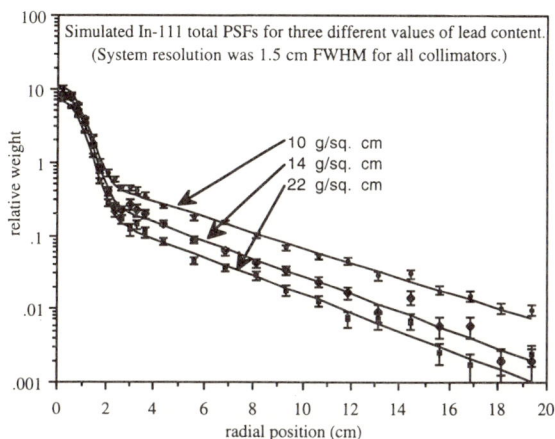

Figure 13.3   *^{111}In radial PSFs from the sum of the two energy windows for different values of lead content, with superimposed fits to equation (13.1). The collimator spatial resolution was the same for all three collimators, and was chosen to provide a system resolution of 1.5 cm FWHM at a distance of 16 cm from the collimator surface.*

10–22 g cm^{-2}. For each collimator, a radial PSF histogram was first accumulated from the sum of the photon weights in both energy windows. In order to assess the possible impact of a patient scatter correction procedure, we also obtained radial PSFs for all photons which did not scatter before entering the collimator, although they may have scattered in the collimator or detector. The pixel size was 0.3125 cm for all simulations.

Because the overall count efficiency obtained by imaging an extended source in a scattering medium would be different from that obtained for a point source in the same attenuator, we also simulated for each collimator a uniform cylindrical activity distribution. The simulated PSFs were each fitted to the sum of a radial Gaussian plus a radially decreasing mono-exponential (see, e.g., figures 13.3 and 13.4)

$$\text{PSF}(r) = A_{\text{Gauss}} \exp\left(\frac{-r^2}{2\sigma^2}\right) + A_{\text{exp}} \exp(-\lambda r). \tag{13.1}$$

For the data which included scatter in the phantom, the fit coefficients (A_{Gauss}, σ, A_{exp} and λ) and extended source count efficiency were then parametrized as products of various functions of lead content and spatial resolution to provide a reasonable means of interpolating between simulated collimator designs in a collimator optimization procedure. (For most of these parametrizations, the spatial resolution and lead content dependences of the parameters were fitted separately and error bars were not used to

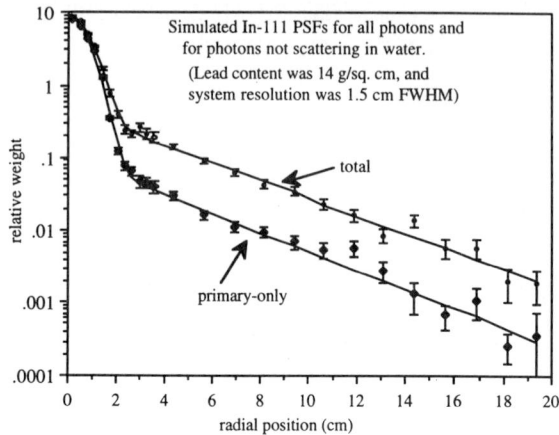

Figure 13.4 ^{111}*In radial PSFs from all photons in both energy windows (total) and from photons which did not scatter in the attenuator (primary only), along with superimposed fits. The collimator had a lead content of 14 g cm^{-2} and provided a system resolution of 1.5 cm FWHM at a distance of 16 cm from its surface.*

weight the contributions from each data point.) The parametrized PSF fit coefficients are shown in formulae (13.2)–(13.5) and figure 13.5. In the following formulae, S is the system spatial resolution in centimetres FWHM, obtained from the quadrature sum of the collimator resolution and the intrinsic spatial resolution, and P is the collimator lead content in grams per square centimetre

$$A_{\mathrm{Gauss}} = 6.81(1.39 - e^{-1.13S})(1.7463 - 0.0663P + 0.0011P^2) \quad (13.2)$$

$$\sigma = (0.0124 + 0.4823S)(1.0315 - 0.002P) \quad (13.3)$$

$$A_{\mathrm{exp}} = (-0.1161 + 0.2831S + 0.0697S^2) \\ \times (8.1073 - 1.1027P + 0.0558P^2 - 0.0001P^3) \quad (13.4)$$

$$\lambda = -2.228(1.06 + e^{-2.04S})(0.116 - e^{-0.413P}). \quad (13.5)$$

The Monte Carlo program was not always set up to simulate the same number of point source decays for each collimator. In general, more decays were simulated for the less efficient collimators to improve the precision with which their PSF parameters could be determined; however, the PSFs were all scaled appropriately to correspond to images acquired for the same scan time from sources of the same activity. (This is why the ordinates of figure

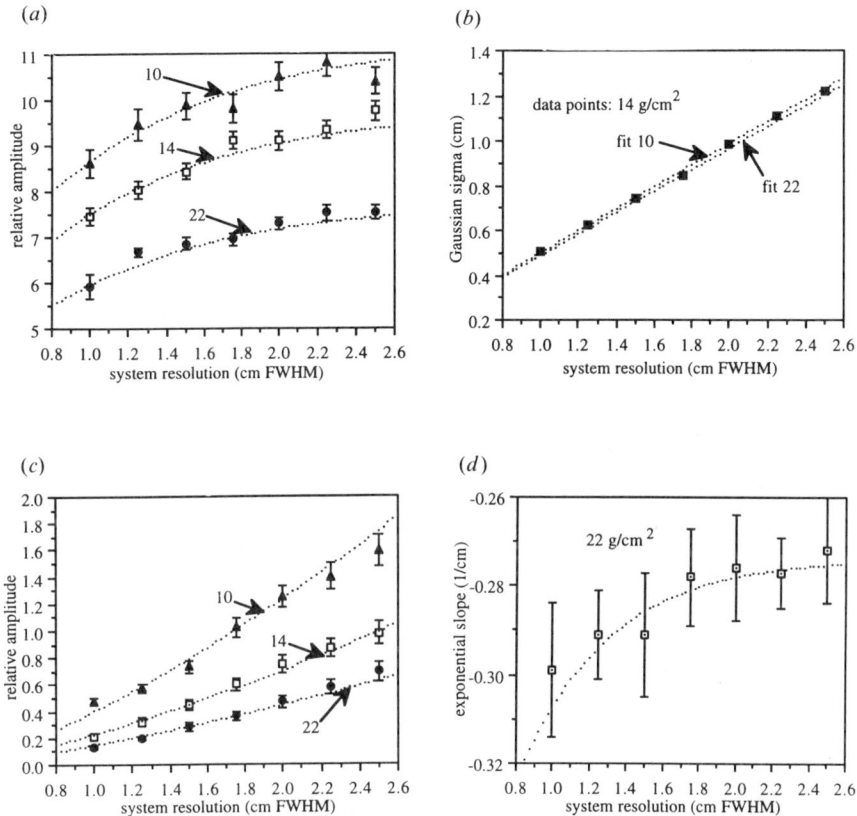

Figure 13.5   *PSF fit coefficients as functions of lead content and system spatial resolution. (a) Gaussian amplitude for three different values of lead content (10, 14 and 22 g cm⁻²) against system spatial resolution, with superimposed parametrized curves. (b) Gaussian σ (cm) from the PSF fits for 14 g cm⁻²; this parameter changed very little over the range of simulated lead content values, as demonstrated by the parametrized curves for 10 and 22 g cm⁻². (c) Exponential amplitude for the same cases as shown in (a). (d) For figure clarity, the exponential slope, λ (cm⁻¹), is shown only for 22 g cm⁻² because this parameter changed very little (compared to the error bars from the PSF fits) for a wide range of lead content values.*

13.5(*a*) and (*c*) are labelled 'relative amplitude'.) The average number of simulated decays was approximately 50 million. For the Bayesian detection procedure described in the next section, we used the modulation transfer function (MTF) of each collimator, rather than its PSF, because our images were modelled in frequency space. To obtain the MTF for a given collimator, the collimator's Gaussian and exponential PSF amplitudes were first renormalized by a constant factor in order that the two-dimensional integral of equation (13.1) would be unity. The MTF was then simply defined by the Hankel transform of (13.1).

13.4 ^{111}In COLLIMATOR OPTIMIZATION USING A BAYESIAN DETECTION PARADIGM

We have previously described the use of a Bayesian parameter estimator as a paradigm for human observer performance in detecting lesions of unknown size in a uniform noisy background [2, 3]. The parameters to be determined from the projection image data were the lesion amplitude, A, the background amplitude, B, and the lesion size or radius, R. This parameter estimation paradigm was based on our previous experience with maximum-likelihood (ML) parameter estimation [4], in which we also demonstrated the mathematical equivalence between estimation from the projection images and from the tomographically reconstructed images. The modified Bayesian 'observer' used knowledge of the possible range of lesion sizes as a prior; its predictions agreed well with the results of a six-observer perceptual study. The average human response to changes in collimator resolution, as measured by the detectability index, d_A, was tracked well by the Bayesian detector's signal-to-noise ratio (SNR). An analytic approximation for the variance of lesion activity estimates (which included the same prior) was shown to predict the variance of the Bayesian estimator over a wide range of collimator resolution values. Because the bias of the Bayesian estimator was small ($<1\%$), the analytic variance estimate permitted a rapid and convenient prediction of the Bayesian detection SNR:

$$\text{SNR}_{\text{Bayes}} = (\langle A_{\text{target}} \rangle - \langle A_{\text{nontarget}} \rangle)/\sqrt{\tfrac{1}{2}(\sigma^2_{\text{target}} + \sigma^2_{\text{nontarget}})}. \quad (13.6)$$

In this formula, $\langle A_{...} \rangle$ represents the expected mean value of the lesion amplitude parameter, estimated from images with many different statistical noise realizations. The subscript 'target' denotes the class of images containing a lesion superimposed on the background, whereas 'nontarget' images contained no lesions. The variances, σ^2, of the estimates of lesion amplitude from the two classes of images were approximated using an analytic calculation based on the Cramer–Rao bound (CRB). For this calculation, with three unknown parameters, the Fisher information matrix is 3×3. We assumed that the lesion and background amplitude parameters, A and B, were completely unknown to the observer, whereas the prior probability density on lesion radius, R, was described by the uniform distribution on some interval $\{R_1, R_2\}$. For a uniform prior, an exact calculation of the CRB is impossible because the contribution of the prior term to the Fisher

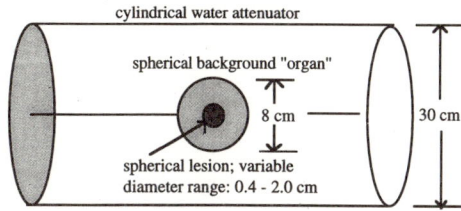

Figure 13.6 *The phantom geometry used for Bayesian lesion detection paradigm.*

matrix is singular. The effect of this prior was, therefore, approximated by adding to the third diagonal element of the Fisher information matrix (corresponding to the lesion radius parameter) the inverse of the variance of the uniform distribution, i.e., $12/(R_2 - R_1)^2$. This modification was made following the exact procedure appropriate for Gaussian priors [5]. The estimated variance of the lesion activity estimates was then obtained from the first diagonal element of the inverse Fisher matrix. This variance approximation agreed well with the measured variance of the Bayesian estimator over a wide range of collimator resolution values [3].

The results of our perceptual studies encouraged us to apply the Bayesian paradigm to optimize the design of a lead collimator for detecting spherical lesions of unknown size, distributed uniformly on the interval 0.4–2.0 cm in diameter, centred in an 8 cm diameter spherical background 'organ' of uniform ^{111}In concentration (figure 13.6). Because the frequency-space image model was equivalent to a real-space image represented by a 2D convolution of the target plus background spheres with a fitted PSF, and because the simulated PSFs were obtained for a centred point source in a 30 cm diameter water cylinder, the spherical background was, therefore, also effectively centred in a 30 cm diameter attenuator. (Note that the assumption of PSF stationarity should be reasonable over the small region containing activity.)

The variance estimation program used the different collimator PSFs and count efficiencies in order to optimize the designs of two collimators for ^{111}In imaging. The first collimator was designed assuming that the sum of the two energy windows would contribute to the image. (For this collimator, the parametrized PSF fit coefficients were used to provide a reasonable means of interpolating between simulated collimator designs.) The second collimator was designed with the assumption that only those photons which did not scatter in the patient contributed to the image, in order to assess any possible effects of an ideal patient scatter correction on the optimal collimator design. The Bayesian SNR (equation (13.6)) was calculated with and without patient scatter for many different values of collimator spatial

(a)

(b)

(c)

(d)

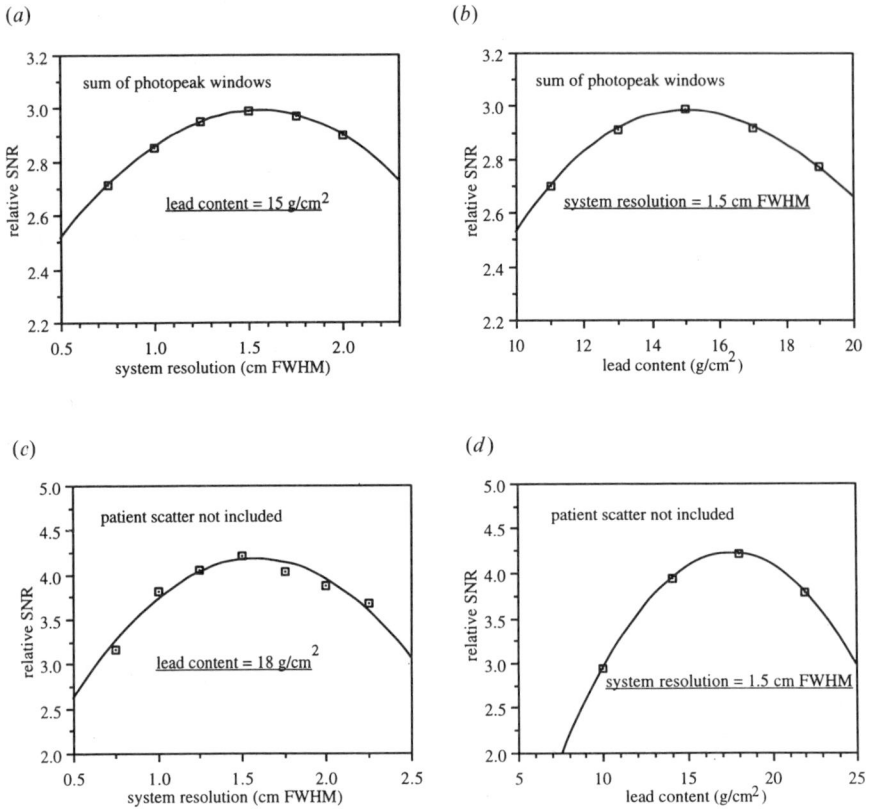

Figure 13.7 *Relative Bayesian SNR calculated from equation (13.6) assuming different values of collimator lead content and system spatial resolution. For the sum of all counts in the two photopeak windows, (a) shows the spatial resolution dependence of the SNR near the optimal lead content (15 g cm^{-2}) and (b) shows the dependence on lead content near the optimal spatial resolution (1.5 cm FWHM). When patient scatter was not included in the image data, (c) shows the spatial resolution dependence near the optimal lead content (18 g cm^{-2}) and (d) shows the dependence on lead content near the optimal spatial resolution (1.5 cm FWHM).*

resolution and lead content (figure 13.7). From these curves, it was possible to find the optimal system resolution (16 cm from the collimator surface) and lead content for this imaging task with ^{111}In. These values are shown in table 13.2, along with the corresponding collimator geometric parameters (a, d and s). The single-septal penetration probability is also listed in order to facilitate comparison with existing collimators.

Table 13.2. Results of ^{111}In collimator design optimization.

	System resolution (cm)	Pb content (g cm^{-2})	SSP (%)	a (mm)	d (mm)	s (mm)
Using all photons in both windows	1.5	15	3.0	34.8	2.47	0.67
With ideal correction for patient scatter	1.5	17	1.9	37.3	2.62	0.77

13.5 CONCLUSIONS

We have shown that, when imaging ^{111}In, the 247 keV contamination detected in the 172 keV window, resulting from patient, collimator and detector scatter, is significant and exceeds the 172 keV scatter component (figure 13.2). ^{111}In collimators were optimized for a useful perceptual task that is equivalent to the task of estimating lesion activity when background activity is unknown and lesion size is known only within a specified range. Our collimator design results (table 13.2) suggest that if the patient scatter 'tails' can be effectively removed from the PSF, image quality may be further improved by adding somewhat more lead to the collimator, thereby also suppressing the collimator penetration and scatter tails.

Monte Carlo simulation was an essential tool for this work because it allowed us to analyse a large number of different collimator designs without having to construct real collimators. Furthermore, our approach of combining a theoretical model (the Bayesian observer) with a numerical tool (the Monte Carlo simulator) allowed us to predict optimal collimator designs for this task without the time, effort and cost of human observer studies.

ACKNOWLEDGMENT

The authors are grateful to Siemens Medical Systems for their support of the work presented here.

REFERENCES

[1] deVries D J, Moore S C, Zimmerman R E, Mueller S P, Friedland B and Lanza R C 1990 Development and validation of a Monte Carlo simulation of photon transport in an Anger camera *IEEE Trans. Med. Imaging* **MI-9** 430–8

[2]   Moore S C, deVries D J, Nandram B, Dardzinski B J, Kijewski M F and
       Mueller S P 1992 Collimator design optimization using Monte Carlo simula-
       tion and Bayesian parameter estimation. *Proc. IEEE Nuclear Science Symp.
       Medical Imaging Conf. (Orlando, FL, 1992)* (Piscataway, NJ: IEEE)
       pp 847-9
[3]   Moore S C, deVries D J, Nandram B, Kijewski M F and Mueller S P 1995
       Collimator optimization for lesion detection incorporating prior informa-
       tion about lesion size *Med. Phys.* **22** 703-13
[4]   Mueller S P, Kijewski M F, Moore S C and Holman B L 1990 Maximum-
       likelihood estimation: a mathematical model for quantitation in nuclear
       medicine *J. Nucl. Med.* **31** 1693-701
[5]   Van Trees H L 1968 *Detection, Estimation, and Modulation Theory* Part 1 (New
       York: Wiley) pp 84-5

CHAPTER 14

ESTIMATION OF THE LUNG REGIONS FROM COMPTON SCATTER DATA IN SPECT*

Tin-Su Pan and Michael A King

14.1 INTRODUCTION

Compton or incoherent scattering occurs when a photon interacts with an electron and its incident energy is considerably greater than the binding energy of the electron [1]. In the energy range of gamma and x-ray photons of 30 keV and 30 MeV, which covers most of the nuclear medicine energy range, Compton scattering is the dominant interaction for the body tissues [1, 2]. However, the major concern regarding Compton scatter photons in nuclear medicine has been in correcting for the degradation in the reconstructed images caused by the partial inclusion of these photons in the photopeak window. As we will show in this chapter, the imaging of Compton scatter photons can provide useful information beyond their use in scatter corrections.

Using Compton scatter data to determine the object outline has been reported in [3–7]. In this research, we explore the possibility of using Compton scatter data for the purpose of estimating the lung regions of the patient data with no assistance from any external source. Reports of similar work requiring usage of an external source can be found in [1] and [8]. Although some preliminary results of this approach have been reported in cardiac perfusion studies using either ^{201}Tl or ^{99m}Tc Sestamibi as a perfusion agent [9, 10], we will focus our discussion only on the application with the ^{99m}Tc-labelled agents. Since the lungs are the primary nonuniform attenuation material in the chest, knowing the regions of the lungs can be important

* This work was supported by the National Heart, Lung and Blood Institute under Grant HL-50349. The contents are solely the responsibility of the authors and do not necessarily represent the official view of the National Heart, Lung and Blood Institute.

for the application of attenuation correction of photo-peak data in SPECT. Correction for nonuniform attenuation is an important step toward making SPECT slices quantitatively accurate so that the incidence of attenuation artifacts can be decreased, and specificity increased [11]. This work was based on the investigation reported in [10] that low-count lung-shaped regions could be observed in, and segmented from, the reconstructed Compton scatter slices in some cardiac ^{99m}Tc Sestamibi perfusion patient studies. Because Compton scattering occurs primarily with loosely bound electrons, and the Compton mass attenuation coefficient is nearly independent of the atomic number of the attenuation medium [12], the lower density of the lungs results in less attenuation in the lung regions and leads to fewer scattered photons being detected from these regions.

If the lung regions and object outlines can be estimated with a reasonable accuracy from Compton scatter data, a patient-specific attenuation map can be derived by assigning appropriate attenuation coefficients to the regions determined. The idea of assigning attenuation coefficients to different density regions has also been applied in transmission imaging to reduce the transmission scan time of a patient in PET and SPECT [13, 14]. Madsen *et al* [15] demonstrated that in the approach of assigning constant attenuation coefficients to the regions of the body and lungs, 20% variation in the assigned coefficients resulted in less than 15% alteration in the relative distribution of counts in the left ventricle. The impact of such a variation on the detection task in a cardiac study has yet to be determined.

If assignment of a constant attenuation coefficient to the segmented regions (with some blurring on the assigned attenuation map) can lead to a satisfactory attenuation correction, then the approach will have great potential for application due to the following factors: (i) dual-energy-window data acquisition (photo-peak and Compton scatter windows) can be performed on most of the current SPECT systems; (ii) the registration of the images acquired from the two windows simultaneously is inherently exact; (iii) no increase in acquisition time or other alteration in the acquisition procedure is required and (iv) the scatter data can be retrieved for the purpose of attenuation compensation when the cause of a defect in a diagnosis is probably due to attenuation.

14.2 PHANTOM STUDY

To verify that Compton scatter window images are useful in estimating the region of the lungs, a phantom consisting of two lung-shaped styrofoam inserts, the Iowa heart phantom and a plastic bottle simulating the liver was first emission imaged and then transmission imaged with a line source at the focal distance of a fan beam collimator. The SPECT system used to do this was a Picker Prism 3000. The photo-peak window was set to be 129–

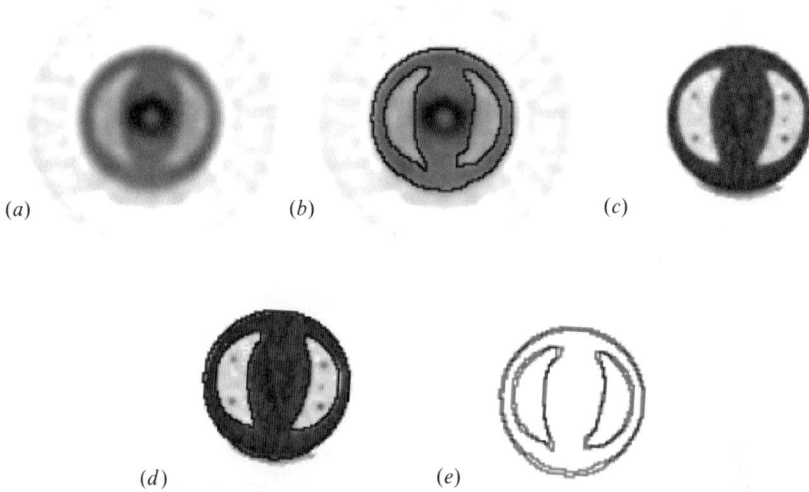

(a)   (b)   (c)

(d)   (e)

Figure 14.1 *A comparison of the outlines of the body and the lungs for the acquisition of a phantom: (a) Compton scatter window image; (c) transmission image; (b) and (d), the segmentation results of (a) and (c) superpositioned on (a) and (c), respectively; (e) superposition of edges in (b) and (d).*

151 keV, and the Compton scatter window was 97–125 keV. The line source was 16 cm in length, and filled with a 296 MBq ^{99m}Tc solution. The edges of the Compton scatter window slices and transmission slices were compared by computing the zero crossings of the second directional derivative in the direction of the count gradient [16, 17]. Figure 14.1 shows that the edges estimated from the Compton scatter window slices are very similar to the ones from the transmission slices.

14.3 ESTIMATION OF THE LUNG REGIONS

The Compton scatter projection data are first two-dimensionally (2D) filtered with a Butterworth filter with an order four and a cut-off at less than half of the Nyquist frequency, and reconstructed with the filtered backprojection. The Compton scatter slices are then used to estimate the outlines of the body and lung regions. The zero crossing of the second directional derivative in the direction of the count gradient is computed in 3D to determine the body outlines [17]. The gradient points in the direction of greatest difference or rate of change in counts at the given location in the image. That is, it tells which direction heads most 'up-hill' in the image counts. The first derivative taken in the gradient direction calculates the magnitude of this difference or rate of change in the direction of greatest change. That is, it tells how steep the 'up-hill' direction is at this location. The second derivative in the direction of the gradient calculates the change in the rate of change in counts, or how steepness is changing at this location. One definition of an edge is that it occurs at the steepest location in an 'up-hill' climb. Below the steepest point, the rate of count change will be increasing and thus have a positive value to its second derivative. Above the steepest point, the rate of count change will be less and thus it will have a negative second derivative. At the point of steepest change, or the edge, the second derivative will be zero. If one connects the locus of all the points whose second derivative in the direction of the gradient is zero, one thus obtains a closed contour which can be used to estimate the location of the boundary of the object. It is closed because it is just a level crossing in a set of values. That is, negative values will be inside, and positive values will be outside the object. We say an estimate because blurring and noise can significantly alter the location of the calculated edge from that of the true boundary.

A counts-threshold-based segmentation strategy is used to segment the regions of the lungs, with the input of the sternum and backbone locations estimated from the photo-peak window slices. The counts threshold method also forms a closed boundary for estimating edge location. Locations whose counts fall below the given threshold count, and are in contact with each other, are placed inside the object (for an object which is lower in counts than its surroundings as is the case of the lungs in this application). Locations which are above the threshold, or not connected, are outside the object. There are several possible definitions of being connected, or in contact, in digital images. In 2D, one can talk of pixels as being four-connected when they share a common side (are located north, south, east or west of the pixel in question), or eight-connected when they share a side or edge (are one of the eight surrounding neighbours of the pixel). Similarly in 3D, one can talk of six-connected, or 26-connected. Herein we have used 26-connected. The user is allowed to interactively mark the locations of the sternum and backbone to constrain the segmented lung regions (see figure 14.2(c)) from getting into the 90° fan regions determined by the sternum and backbone locations and the body outline (see figure 14.2(d)). That is, we look to the left or right of the mid-plane of the patient, within the body as determined by the zero crossing of the second directional derivative in the direction of the count gradient, and outside the regions in front of the sternum, and behind the backbone. The user adjusts two separate thresholds for segmenting the right and left lung regions. A centre line which evenly splits the width of the body region in a slice is employed to divide the region of the same slice into left-half and right-half body regions for the segmentation of the left lung and right lung, respectively. Two user-defined thresholds are used to select the voxels which are considered as making up the lung regions. The thresholds for segmenting the right and left lung regions can be different

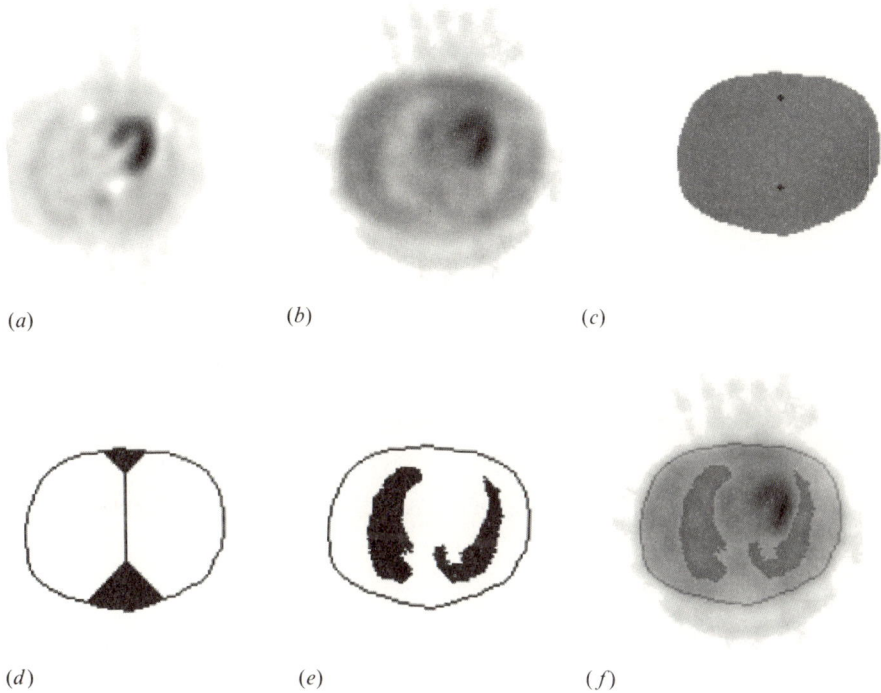

(a)   (b)   (c)

(d)   (e)   (f)

Figure 14.2 *Steps in segmentation of the body outline and lungs from a clinical study: (a) photo-peak window slice; (b) Compton scatter window slice; (c) segmented body region and positions of backbone and sternum; (d) the mask constraining the lung region; (e) segmented lung regions and (f) superposition of (e) on (b).*

due to the nonuniform activity distribution in the object, for example, the heart muscles and the liver take up much more activity in a cardiac perfusion study than the other soft tissues and are not sitting at the centre of a cross section of the body. User interaction in this process is currently necessary; however, ways to further automate the process are under investigation. In addition to thresholding the left and right lung regions separately, the largest grouping of voxels which are 3D connected is taken as being the lung regions.

14.4 PATIENT STUDY

Figure 14.3 shows two examples of patient data: one male and one female. The photo-peak window was set to be 129–151 keV, and the Compton scatter window was 97–125 keV. A Picker Prism three-head camera was used for

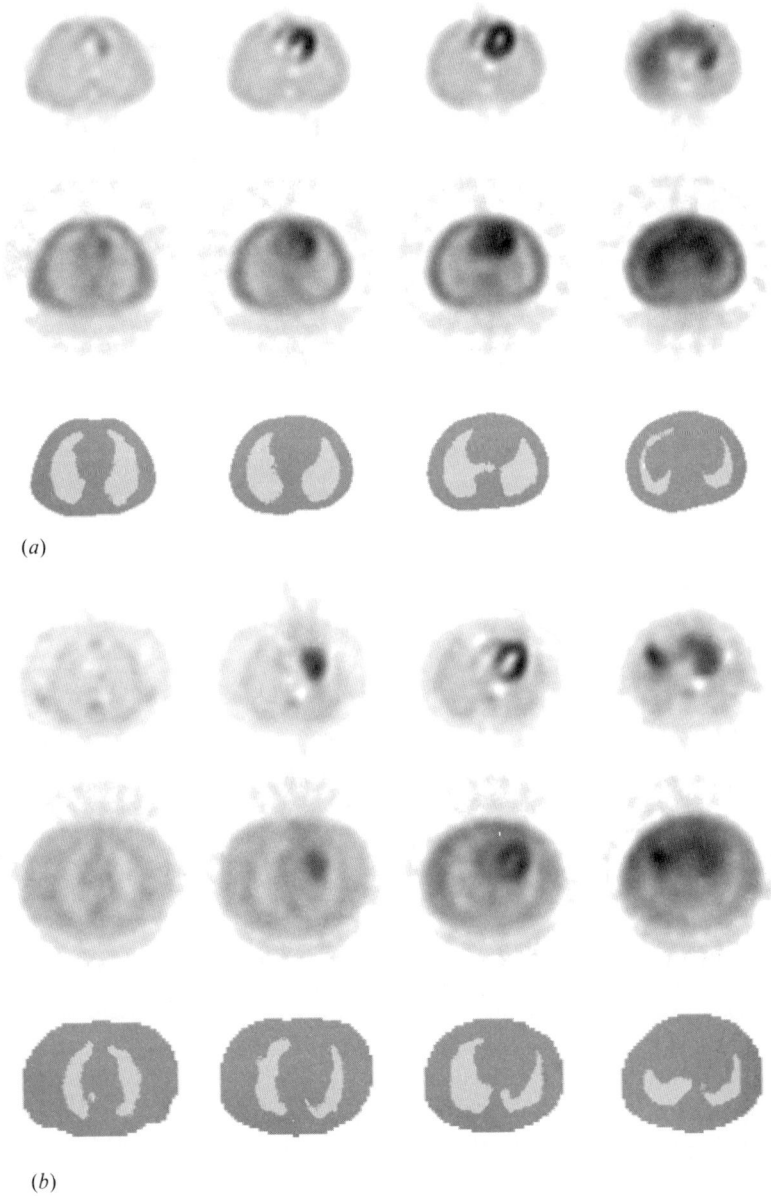

Figure 14.3 *Examples of (a) male and (b) female patient data. From the top to the bottom rows in (a) or (b) are the selected slices of the photo-peak, Compton scatter and estimated body and lung regions. There is 2.85 cm between any two displayed adjacent slices.*

Table 14.1 *Relative activity per voxel in the various organs of the phantom. Note that everything in the chest other than the organs listed is categorized as 'other soft tissues'.*

Organs	Activity/voxel
Liver	5.32
Spleen	5.32
Heart	3.99
Bone marrow	3.99
Spine, rib cage and long bones	1.32
Lungs	0 or 1.00 or 2.00
Other soft tissues	1

the imaging of the patients. The total acquisition time was 10 and 13.3 min for the acquisition of the male and female patient data, respectively, which corresponded to 15 and 20 s/step for a total of 40 steps (three degrees per step) per head. The difference in the acquisition time was primarily determined by the comfort level of a patient lying down on the patient supporting table. Although the Compton scatter data were blurry, we can still identify the regions of the lungs as the low-count regions in the Compton scatter slices.

14.5 COLD LUNG REGIONS

Because the proposed segmentation approach is based on segmenting the low-activity lung regions from the Compton scatter data, it is important to demonstrate that the cold or low-activity regions of the lungs can still be maintained even with the increase of uptake in the lungs. Through the use of Monte Carlo simulation, we can show the cold lung regions can be maintained with different uptake in the lung regions. The Monte Carlo software used in the simulation of SPECT imaging was the SIMIND code [18].

An anthropomorphic phantom [19] was used to define the activity and attenuation maps. Relative activities per unit volume in various organs of the phantom were derived from patient data and are listed in table 14.1. Figure 14.4 shows the reconstructed photo-peak window (130–151 keV) and Compton scatter window (99–120 keV) slices of the three different activity uptakes in the lungs. The selection of the 99–120 keV Compton scatter window was based on the observation that a better contrast between the lungs and the other soft tissues can be obtained from among the various windows simulated in [20]. The energy gap between the photo-peak and Compton scatter windows can reduce the influence of the photo-peak data on the Compton scatter data. Although the lung uptake increases as shown in the

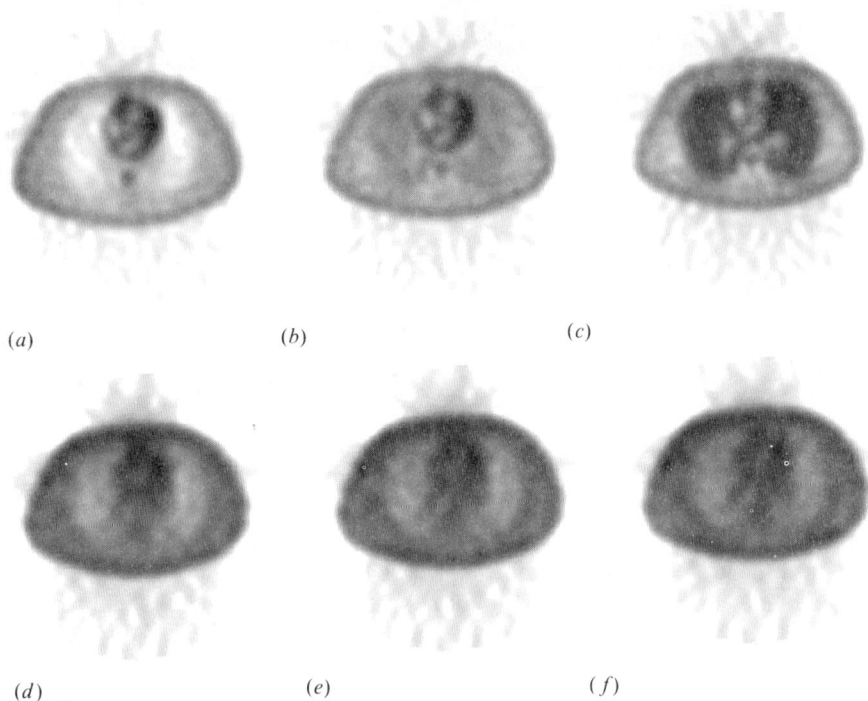

(a)   (b)   (c)

(d)   (e)   (f)

Figure 14.4 *Photo-peak (130–151 keV) and Compton scatter (99–120 keV) window slices with simulated activity ratios of zero, one and two between the lungs and other soft tissues are shown in (a), (b) and (c) for the photo-peak and (d), (e) and (f) for Compton scatter, respectively. The activity ratios calculated between the lungs and other soft tissues in the photo-peak slices are 0.36, 1.2 and 1.99 for (a), (b) and (c), respectively, while for Compton scatter slices they are 0.72, 0.77 and 0.82 for (d), (e) and (f), respectively. It is clear that the Compton scatter slices are not as dependent on the activity ratio between the lungs and other soft tissues as the photo-peak window slices are.*

photo-peak window slices, the same lung regions appear to be relatively cold in the Compton scatter window slices. This further demonstrates that the segmentation based on the cold lung regions in the Compton scatter window slices is reasonable and is not very dependent on the activity in the lungs.

Two simulations were also performed to examine the spatial distribution of scatter events for photons that were detected in the 99–120 keV window. The objective was to demonstrate that these photons contain the contrast information for distinguishing the lungs (lower density) from the surrounding bones and soft tissues (higher density). A point source was placed in the heart of the phantom, and was located in the slice chosen for display. Up to third-order scattering was simulated, since photons that scattered more than three times would have a negligible contribution [21] in this window.

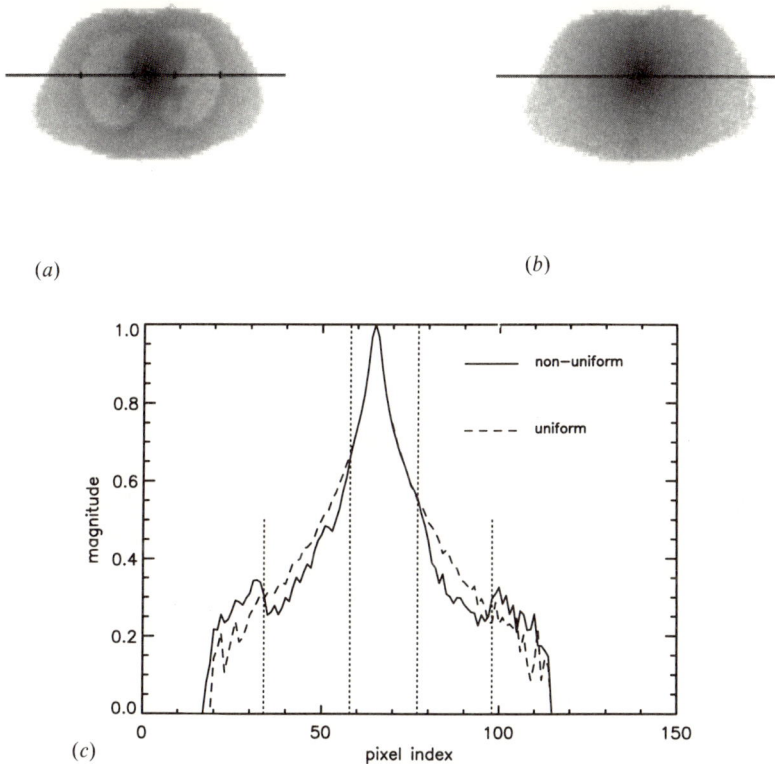

(a)

(b)

(c)

Figure 14.5 *Probability maps of the scatter interactions (up to third order and with detected photon energy in 99–120 keV) in the simulations of (a) nonuniform and (b) uniform media. The marked profiles are compared in (c), where four vertical dotted lines correspond to the edges of the lungs (shown as the four crosses in (a)). There were ten million emitted counts simulated in both experiments.*

The coordinates of the last scatter interaction as well as its probability of occurrence on a logarithmic scale in the phantom were used to map the origin of photons detected in the window.

The first simulation used a nonuniform density map, while the second used a map with uniform density equal to that of water. There were ten million emitted counts simulated at the point source in both simulations. Figure 14.5(a) and (b) shows the maps of the probability of scatter interactions (up to third order with a detected energy in the window 99–120 keV) of the nonuniform and uniform density maps, respectively. The contrast for dfferentiating the lungs from the surrounding tissues can be appreciated in figure 14.5(a). The profiles in figure 14.5(c) clearly indicate that there is less scattering (colder) in the lung regions than in the surrounding (hotter) soft-tissue region. The inner edges of the lungs in the map correspond to the locations where the profiles from the two different attenuation cases deviate from each other. The outer edges of the lungs in the map correspond to a significant change in the magnitude of the nonuniform attenuation case.

14.6 CONCLUSIONS AND FUTURE WORK

We propose to estimate the lung regions mainly using the Compton scatter data in the ^{99m}Tc Sestamibi cardiac perfusion study. No external source of radiation is required in this approach. The estimation of the body outline can be derived from the zero crossings of the second directional derivative in the direction of the count gradient of Compton scatter slices. To estimate the regions of the left and right lungs, we use a count threshold segmentation on the scatter data with the constraints of the anatomical information of the backbone and sternum locations from the photo-peak data. To verify the feasibility of this method, the emission and transmission data of a phantom with a composition similar to that of the human thorax were acquired and compared. The edges from both emission and transmission studies were very similar. The applicability of this approach has been supported by some patient data. Through Monte Carlo simulations, we demonstrated that the cold lung regions can be maintained even with the activity ratio of two to one between the lungs and the soft tissues, and a contrast differentiating the lung regions from the other tissues can be derived from a point source scattering in the heart.

For further research, a segmentation strategy of incorporating the heart boundary from photo-peak data to help define the inner edges of the lungs should be employed. The motion of the diaphragm and the heart, as well as the lung density change in a patient and between patients, deserve more investigations. One logical extension of this research is to compare the results of using the attenuation maps based on (i) segmentation on the Compton scatter data and (ii) transmission imaging for attenuation correction of patient studies. The important questions of how accurate the segmentation has to be to be clinically useful and whether this accuracy can be achieved with clinical images will be addressed in future investigations.

REFERENCES

[1] Guzzardi R and Licitra G 1988 A critical review of Compton imaging *Crit. Rev. Biol. Eng.* **15** 237–68
[2] Sorenson J A and Phelps M E 1987 *Physics in Nuclear Medicine* 2nd edn (Orlando, FL: Frune and Stratton)
[3] Jaszczak R J, Chang L-T, Stein N A and Moore F E 1979 Whole-body single-photon emission computed tomography using dual large-field-of-view scintillation cameras *Phys. Med. Biol.* **24** 1123–43

[4] Hosoba M, Wani H, Toyama H, Murata H and Tanaka E 1986 Automatic body contour detection in SPECT: effects on quantitative studies *J. Nucl. Med.* **27** 1184–91

[5] Tomitani T 1987 An edge detection algorithm for attenuation correction in emission CT *IEEE Trans. Nucl. Sci.* **NS-34** 309–12

[6] Younes R B, Mas J and Bidet R 1988 A fully automated contour detection algorithm: the preliminary step for scatter and attenuation compensation in SPECT *Eur. J. Nucl. Med.* **14** 586–9

[7] Macey D J, DeNardo G L and DeNardo S J 1988 Comparison of three boundary detection methods for SPECT using Compton scattered photons *J. Nucl. Med.* **29** 203–7

[8] Chang W, Huang G, Al-Doohan S, Pawlowski J and Loncaric S 1993 Scatter imaging with external source for attenuation correction of cardiac SPECT *J. Nucl. Med.* **34** 195P (abstract)

[9] Pan T-S, Seldin D W, Dahlberg S T and King M A 1994 An approach for reduction of attenuation artifacts from using the emission data in SPECT ^{201}Tl cardiac perfusion imaging *J. Nucl. Med.* **35** 82P (abstract)

[10] Pan T-S, King M A, Penney B C and Rajeevan N 1993 Segmentation of the body lungs and patient table from scatter and primary window images in SPECT *J. Nucl. Med.* **34** 195P (abstract)

[11] Schwaiger M 1994 Myocardial perfusion imaging with PET *J. Nucl. Med.* **35** 693–8

[12] Hendee W R 1979 *Medical Radiation Physics* second edn (Chicago: Year Book)

[13] Huang S-C, Carson R E, Phelps M E, Hoffman E J, Schelbert H R and Kuhl D E 1981 A boundary method for attenuation correction in positron computed tomography *J. Nucl. Med.* **22** 627–37

[14] Galt J R, Cullom J and Garcia E V 1992 SPECT quantification: a simplified method of attenuation and scatter correction for cardiac imaging *J. Nucl. Med.* **33** 2232–7

[15] Madsen M T, Kirchner P T, Edlin J P, Nathan M and Kahn D 1993 An emission-based technique for obtaining attenuation correction data for myocardial SPECT studies *Nucl. Med. Commun.* **14** 689–95

[16] Canny J F 1986 A computational approach to edge detection *IEEE Trans. Pattern Anal. Machine Intell.* **PAMI-6** 679–98

[17] Pan T-S, Gennert M A, Gauch J M and King M A 1992 Comparison of second directional derivative boundary detection methods for SPECT *IEEE Medical Imaging Conf. Record (Orlando, FL, 1992)* ed G T Alley (Piscataway, NJ: IEEE) pp 1080–2

[18] Ljungberg M and Strand S-E 1989 A Monte Carlo program for the simulation of scintillation camera characteristics *Comput. Methods Programs Biomed.* **29** 257–72

[19] Zubal I G, Harrell C R, Smith E O, Rattner Z, Gindi G and Hoffer P B 1994 Computerized three-dimensional segmented human anatomy *Med. Phys.* **21** 299–302

[20] Pan T-S, King M A, deVries D J and Ljungberg M 1996 Segmentation of the body and lungs from Compton scatter and photo-peak window data in SPECT: a Monte Carlo investigation *IEEE Trans. Med. Imaging* **MI-15** 13–24

[21] Floyd C E, Jaszczak R J, Harris C C and Coleman R E 1984 Energy and spatial distribution of multiple order Compton scatter in SPECT: a Monte Carlo investigation *Phys. Med. Biol.* **29** 1217–30

CHAPTER 15

THE MONTE CARLO METHOD APPLIED IN OTHER AREAS OF SPECT IMAGING

Michael Ljungberg

15.1 INTRODUCTION

This chapter summarizes some topics that have been successfully investigated by Monte Carlo simulations. One of the key features in these works is the ability to distinguish between primary photons and photons scattered in the object.

15.2 ENERGY PULSE-HEIGHT DISTRIBUTION ANALYSIS

Spectral analysis is very important in different topics. In general detector research such studies have been carried out by several groups, for example [1]. An early Monte Carlo study of the spectral distribution was made by Anger and Davis [2] that calculated the intrinsic efficiency and the intrinsic spatial resolution for NaI(Tl) crystals of different thicknesses and for various photon energies. Many investigators have studied the characteristics of the scattered photons in single-photon emission computed tomography (SPECT). These studies have included measurements of the scatter to primary ratios, the shape of the scatter response function, the shape of the energy spectrum, proportion of photons undergoing various numbers of scattering events and the effect of attenuator shape and composition in addition to camera parameters such as energy resolution and window size. Based on knowledge gained in these studies several methods for modelling the scatter response function have been proposed. Floyd *et al* [3] have studied the components of the energy spectra, expressed as first, second and third or more scattered orders for different point source locations for planar

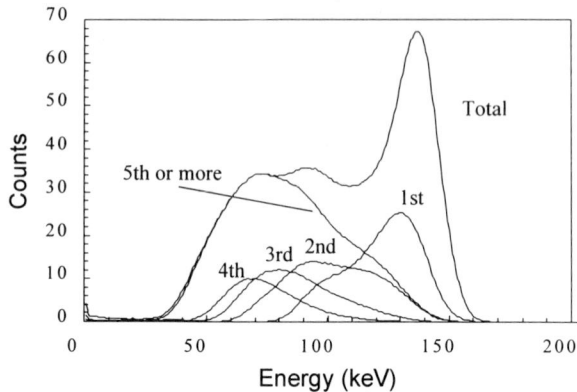

Figure 15.1   *Example of an energy pulse-height distribution, separated into components of scatter orders.*

imaging as well as for SPECT. An example of such a simulation, performed with the code SIMIND, is shown in figure 15.1.

Simulation has been performed for an 11 cm radius × 20 cm cylindrical water phantom with a ^{99m}Tc point source in the centre of the phantom. Similar studies have also been published by Zubal *et al* [4]. These types of study give important information about how the amount and distribution of scatter depend on source location, energy window width and photon energy [5]. From the differentiated energy pulse-height distribution, the scatter fraction (the ratio of scattered events to primary events) and scatter distribution can be calculated. Work on this topic has also been done by Manglos *et al* [6], Kojima *et al* [7] and Smith *et al* [8].

15.3 MONTE CARLO SIMULATION IN THE DEVELOPMENT AND SIMULATION OF SCATTER MODELS FOR SPECT†

While Monte Carlo simulation is a gold standard for simulating SPECT data, there are situations when it is not suitable due to the large computational requirements. In these cases it is desirable to develop quick, accurate methods for modelling the projection data. In the following discussion we refer to a method for analytically estimating the scatter component of the projection data for a known source and attenuator as a scatter model. Two such applications are the use of iterative reconstruction methods for scatter compensation and the generation of simulated projection data for applications requiring large numbers of sets of different projection data such as

†Written by Dr Eric C Frey, Department of Biomedical Engineering, University of North Carolina at Chapel Hill, NC, USA.

observer studies or for training neural networks. However, Monte Carlo simulation is invaluable in developing and evaluating these scatter models.

The reason scatter models are desirable is evident when one considers the requirements of these applications. For example, for iterative reconstruction scatter compensation it is necessary to compute the scatter response function (SRF) at each point in the attenuator with respect to each projection view. Both the time required to compute this and the memory required to store these are impractical, especially for three-dimensional (3D) reconstruction.

A second application where scatter models is important is for use in generating projection data for applications requiring large numbers of such data. Two such applications are receiver operating characteristic (ROC) analysis and developing training sets for neural networks. In both these applications it is desirable to generate projection data for large numbers of projection data sets from different source and attenuator configurations. In addition, these data sets must have Poisson-distributed noise. Using Monte Carlo simulation to generate these is prohibitive. For example, simulating a noise-free 3D data set with Monte Carlo typically takes of the order of weeks of computer time.

While Monte Carlo simulation is not practical as a method for simulating projection data in these cases, it is an indispensable tool in developing and evaluating scatter models. The process of developing scatter models occurs in four stages: characterization, development, validation and evaluation. First, Monte Carlo is used to study the scatter response functions in various phantoms to obtain an idea of how parameters such as source position, attenuator shape and composition, source energy, detector energy resolution and collimator hole length and shape affect the SRF. Next, based on the insight gained in these studies, a scatter model is developed. In this stage Monte Carlo may again prove valuable in estimating parameters necessary for the scatter model. After developing the model it is important to validate it. This validation process involves comparing the predictions of the model, for example the SRF and projection data for various phantoms, with those from either Monte Carlo simulations or direct experimental measurements. As a result of the ability to simply and accurately control and know the phantom geometry and source position, as well as the ability to separate scattered and unscattered photons, Monte Carlo simulations are often preferable to direct measurements. The final step is evaluation of the scatter model in terms of the purpose for which it was developed. For example, for the case of iterative reconstruction scatter compensation this includes evaluating the effectiveness of the method in comparison with other scatter compensation methods.

The slab phantom method, slab-derived scatter estimation (SDSE), can be used to estimate the SRF in uniform objects. It is based on a parametrization of the SRF of sources in slab phantoms and has been developed independently by two groups. Two variations of this method have been developed

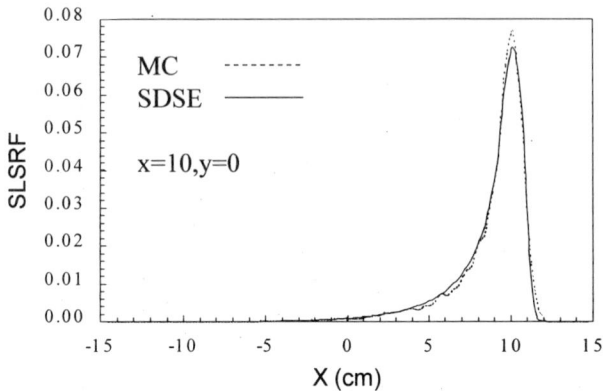

Figure 15.2 *A comparison of scatter line source response function computed using Monte Carlo simulation (MC) and SDSE. The response function was simulated for a ^{99m}Tc line source in a 22 cm diameter water-filled cylinder. The source was placed parallel to the axis of the cylinder lying on the diameter parallel to the detector at a distance 10 cm from the axis of the cylinder.*

by Frey *et al* [9–11] and Beekman *et al* [12]. A slab phantom is defined by a plane dividing a three-space into regions filled with an attenuating medium and regions filled with air. A point source is placed at a given distance below the plane in the attenuating medium and a gamma camera is placed in air at a specific distance from the source with the detection plane parallel to the plane dividing the two regions. For a given collimator and detector, the only parameters that determine the SRF in a slab phantom are the source depth and the distance to the collimator. There have been some recent investigations of extensions of this method to nonuniform attenuators [11].

Since the SRF and detector response function (DRF) are stationary in the plane parallel to the collimator, it is possible to use the series-equivalent formulation. In this formulation the SRF is given by the convolution of the series-equivalent kernel with the DRF. Another way of thinking of this is to consider the kernel to be the SRF for a detector having perfect resolution and it is thus independent of the distance from the face of the collimator. Thus, it is relatively easy to parametrize the shape and magnitude of the SRF as a function of depth. This can be done by fitting empirical relationships to either measured or Monte Carlo simulation total response functions (TRFs) or SRFs. Fitting experimentally measured SRFs has the advantage that the resulting parametrization will more accurately model the parameters of the imaging system. Fitting Monte Carlo-simulated SRFs on the other hand is important in determining the empirical shape of the SRFs. As mentioned above, Monte Carlo simulation is useful in validating these methods. Figure 15.2 shows a comparison of the line source response function computed

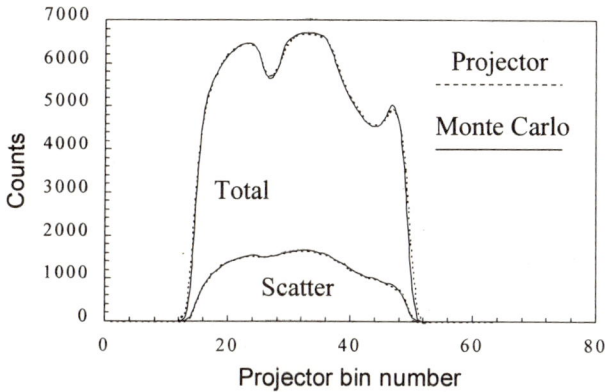

Figure 15.3 *Comparison of MC-simulated total and scatter projection data for a 22 cm diameter uniform cylinder filled with ^{99m}Tc. Three cold rods with diameters of 2, 4 and 6 cm were placed in the phantom parallel to the axis. The centres of these rods lay on a circle 12 cm in diameter and the centres of the rods were placed at equiangular intervals.*

using Monte Carlo simulation methods and the SDSE method for a ^{99m}Tc line source in a 22 cm diameter cylindrical phantom. The line source was placed on the diameter parallel to the detector 10 cm from the axis of the cylinder. Figure 15.3 shows a comparison of projection data for an extended source distribution. This simulation was for a 22 cm water-filled attenuator with a uniform background of ^{99m}Tc and three cold rods with diameters of 2, 4 and 6 cm. The rods were placed with their centres on a 6 cm radius circle concentric with the centre of the phantom and distributed at equiangular intervals.

The second major method proposed for modelling the SRF is based on numerically evaluating the transport equation assuming first-order scatter [13–16]. This method has the advantage of being immediately applicable to nonuniform attenuators or converging beam geometries but requires much more computation. Also, these methods may be less accurate since they often neglect the effects of multiple scatter. Multiple scatter can give rise to a significant portion of the scatter. For example, for a ^{99m}Tc line source at the centre of a 22 cm diameter cylindrical phantom approximately 20% of the scatter measured in a 20% photo-peak energy window is multiply scattered. One method used to overcome this difficulty is to use empirical cross sections or attenuation coefficients [15]. Monte Carlo simulations are invaluable in determining these. Again, Monte Carlo simulations are important for validating these methods. Figure 15.4 shows a plot comparing a model-based estimation of the SRF for a ^{99m}Tc line source in a 22 cm diameter water-filled cylinder. The calculations were performed for a source located on the diameter parallel to the detector and 5 cm from the axis of the

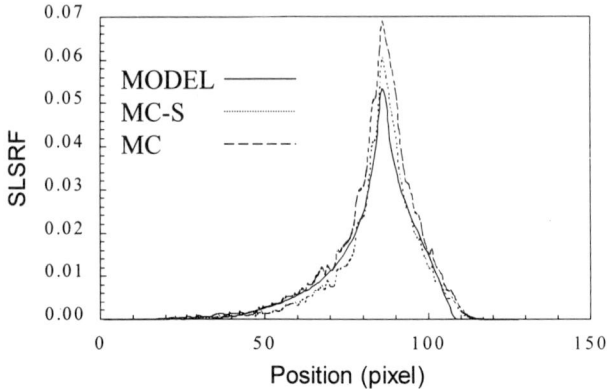

Figure 15.4 *Comparison of the line source response function obtained using Monte Carlo simulation and an analytical scatter model for a cone beam collimator. These data are for a line source placed parallel to the phantom and collimator axis in a 22 cm diameter water-filled cylinder. The source was positioned 5 cm from the centre of the phantom along the diameter parallel to the detector. For the Monte Carlo simulation, the calculation was performed for both singly scattered photons (MC-S) and photons with all scatter orders (MC). The analytical model is based on a single-scatter approximation.*

phantom. Particularly valuable for this application is the ability of the MC simulation to provide separate information about first-order scatter (which is all that is included in this model) and all orders of scatter.

A final method for modelling scatter is the effective source method. In the previous two methods the scatter distribution in the projection data is estimated directly from the activity distribution. In contrast, in the effective source method, an effective scatter source distribution is computed from the activity distribution and the attenuated projection of these distributions gives the scatter component of the projection data. A major advantage of this method is that it is more accurate for isotopes such as ^{201}Tl that have lower energy and more complex emission spectra. It should also be easier to generalize to converging beam geometries and nonuniform attenuation distributions. This method was first proposed in [17] where various approximations were made to derive a closed-form expression for the relationship between the activity and effective scatter source distributions. More recently, a more accurate expression has been proposed in [18]. In this method, the effective source is obtained using two distributions that are obtained using a modified Monte Carlo simulation program. Monte Carlo simulations were also used to validate the algorithm, as shown in figure 15.5, which shows a comparison between the estimate SRF for a line source computed using the SDSE method described, the new effective source method (ESSE) and Monte Carlo (MC) simulation.

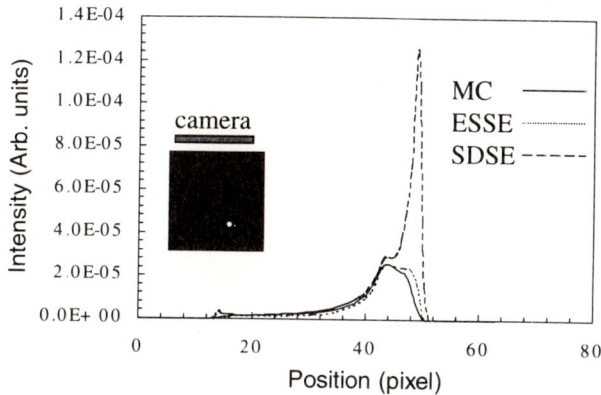

Figure 15.5 *A comparison of projection data computed using Monte Carlo simulation (MC) and the SDSE and effective scatter source estimation methods for a ^{201}Tl line source at the position indicated in the elliptical phantom. The major and minor axes of the elliptical cylinder were 32 and 22 cm, respectively. The phantom was positioned so the major axis was perpendicular to the detector. The line source was placed 9.92 cm from the centre of the phantom, away from the detector, in the direction of the major axis and 7.62 cm from the centre of the phantom along the minor axis.*

15.4 MONTE CARLO SIMULATION OF MULTI-WINDOW IMAGING†

In dual-isotope imaging, the same problem with cross talk occurs as described in the transmission section in chapter 11. This is of course a limitation since it affects both the ability to quantify activity uptake and also the image quality. In these cases, the Monte Carlo technique is also widely used to determine the effect and how to best correct for it. Dual-radionuclide imaging would allow simultaneous emission and transmission imaging and viewing of paired tracers during the same clinical and physiologic state and eliminate problems in image registration incurred in comparing images from sequential imaging studies. Several approaches for simultaneous dual-radionuclide imaging with ^{201}Tl and ^{99m}Tc have been reported [19–22]. The cross-talk components present in one or both energy windows (EWs) in a dual-radionuclide study can influence the detectability of cold defects and overestimate areas of hyperperfusion [23, 24], and the effects are dependent on the comparative administered activity doses. Monte Carlo is here useful to evaluate the feasibility and validity of simultaneous

†Written by Marija Ivanovic, Department of Nuclear Medicine, University of California at Davis Medical Center, CA, USA.

dual-radionuclide imaging of tracers (and ^{133}Xe) using multi-window acquisition.

An example is the work of Ivanovic *et al* who adapted an EGS4 simulation code for use with scintillation camera imaging of uniform and nonuniform source and media geometries. Energy spectra, planar images and SPECT images were obtained for a point source placed (i) in the centre of a water-filled cylindrical phantom and at eight different positions in a brain phantom and (ii) in a brain phantom with and without lesions. Simulations of the grey- and white-matter compartments and each lesion were run individually to allow greater flexibility in evaluating different patterns of radionuclide uptake distributions. The uptake ratio between grey and white matter could be varied, and each simulated lesion could be used as a cold or hot defect in the image by subtracting or adding the lesion images from the brain phantom images.

The system energy resolution was set at 1% intervals between 6 and 14%, to allow evaluation of the influence of energy resolution on peak separation and cross-talk components. For each radionuclide, images in 16 energy windows were recorded and selected according to the baseline settings and energy window widths commonly used for imaging all three radionuclides.

The spatial and energy distributions (figure 15.6) of the ^{99m}Tc cross-talk photons in Tl and Xe windows differ from the spatial and energy distributions of photo-peak photons. The Tc photo-peak image cannot therefore be used to obtain a good estimate of the cross-talk component. The eight 5 keV EWs were summed, varying the number and the starting energy in order to determine the EW that gives the best estimate of the ^{99m}Tc cross-talk components in the ^{201}Tl and the ^{133}Xe windows. The line spread functions of the simulated ^{99m}Tc point sources at different positions in the brain phantom geometry for cross-talk (^{201}Tl and ^{133}Xe) and cross-talk correction windows (windows in 91–120 keV intervals) were used to determine a blurring function. The image from cross-talk correction windows was convolved with this function before correction to account for a small difference in spatial and energy distribution of the photons in corresponding windows.

Results show that cross-talk in simultaneous dual-radionuclide imaging with Tc/Tl and Tc/Xe is very high and cannot be lowered by improving energy resolution. The number of Tc cross-talk photons present in ^{201}Tl and ^{133}Xe depends on imaging and scattering media geometry and comparative administered activity doses. It ranged from 39 to 48% of the total count and from 64 to 92% of the ^{201}Tl or ^{133}Xe counts recorded in corresponding windows for evaluated phantom geometries and a Tc/Tl activity ratio of 5 : 1 and a Tc/Xe activity ratio of 4 : 1. This high cross talk degrades the image and lesion contrast. Particularly large changes in lesion contrast were observed when a mixture of lesions with increased and decreased activity were used in the same study, especially if cold lesions on ^{133}Xe and ^{201}Tl images appear as regions with increased activity uptake on the ^{99m}Tc study.

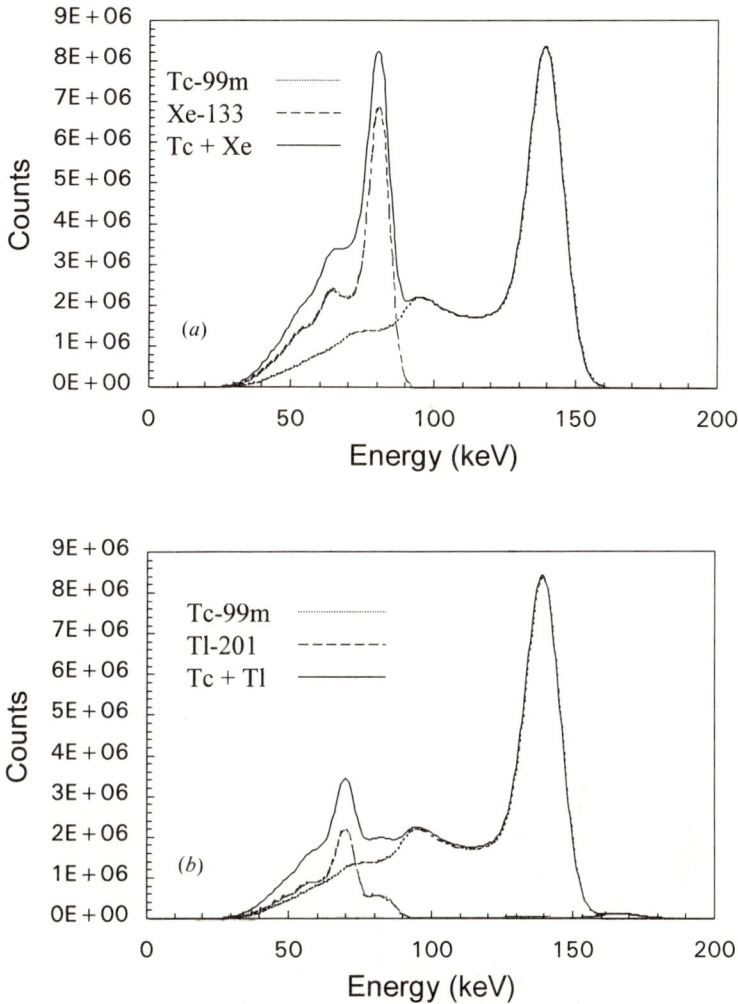

Figure 15.6 *(a) Simulated energy spectra for ^{99m}Tc and ^{133}Xe; (b) simulated energy spectra for ^{99m}Tc, and ^{99m}Tc + ^{201}Tl. Both studies were performed with a filled single-slice brain phantom (21 cm major axes, 16 cm minor axes and 8.0 cm thick). System energy resolution was set at 10%. The ^{99m}Tc to ^{133}Xe activity ratio was 2:1 and the ^{99m}Tc to ^{201}Tl activity ratio 5:1.*

Lesion contrast that ranged from −0.25 to −0.65 for single-radionuclide imaging with Tl changed to the range of −0.07 to 0.34 in dual-radionuclide imaging with a Tc to Tl activity ratio of 5:1. The corresponding lesion in the Tc study had increased uptake that ranged from 0.68 to 1.13. Similar changes were observed for Tc and Xe dual-radionuclide imaging. Image contrast improved when images recorded in the 95–110 keV energy window and blurred with a 2D Gaussian filter were used to estimate and correct for Tc cross talk in the Tl and Xe studies.

15.5 EVALUATION OF LESION DETECTION BY ROC ANALYSIS USING THE MONTE CARLO METHOD†

An application of Monte Carlo simulation is to produce images for lesion detection studies. These studies seek to determine the impact on detection of some perturbation in the acquisition or processing strategies. Lesion detection can be assessed by human observers using ROC, localization ROC (LROC) or free-response ROC (FROC) methods [25–28], or by numerical observers [29–31]. The major advantage of Monte Carlo simulation over phantom or clinical acquisitions is that it uniquely provides images formed with full control of the sources of degradation during imaging. For example, the primary and scatter photons can be recorded separately, allowing images in which the truth is known as to the presence or absence of scatter. Its disadvantages are that real clinical images are not used in the investigation, that questions can be raised about how closely the simulation approximates the real imaging system and the time it takes to form the images. A problem with using actual clinical images is the need to establish the truth of lesion presence. Monte Carlo methods can be used to simulate lesions which are then added to otherwise normal clinical studies [32].

The recent investigation of deVries *et al* [33, 34] into the impact of scatter and scatter correction in SPECT imaging upon the detection of tumours by human observers can serve as an example of the use of Monte Carlo methods with ROC investigations. The experiments approximated imaging of the liver with radiolabelled monoclonal antibodies with a relatively uniform attenuating medium. The SIMIND Monte Carlo package was used to produce high-count projection images, with primary and scattered photons stored separately. Separate Monte Carlo simulations of 2.5 cm spherical tumours at each of three locations in the liver were created. The resulting projection images were scaled to provide the desired tumour-to-background ratio and either added to or subtracted from the projection images of the normal distribution to produce 'hot' and 'cold' lesions, respectively. Case variation was obtained by using three different tumour sites, and different Poisson noise realizations. The slices were reconstructed with filtered back-projection (FBP), using a 2D Butterworth filter for noise suppression, and zeroth-order Chang attenuation compensation. The detection experiments

†Written by Dr Michael A King, University of Massachusetts Medical Center, Worcester, MA, USA.

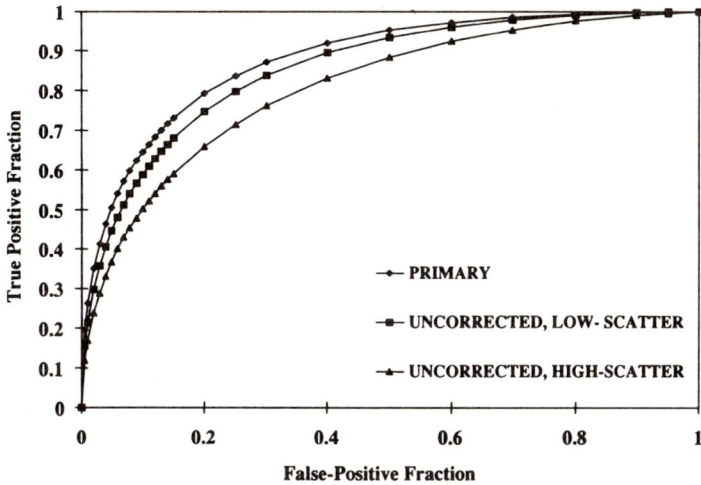

Figure 15.7 *A graph of ROC curves obtained for primary photons only, an uncorrected low-scatter case and an uncorrected high-scatter case.*

were conducted under signal-known-exactly (SKE) conditions. That is, lesion size was constant and potential location was indicated by cross hairs. Each slice presented to the observer contained either one lesion or none. For image display and scoring, software developed at UNC at Chapel Hill [35] was used. This software provides a continuous scale of confidence ratings [36]. Analysis was performed using the LABROC 1 program of Metz [36] to compute the areas under the binomial ROC curves (A_z). A paired t test is used to compare A_z values for the multiple readers. The processing methods compared were:

(i) reconstruction of primary only counts (this is 'ideal' scatter subtraction, or image acquisition with a camera with ideal energy resolution);

(ii) reconstruction of primary plus scatter (this is the no-correction case and the scatter fraction was 0.4–0.5 in the projections);

(iii) reconstruction of dual-photo-peak window (DPW) [37] scatter-corrected projections (DPW was selected to serve as an example of what a clinically feasible scatter subtraction method could achieve);

(iv) reconstruction of primary plus segmented scatter (in these images the amount of scatter was increased by a factor of 2.5 to investigate tumour detection in a significantly higher-scatter situation which is in magnitude, but not spatial distribution, similar to that of ^{201}Tl or ^{111}In studies); and

(v) DPW correction of the augmented scatter images.

In each case the number of primary counts was kept the same so that when scatter is present it adds the corresponding number of counts to the images.

Figure 15.7 shows a graph of ROC curves obtained for primary photons only, an uncorrected low-scatter case and an uncorrected high-scatter case. For both the true and augmented scatter cases, 'ideal' scatter subtraction resulted in a statistically significant improvement in area under the ROC curve (detection accuracy) for both 'cold' and 'hot' tumours. Also a trend toward (but not statistically significant at $p = 0.05$) improved detection with DPW correction was observed in the case of augmented scatter.

REFERENCES

[1]   Beattie R J D and Byrne J 1972 A Monte Carlo program for evaluating the response of a scintillation counter to monoenergetic gamma rays *Nucl. Instrum. Methods* **104** 163–8

[2]   Anger H O and Davis D H 1964 Gamma-ray detection efficiency and image resolution in sodium iodide *Rev. Sci. Instrum.* **35** 693–7

[3]   Floyd C E, Jaszczak R J, Harris C C and Coleman R E 1984 Energy and spatial distribution of multiple order Compton scatter in SPECT: a Monte Carlo investigation *Phys. Med. Biol.* **29** 1217–30

[4]   Zubal I G, Harrell C R and Esser P D 1990 Monte Carlo determination of emerging energy spectra for diagnostically realistic radiopharmaceutical distribution *Nucl. Instrum. Methods Phys. Res.* A **299** 544–7

[5]   Floyd C E, Jaszczak R J and Coleman R E 1988 Scatter detection in SPECT imaging: dependence on source depth, energy and energy window *Phys. Med. Biol.* **33** 1075–81

[6]   Manglos S H, Floyd C E, Jaszczak R J, Greer K L, Harris C C and Coleman R E 1987 Experimentally measured scatter fractions and energy spectra as a test of Monte Carlo simulations *Phys. Med. Biol.* **22** 335–43

[7]   Kojima A, Matsumoto M, Takahashi M and Uehara S 1993 Effect of energy resolution on scatter fraction in scintigraphic imaging: Monte Carlo study *Med. Phys.* **20** 1107

[8]   Smith M F and Jaszczak R J 1994 Three-dimensional photon scatter in nonuniform media for a simulated ^{99m}Tc SPECT myocardial perfusion study *J. Nucl. Med.* **35** 17P (abstract)

[9]   Beekman F J, Frey E C, Tsui B M W and Viergever M A 1993 A new phantom for fast determination of the scatter response of a gamma camera *IEEE Medical Imaging Conf. Records (San Francisco, CA, 1993)* (Piscataway, NJ: IEEE) pp 1847–51

[10]   Frey E C, Ju Z-W and Tsui B M W 1993 A fast projector–backprojector pair modeling the asymmetric, spatially varying scatter response function for scatter compensation in SPECT imaging *IEEE Trans. Nucl. Sci.* **NS-40** 1192–7

[11]   Frey E C and Tsui B M W 1993 A practical method for incorporating scatter in a projector–backprojector for accurate scatter compensation in SPECT *IEEE Trans. Nucl. Sci.* **NS-40** 1107–16

[12]   Beekman F J, Eijman E, Viergever M A, Borm G and Slijpen E 1993 Object shape dependent PSF model for SPECT imaging *IEEE Trans. Nucl. Sci.* **NS-40** 31–9

[13] Cao Z J, Frey E C and Tsui B M W 1994 A scatter model for parallel and converging beam SPECT based on the Klein–Nishina formula *IEEE Trans. Nucl. Sci.* **NS-41** 1594–600

[14] Riauka T A and Gortel Z W 1994 Photon propagation and detection in single-photon emission computed tomography—an analytic approach *Med. Phys.* **21** 1311–21

[15] Welch A, Gullberg G T, Christian P E, Datz, F L and Morgan H T 1995 A transmission-map-based scatter correction technique for SPECT in inhomogeneous media *Med. Phys.* **22** 1627–35

[16] Riauka T A, Hooper H R and Gortel Z W 1997 Experimental and numerical investigation of the 3D SPECT photon detection kernel for nonuniform attenuating media *Phys. Med. Biol.* **41** 1167–90

[17] Clough A V 1986 A mathematical model of single-photon emission computed tomography (abstract) *PhD Dissertation* (University of Arizona)

[18] Frey E C and Tsui B M W 1996 A new method for modeling the scatter response function in SPECT *IEEE Medical Imaging Conf. Records (Anaheim, CA)* vol 2 (Piscataway, NJ: IEEE) pp 1082–6

[19] Moore S C 1993 Simultaneous SPECT imaging of ^{201}Tl and ^{99m}Tc using four energy windows *J. Nucl. Med.* **34** 188P (abstract)

[20] Knesaurek K, Ivanovic M and Machac J 1994 A new dual-isotope convolution cross-talk correction technique: a ^{201}Tl/^{99m}Tc SPECT phantom study *Eur. J. Nucl. Med.* **21** 793 (abstract)

[21] Yang D C 1993 Radionuclide simultaneous dual-isotope stress myocardial perfusion study using the three window technique *Clin. Nucl. Med.* **18** 852–7

[22] Moore S C, English R J, Syravanh C, Tow D E, Zimmerman R E, Chan K H and Kijewski M F 1994 Simultaneous ^{99m}Tc/^{201}Tl imaging using energy-based estimation of the spatial distribution of contaminant photons *IEEE Medical Imaging Conf. Records (Norfolk, VA)* vol 3 (Piscataway, NJ: IEEE) pp 1507–11

[23] Ivanovic M, Esser P, Van Heertum R L and Weber D A 1994 Evaluation of lesion detectability and cross-talk correction methods in dual radioisotope imaging *J. Nucl. Med.* **35** 61P (abstract)

[24] Ivanovic M, Esser P, Van Heertum R L and Weber D A 1993 Monte Carlo simulations of the cross-talk component in dual-isotope imaging *J. Nucl. Med.* **34** 60P (abstract)

[25] Green D and Swets J 1966 *Signal Detection Theory and Psychophysics* (New York: Wiley)

[26] Starr S J, Metz C E, Lusted L B and Goodenough D J 1975 Visual detection and localization of radiographic images *Radiology* **116** 533–8

[27] Chesters M S 1992 Human visual perception and ROC methodology in medical imaging *Phys. Med. Biol.* **37** 1433–76

[28] Swensson R G 1996 Unified measurement of observer performance in detecting and localizing target objects on images *Med. Phys.* **23** 1709–25

[29] Fiete R D, Barrett H H, Smith W E and Myers M J 1987 Hotelling trace criterion and its correlation with human-observer performance *J. Opt. Soc. Am.* **4** 945–53

[30] Myers M J and Barrett H H 1987 Addition of a channel mechanism to the ideal observer model *J. Opt. Soc. Am.* **12** 2447–57

[31] Barrett H H, Yao J, Rolland J P and Myers M J 1993 Model observers for assessment of image quality *Proc. Natl Acad. Sci.* **90** 9758–65

[32] Seltzer S E, Swensson R G, Nawfel R D, Lentini J F, Kazda I and Judy P F 1991 Visualization and detection—localization on computed tomographic images *Invest. Radiol.* **26** 285–94

[33] deVries D J, King M A, Tsui B M W and Metz C E 1997 Evaluation of the effect of scatter correction on lesion detection in hepatic SPECT imaging *IEEE Trans. Nucl. Sci.* **44** 1733–40

[34] deVries D J 1996 Development and evaluation of scatter subtraction for SPECT imaging *PhD Thesis* (Worcester Polytechnic Institute, Worcester, MA 1995)

[35] Tsui B M W, Terry J A and Gullberg G T 1993 Evaluation of cardiac cone-beam single photon emission computed tomography using observer performance experiments and receiver operating characteristic analysis *Invest. Radiol.* **28** 1101–12

[36] Rochette H E, Gur D and Metz C E 1992 The use of continuous and discrete confidence judgment in receiver operating characteristic studies of diagnostic imaging techniques *Invest. Radiol.* **27** 169–72

[37] King M A, Hademenos G J and Glick S J 1992 A dual photo-peak window method for scattering correction *J. Nucl. Med.* **33** 605–13

CHAPTER 16

POSITRON EMISSION TOMOGRAPHY—BASIC PRINCIPLES

Kjell Erlandsson and Tomas Ohlsson

16.1 INTRODUCTION

Positron emission tomography (PET) has advanced the development of modern nuclear medicine through the introduction of new radiopharmaceuticals making it possible to study physiological and biochemical processes inaccessible with standard nuclear medicine techniques. The PET technique is complex, requiring radionuclide production facilities, advanced radiochemistry, PET scanners, mathematical modelling of biochemical processes and trained personnel.

The most commonly used radionuclides in PET are ^{11}C, ^{13}N, ^{15}O and ^{18}F (table 16.1). There are several reasons for using these positron emitters. Firstly, a broad range of compounds that trace physiological and biochemical processes can be labelled with only a change of isotope in the case of ^{11}C, ^{13}N and ^{15}O, whilst ^{18}F can be used as a hydrogen substitute. Secondly, the use of isotopes of these biologically ubiquitous elements makes it possible to label molecules that trace biochemical processes precisely, which is rarely possible with conventional radiopharmaceuticals for scintillation camera imaging. Thirdly, the short half-life of these positron-emitting radionuclides results in a low absorbed dose to the patient, and allows for several PET studies to be made on the same patient on the same day.

The interest in PET has increased considerably, from being only a research tool for brain imaging, to being a clinical whole-body imaging modality. Several specific clinical indications have been defined in cardiology [2], neurology [3–6] and oncology [7].

221

Table 16.1 *Physical characteristics of the most commonly used positron-emitting radionuclides [1].*

Radionuclide	$T_{1/2}$ (min)	β^+ decay (%)	Max. positron energy (MeV)	Positron range (mm)
^{11}C	20.4	99.8	0.96	0.9
^{13}N	9.96	100	1.19	1.3
^{15}O	2.04	99.9	1.72	2.4
^{18}F	109.8	96.9	0.64	0.6

16.2 THE β^+ DECAY

Radionuclides with excess protons may decay in two ways to reduce the number of protons and achieve a more stable nuclear configuration. The nucleus may either capture an orbital electron, or emit a positron (β^+ particle) and a neutrino. The excess energy from the β^+ decay is divided between the positron and the neutrino. The kinetic energy of the positrons, for a number of disintegrations, therefore exhibits a continuous distribution from zero up to a maximum energy E_{max} dependent on the radionuclide (table 16.1). After losing almost all of its kinetic energy, the positron interacts with an atomic electron. The two particles form a short-lived hydrogen-like structure called positronium, and then undergo a process called annihilation, in which the masses of the two particles are converted into electromagnetic radiation, usually in the form of two gamma photons. These annihilation photons are emitted in nearly opposite directions, each one carrying an energy of 511 keV. The sum of the two photon energies corresponds to the rest mass of the two particles. The opposite directions of the photon trajectories is demanded by the law of momentum conservation and is exact only in the centre of mass system. A residual kinetic energy of the positron at the instant of annihilation results in a slight deviation from 180° in the laboratory system, with a mean value of approximately 0.5°.

16.3 PET SCANNERS

The suggestion of using positron emitters for medical imaging purposes was first made in the early 1950s, the basic principle being that of observing the emission of positrons in tissue through the detection of annihilation radiation. A PET scanner is composed of at least two radiation detectors operating in time coincidence mode. When two 511 keV photons are detected in two different detectors simultaneously, a β^+ decay is assumed to have occurred somewhere on the line connecting the two detection points. In this case, 'simultaneously' means within a time window of 10–20 ns. If an annihilation occurs outside the region between the detectors, only one annihilation photon can be detected and, since this does not satisfy the coincidence condition, the event is rejected. The spatial resolution in PET images is primarily dependent on the physical size and cross-sectional geometry of the detectors, but is also affected by the positron range before annihilation and by the deviation from 180° of the photon trajectories. The influence of the latter effect depends on the detector separation.

PET scanners can be divided into two groups; those based on a small number of large-area position sensitive detectors, and those based on a large number of small individual detectors (ring scanners). In table 16.2 the design parameters and performance characteristics of a few different PET scanners are presented.

16.3.1 Area detector scanners

A PET scanner can be constructed with two opposed large-area detectors (figure 16.1(*a*)). For each coincidence detection event a two-dimensional (2D) coordinate is obtained from each detector, and coincidence lines will thus be registered for a number of different angles. Such a system can be used either in stationary mode for limited-angle tomography, or in a rotating mode to obtain a complete tomographic data set. Two different types of area detector have been used for PET scanners: scintillation cameras (see Chapter 6) and multiwire proportional chambers.

Brownell and Sweet [20] and Anger and Rosenthal [21] described the first PET imaging devices. With two scintillation cameras in coincidence mode, images of focused longitudinal planes were produced. Muehllehner *et al* [22] developed a positron camera based on two scintillation cameras, which could be used in both stationary and rotational mode. Muehllehner and Karp [23] later developed another PET scanner based on six position-sensitive NaI(Tl) scintillation detectors in a hexagonal arrangement, as shown in figure 6.1(*b*), giving a virtually complete tomographic data set without the need for rotation. Rotating dual-headed scintillation cameras have for some time now been used for single-photon emission computed tomography (SPECT). Recently, several such commercial systems have been modified for coincidence measurements [24, 25]. However, a crystal thickness of 1 cm, which is optimal for the photon energies used in SPECT, gives quite a low efficiency for annihilation photons. The main drawbacks of a PET scanner developed by Sandell *et al* [26] based on two rotating scintillation cameras, were low sensitivity and poor count rate capability. The spatial resolution, however, was comparable to that of most commercial PET scanners.

The multiwire proportional chamber (MWPC) is another type of large-area detector which has been used in PET [27–29]. These detectors have the advantage of good spatial resolution and low cost, but suffer from low

Table 16.2. *Comparison of design parameters and performance characteristics for some PET scanners.*

Scanner	Crystal material	Crystal dimension[a]	No rings/No slices	No crystals/ring	Ring diameter[c]	Spatial resolution[b] In-plane	Spatial resolution[b] Axial	Sensitivity[d]	Operation mode	Refs
PC384-7B	BGO	12×20×30	4/7	96	48	7.6	11.6	595	2D	[8]
TOFPET	CsF	180×45	5/9	144	99	9.0	10.8	366	2D/3D	[9]
ECAT III	BGO	5.6×30×30	2/3	512	100	4.4	10	405	2D	[10]
PC2048-7WB	BGO/GSO	6×20×30	4/7	2×256	107	5.0	11.2	335	2D	[11]
Donner-600	BGO	3×10×23–30	1/1	600	60	2.6	5.0	216	2D	[12]
ECAT 931	BGO	5.6×12.9×30	8/15	512	102	5.5	6.2	162	2D	[13]
PC4096-15WB	BGO	6×12×30	8/15	512	101	4.9	6.0	132	2D	[14]
PENN-PET	NaI(Tl)	500×150×25	1/≤50	6	84	5.5	5.5	126	3D	[15]
ECAT 953B	BGO	5.6×6.15×30	16/31	384	76.5	4.6	4.3	128	2D/3D	[16]
ECAT EXACT HR	BGO	2.9×5.9×30	24/47	784	82.0	3.6	4.0	102	2D/3D	[17]
GE Advance	BGO	4.0×8.1×30	18/35	672	92.7	3.8	4.0	172	2D/3D	[18]
MicroPET	LSO	2.0×2.0×10	8/15	240	17.2	2.0	2.0	—	3D	[19]

[a] Dimensions in millimetres.
[b] FWHM in millimetres.
[c] Ring diameter in centimetres.
[d] cps $(kBq\,ml^{-1})^{-1}$ per slice for a 20 cm phantom.

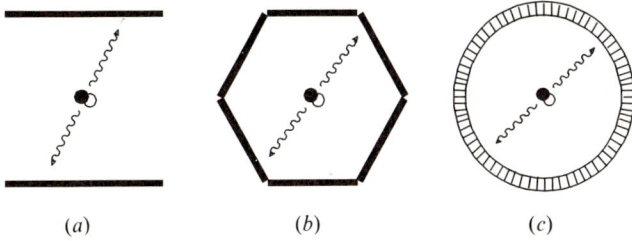

Figure 16.1 *PET scanners; (a) planar detectors; (b) hexagonal geometry; (c) detector ring.*

intrinsic efficiency. To address this problem, a new positron camera is currently being developed with hybrid detectors that combine MWPCs and BaF$_2$ scintillation crystals [30].

16.3.2 Ring detector scanners

The most common design for PET scanners today is a large number of small detectors in a ring surrounding the patient (figure 16.1(*c*)). Ring scanners have evolved through several generations; from single-ring to multiring systems, from NaI(Tl) to Bi$_4$Ge$_3$O$_{12}$ (BGO) crystals, from individual detectors to block detectors and from 2D to 3D scanners.

   The first positron tomograph that functioned on a practical clinical level was built by Hoffman *et al* [31], consisting of 48 cylindrical NaI(Tl) detectors in a hexagonal array. Bohm *et al* [32] built a PET scanner with 95 NaI(Tl) detectors in a ring geometry, and introduced a technique called 'wobbling' to improve radial sampling. This was a movement of the detector system, its centre describing a small circle with a diameter equal to the detector spacing. A multiring PET scanner was developed by Eriksson *et al* [8], consisting of four axial rings of detectors. A total of seven transaxial image planes were obtained by utilizing coincidences between adjacent rings to generate three 'cross-planes' in addition to the four 'direct planes' (figure 16.2(*a*)). The cross planes were treated as transaxial planes parallel to and midway between the direct planes. Between the detector rings, lead shields called septa were placed, acting as a collimator to reduce the number of unwanted scattered and random events (see below). The scintillation material BGO was used instead of NaI(Tl) due to its higher intrinsic efficiency for 511 keV photons. Table 16.3 summarizes the properties of different scintillation materials used in PET scanners. BGO has become the most commonly used material in conventional scanners despite some drawbacks. Its relatively low light output leads to poor energy resolution, and its long fluorescent decay time to limited count rate capability. BaF$_2$ and CsF have rapid decay times and have been used in systems designed for high count

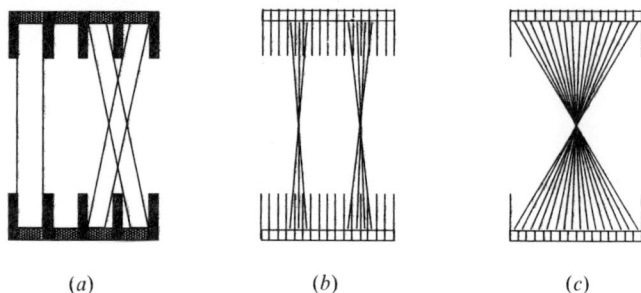

(a)   (b)   (c)

Figure 16.2 *Axial ring scanner geometries; (a) four rings with septa; (b) 16 rings with septa 2D mode; (c) 16 rings without septa, 3D mode. The axial dimension has been exaggerated.*

Table 16.3 *Physical properties of inorganic scintillators used in PET scanners [33, 34].*

	NaI(Tl)	BGO	GSO	BaF$_2$	CsF	LSO
Density (g cm^{-3})	3.67	7.13	6.70	4.88	4.64	7.4
Effective atomic no	51	75	59	51	52	66
Index of refraction	1.85	2.15	1.85	1.50	1.48	1.82
Relative emission intensity	100	15	25	10	5	75
Peak wavelength (nm)	410	480	440	310 (slow) 220 (fast)	390	420
Decay constant (ns)	230	300	56, 600	430, 620 (slow) 0.6, 0.79 (fast)	5	40
Hygroscopic	yes	no	no	no	yes	no

rates, and also in time-of-flight (TOF) systems [9], which measure the time difference between the detection of the two annihilation photons. The time resolution is at best 200 ps, corresponding to a position uncertainty of 3 cm. The incorporation of TOF information into the image reconstruction process increases the signal-to-noise ratio in the images. The low density and atomic number of CsF and BaF$_2$ results in a low efficiency as compared to BGO and makes these materials less suitable in conventional PET systems. However, cerium-doped lutetium oxyorthosilicate (LSO) has a unique combination of high density, atomic number, light emission intensity and speed that makes it a very promising scintillation crystal for PET scanners. The one drawback is its 2.6% content of radioactive ^{176}Lu. LSO crystals have been used by Cherry *et al* [19] to build a high-resolution PET scanner for small animal imaging.

The quest for higher spatial resolution without increased complexity of the scanner has led to the development of the block detector, consisting of a block of BGO which is cut into, for example, 8×8 elements viewed by four photomultiplier tubes (PMTs). Position logic, similar to that in the scintillation camera, can be used for identification of the different elements. The number of resolution elements per PMT then increases by a factor of 16 compared to the previously used one-to-one coupling design. A 16-ring system based on block detectors was described by Spinks *et al* [35]. In order to obtain sufficient sensitivity in each plane, the number of ring combinations used for coincidence detection was increased (figure 16.2(*b*)). Block detectors have also been used for building rotating PET scanners [36].

To increase the sensitivity, modern PET scanners offer the possibility of removing the inter-plane septa and operating in 3D mode, utilizing all possible ring combinations (figure 16.2(*c*)). The sensitivity increases by a factor of five to six, partly due to more coincidence lines and partly due to less shielding from the septa. However, the fraction of unwanted scattered and random events (see below) increases as well [37]. In 3D mode the sensitivity varies in the scanner's axial direction due to geometric factors. The count rate capability of the system will remain the same, reaching a maximum value with a lower amount of activity in the field of view (FOV). 3D PET offers the possibility of lowering the absorbed dose of radiation to the patient, or of acquiring a higher number of counts in low-count-rate applications. This was utilized by Cherry *et al* [38] to develop a whole-body PET technique that generates tomographic images of the entire patient by moving the bed through the FOV of a scanner operating in 3D mode. This technique is very useful in oncology for localization of occult metastases and unknown primary tumours. Examples of modern state-of-the-art PET scanners that operate in both 2D and 3D modes are the ECAT EXACT HR and the GE Advance (table 16.2).

16.4 DATA CORRECTIONS

In order to obtain correct quantitative images of the activity distribution it is necessary to correct the acquired PET data for the following effects: efficiency variations, dead time, random coincidences, scatter and attenuation. In a scanner composed of many individual detectors, the intrinsic efficiency for each detector will not be the same. Also, for large-area detectors, the efficiency may vary over the detector plane. To correct for this effect, the acquired data are normalized using a measurement with a uniform plane source or a rotating rod source. For high count rates, the rate of data acquired will not be a linear function of the incoming photon fluence rate due to electronic dead time in the system. This can be corrected for using a model of the count rate performance of the system (see Chapter 19).

There is no natural background in PET measurements due to the coincidence requirement. However, there are two effects that create background

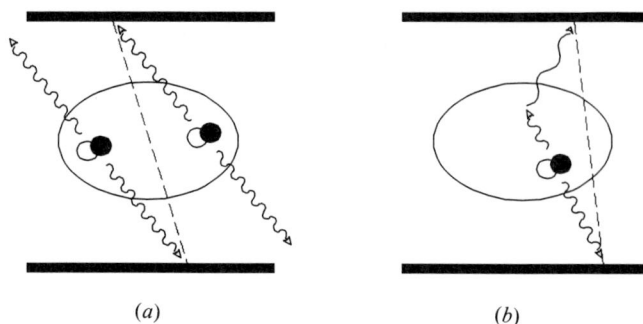

(a)   (b)

Figure 16.3 *Background events: (a) random coincidences; (b) the scattered photon.*

events during a PET scan: random coincidences and Compton scattering. Since the coincidence detection technique is based on a finite time window of 10–20 ns, there is always a chance that two photons originating from different annihilations may cause a coincidence event, as shown in figure 16.3(a). The number of these accidental or 'random' coincidences can be calculated using the equation $n_R = n_1 n_2 2\tau$, where n_R is the count rate of random coincidences between detectors 1 and 2, n_1 and n_2 are the single count rates in each detector and 2τ is the time window (the maximum time difference between the two registrations is τ). It is also possible to measure the random count rate using a delayed coincidence window, which means that one signal is delayed so that no true coincidences can be detected.

The annihilation photons may interact with atomic electrons in the patient either by photo-electric absorption or by Compton (incoherent) scattering. In the latter process (figure 16.3(b)), a photon of lower energy and different direction is produced. If detected, the two photons will still be in coincidence. The energy resolution of the detectors used in PET is not good enough to allow efficient discrimination of scattered photons. Also, a wide energy window is often used in order to accept events in which a photon does not deposit its entire energy in the detector. Scatter correction is essential for accurate quantification, especially in 3D mode. Several correction methods have been developed for both 2D and 3D PET based on either calculations of measurements. These are described in more detail in Chapter 18.

Attenuation correction is also essential for an accurate reconstruction [39]. Some of the emitted photons will never reach the detectors due to absorption or scattering in the patient. The total interaction probability for the two annihilation photons along a certain coincidence line is independent of the annihilation position along the line. The correction factors for each coincidence line can be calculated if the object is uniform and its contour is known [40]. Alternatively, the factors can be derived from a transmission measurement using an external radioactive source. The latter method is more accurate, but requires an extra measurement and increases the noise in the reconstructed image. Huang *et al* [41] developed a hybrid method based on a short transmission scan, segmentation of the body into uniform regions and calculation of the attenuation. For simplicity transmission scanning is preferably done before the administration of the radiopharmaceutical to the patient. However, this is not always convenient, in which case it can be done after administration provided that account is taken of the photons emitted from inside the patient [42].

16.5 MONTE CARLO SIMULATIONS FOR PET

Monte Carlo (MC) simulation is a very useful tool for studying the performance of a PET system. A number of physical factors, such as source distribution, positron range, noncollinearity of the annihilation photons, scanner configuration and photon interactions in the object as well as in active and passive detector components, can be included in the simulation. MC can be used for evaluation of sensitivity, resolution and, in conjunction with other mathematical modelling, also count-rate performance. This is useful both for evaluating existing PET scanners and in designing new ones (Chapters 17 and 19). The possibility of distinguishing between scattered and true events in MC simulations is valuable for development of scatter correction techniques (Chapter 18). MC can also be used for generation of realistic data for testing new correction or reconstruction methods.

REFERENCES

[1]   1978 *Table of Isotopes* 7th edn (New York: Wiley)
[2]   Schwaiger M and Hutchins G D 1992 Evaluation of coronary artery disease with positron emission tomography *Semin. Nucl. Med.* **22** 210–23
[3]   Broich K, Alavi A and Kushner M 1992 Positron emission tomography in cerebrovascular disorders *Semin. Nucl. Med.* **22** 224–32
[4]   Chugani H T 1992 The use of positron emission tomography in the clinical assessment of epilepsy *Semin. Nucl. Med.* **22** 247–53
[5]   Volkow N D and Fowler J S 1992 Neuropsychiatric disorders: investigation of schizophrenia and substance abuse *Semin. Nucl. Med.* **22** 254–67
[6]   Maziotta J C, Frackowiak R S J and Phelps M E 1992 The use of positron emission tomography in the clinical assessment of dementia *Semin Nucl. Med.* **22** 233–46
[7]   Hawkins R A and Hoh C K 1994 PET FDG studies in oncology *Nucl. Med. Biol.* **21** 739–47
[8]   Eriksson L, Bohm C, Kesselberg M, Blomqvist G, Litton J, Widen L, Ericson K and Greitz T 1982 A four ring positron camera system for emission tomography of the brain *IEEE Trans. Nucl. Sci.* **NS-29** 539–43

[9]   Mullani N A Wong W-H, Hartz R, Yerian K, Philippe E A, Gaeta J M and Gould K L 1983 Preliminary results with TOFPET *IEEE Trans. Nucl. Sci.* **NS-30** 739–43

[10]   Hoffman E J, Ricci A R, van der Stee L M A and Phelps M E 1983 ECAT III-basic design considerations *IEEE Trans. Nucl. Sci.* **NS-30** 729–33

[11]   Holte S, Ostertag H and Kesselberg M 1987 A preliminary evaluation of a dual crystal positron camera *J. Comput. Assist. Tomogr.* **11** 691–697

[12]   Derenzo S E, Huesman R H, Cahoon J L, Geyer A, Uber D, Vuletich T and Budinger T F 1987 Initial results from the Donner 600 crystal positron tomograph *IEEE Trans. Nucl. Sci.* **NS-34** 321–5

[13]   Spinks T, Jones J, Gilardi M C and Heather J D 1988 Physical performance of the latest generation of positron scanners employing BGO detectors *IEEE Trans. Nucl. Sci.* **NS-35** 721–5

[14]   Kops E R, Herzog H, Schmid A, Holte S and Feindegen L E 1990 Performance characteristics of an eight-ring whole body PET scanner *J. Comput. Assist. Tomogr.* **14** 437–5

[15]   Karp J S, Muehllehner G, Mankoff D A, Ordonez C E, Ollinger J M, Daube-Witherspoon M E, Haigh A T and Beerbohm D J 1990 Continuous-slice PENN PET: a positron tomograph with volume imaging capability *J. Nucl. Med.* **31** 617–27

[16]   Mazoyer B, Trebossen R, Deutch R, Casey M and Blohm K 1991 Physical characteristics of the ECAT 953B/31: A new high resolution brain positron tomograph *IEEE Trans. Med. Imaging* **MI-10** 499–504

[17]   Wienhard K, Dahlbom M, Eriksson L, Bruckbauer M T, Pietrzyk U and Heiss W-D 1994 The ECAT EXACT HR: performance of a new high resolution positron scanner *J. Comput. Tomogr.* **18** 110–8

[18]   DeGrado T R, Turkington T G, Williams J J, Stearns C W, Hoffman J M and Coleman R E 1994 Performance characteristics of a whole-body PET scanner *J. Nucl. Med.* **35** 1398–1406

[19]   Cherry *et al* 1997 MicroPET: A high resolution PET scanner for imaging small animals *IEEE Trans. Nucl. Sci.* **NS-44** 1161–6

[20]   Brownell G L and Sweet W H 1953 Localization of brain tumors with positron emitters *Nucleonics* **11** 40–5

[21]   Anger H O and Rosenthal D J 1959 Scintillation camera and positron camera, in medical radioisotope scanning *International Atomic Energy Agency* pp 59–82

[22]   Muehllehner G, Atkins F and Harper P V 1977 Positron camera with longitudinal and transverse tomographic capabilities *Medical Radionuclide Imaging*, SM-210/84, vol 1 (Vienna: IAEA) pp 291–301

[23]   Muehllehner G and Karp J 1986 A positron camera using position-sensitive detectors: PENN-PET *J. Nucl. Med.* **27** 90–8

[24]   Nellemann P, Hines H, Braymer W, Muehllehner G and Geagan M 1996 Performance characteristics of a dual-headed SPECT scanner with PET capability *IEEE Nuclear Science Symp. Medical Imaging Conf. Record (San Francisco, CA, 1995)* ed P A Moonier (Piscataway, NJ: IEEE) pp 1751–5

[25]   Miyaoka R S, Lewellen T K, Kim J S, Kaplan M S, Kohlmyer S K, Costa W and Jansen F 1996 Performance of a dual headed SPECT system modified for coincidence detection *IEEE Nuclear Science Symp. Medical Imaging Conf. Record (San Francisco, CA, 1995)* ed P A Moonier (Piscataway, NJ: IEEE) pp 1348–52

[26]   Sandell A, Ohlsson T, Erlandsson K, Hellborg R and Strand S-E 1992 A PET system based on 2-^{18}FDG production with a low energy electrostatic proton accelerator and a dual-headed PET scanner *Acta Oncol.* **31** 771–6

[27] Flower M A *et al* 1984 A clinical evaluation of a prototype positron camera for longitudinal emission tomography *Br. J. Radiol.* **57** 1103–17

[28] Townsend D, Frey P, Jeavons A, Reich G, Tochon-Danguy H J, Donath A, Christin A and Schaller G 1987 High density avalanche chamber (HIDAC) positron camera *J. Nucl. Med.* **28** 1554–62

[29] Cherry S R, Marsden P K, Ott R J, Flower M A, Webb S and Babich J W 1989 Image quantification with a large area multiwire proportional chamber positron camera (MUP-PET) *Eur. J. Nucl. Med.* **15** 694–700

[30] Visvikis D, Wells K, Ott R, Stephenson R, Bateman J E, Connolly J and Tappern G 1995 Preliminary experimental results from the first full size detector and dead time simulation of the count rate performance of a unique whole body PET camera *IEEE Trans. Nucl. Sci.* **NS-42** 1031–7

[31] Hoffman E J, Phelps M E, Mullani N A, Higgins C S and Ter-Pogossian M M 1976 Design and performance characteristics of a whole-body positron transaxial tomograph *J. Nucl. Med.* **17** 493–502

[32] Bohm C, Eriksson L, Bergström M, Litton J, Sundman R and Singh M 1978 A computer assisted ring-detector positron camera system for reconstruction tomography of the brain *IEEE Trans. Nucl. Sci.* **NS-25** 624–37

[33] Koeppe R A and Hutchins G D 1992 Instrumentation for positron emission tomography: Tomographs and data processing and display systems *Semin. Nucl. Med.* **22** 162–81

[34] Melcher C L and Schweitzer J S 1992 Cerium-doped lutetium oxyorthosilicate: A fast, efficient scintillator *IEEE Trans. Nucl. Sci.* **NS-39** 502–5

[35] Spinks T J, Jones T, Bailey D L, Townsend D W, Grootoonk S, Bloomfield P M, Gilardi M C, Casey M E, Sipe B and Reed J 1992 Physical performance of a positron tomograph for brain imaging with retractable septa *Phys. Med. Biol.* **37** 1637–55

[36] Townsend D W, Wensveen M and Byars L G 1993 A rotating PET scanner using BGO block detectors: Design, performance and applications *J. Nucl. Med.* **34** 1367–76

[37] Dahlbom M, Eriksson L, Rosenqvist G and Bohm C 1989 A study of the possibility of using multi-slice PET systems for 3D imaging *IEEE Trans. Nucl. Sci.* **NS-36** 1066–71

[38] Cherry S R, Dahlbom M and Hoffman E J 1992 High sensitivity, total body PET scanning using 3D data acquisition and reconstruction *IEEE Trans. Nucl. Sci.* **NS-39** 1088–92

[39] Huang S-C, Hoffman E J, Phelps M E and Kuhl D E 1979 Quantitation in positron emission computed tomography: 2. Effects of inaccurate attenuation correction *J. Comput. Assist. Tomogr.* **3** 804–14

[40] Bergström M, Litton J, Eriksson L, Bohm C and Blomqvist G 1982 Determination of object contour from projections for attenuation correction in cranial positron emission tomography *J. Comput. Assist. Tomogr.* **6** 365–72

[41] Huang S-C, Carson R E, Phelps M E, Hoffman E J, Schelbert H R and Kuhl D E 1981 A boundary method for attenuation correction in positron computed tomography *J. Nucl. Med.* **22** 627–37

[42] Carson R E, Daube-Witherspoon M E and Green M V 1988 A method for postinjection PET transmission measurements with a rotating source *J. Nucl. Med.* **29** 1558–67

CHAPTER 17

PETSIM: MONTE CARLO SIMULATION OF POSITRON IMAGING SYSTEMS

Chris J Thompson and Yani Picard

17.1 INTRODUCTION

Monte Carlo simulation is a very realistic way to evaluate the performance of a positron imaging system. All the factors affecting sensitivity, count rate performance and resolution can be represented by random processes and geometric properties of the scanner. Since positron emission tomography (PET) systems are used in a wide variety of count rate and source distribution imaging situations, a knowledge of how the system will handle these conditions is essential to manufacturers, and very useful to buyers trying to acquire the 'best' machine for their needs. Inevitably, when faced with a comparison of several machines, one uses extreme situations. Good high-count-rate performance of a scanner is required in transmission scans, bolus water blood flow emission scans and rubidium-82 cardiac studies. Under these conditions, because the patient is exposed to the radiation for a short time (because of the short scan time or isotope half-life) and high-activity concentrations are easily obtained, the highest count rates are encountered.

At the opposite end of the count rate spectrum are images of more exotic, difficult to produce radiopharmaceuticals. These would either be available from low-yield chemical synthesis, have low specific activity, or perhaps be toxic in larger quantities. Sensitivity will be much more important in these cases. If the conflicting requirements of mixing low-dose and high-dose scans on the same machine must be met, an accurate model of the dead time, interchangeable collimators or both will be required. Another reason for wanting maximum sensitivity is that PET is often used to understand normal physiology in control subjects. High scanner sensitivity can be used to reduce the radiation risk in these subjects, by reducing either the dose per subject or the number of subjects required to validate the hypothesis under investigation.

The evaluation of the useful dynamic range of a PET system is well demonstrated in terms of noise-effective counts (NECs) in a particular imaging situation. The NEC is given by

$$NEC = \frac{T}{1 + S/T + nR/T} \qquad (17.1)$$

where T is the true count rate, S the scattered count rate and R the random count rate before correction for live time. This form allows the random counts to be estimated from the singles counts (in this case $n = 1$) or from a delayed coincidence window (in this case $n = 2$). The distribution of scattered counts for a given imaging situation is best obtained by Monte Carlo simulation. However it is important that the source distribution simulated be comparable with the imaging situation. Very few PET imaging situations correspond to uniform-activity test phantoms 20 cm in diameter.

The sensitivity and resolution of PET systems are not isolated from one another. One tends to associate higher-resolution PET studies with ^{18}F-labelled fluorodeoxyglucose (FDG) and ^{11}C CO blood volume studies [1, 2]. The intrinsic resolution of these studies will be slightly higher than ^{15}O-water studies because of the shorter positron range. However the fact that images from the former compounds are usually displayed with higher resolution reflects the long counting times possible while acquiring these images rather than the improved intrinsic resolution. Older PET systems (PETT VI [3] and AECL Therascan 3128 [4]) used removable dense bars to alternate between higher sensitivity and higher resolution. This method of improving spatial resolution has been rendered obsolete by multielement block detectors [5]. These maintain most of their good spatial resolution even at quite high count rates, and their live time falls in a very predictable way [6].

While these bars do improve resolution, they do so with a very substantial drop in sensitivity. This sensitivity cannot be made up for by increasing the activity concentration, or reducing the detector separation. Cho *et al* [7, 8] recently suggested that the sensitivity lost by blocking off a large part of the crystal face could be regained by using a smaller detector array diameter. Unfortunately their simulations neglected the effects of increased random and scattered counts and live time and greater angular acceptance required with smaller diameters. It is thus important to try and include *all* parameters in the simulation in order to make it realistic. In practice no simulation will include all effects. Some must be assumed or ignored, otherwise the simulation time would be unreasonably long, or so complicated it would be difficult to validate. Compromise must be made in neglecting small effects and running the simulations in a reasonable time.

The programs used for these simulations are derived from those of Lupton and Keller [9]. We obtained copies of the original programs and documentation [9], and have modified them substantially to permit more realistic simulation of modern PET scanners [10–13].

Table 17.1 *Information saved on each ray in the GRH file.*

Parameter	Storage	Range	Precision
X, Y, Z origin[a]	6 bytes	±100 cm	0.01 mm
X, Y, Z current	6 bytes	±100 cm	0.01 mm
A, B, C direction cosines	6 bytes	180°	0.006°
Gamma ray status	2 bytes decomposed as:		
Present energy	9 bits	0–511	1 keV
Number of scatters	3 bits	0–7	
Number of interactions in detector	2 bits	0–3	
Identifying code[b]	2 bits	0–3	

[a]First photon has positron creation coordinates. Second photon has positron annihilation coordinates.
[b]First or second photon, singled in or after phantom.

In this chapter we present the methods used for simulating various parameters and we will show typical results based on scanner geometries which represent variations on the GE Advance as a demonstration of the application of these simulation techniques.

17.2 STRUCTURE OF THE PET SIMULATION

The PET simulations are performed as a series of steps, for the source distribution, collimator and detector array. Each of these major programs is run independently. Each (except the first) reads an input gamma ray history (GRH) file and creates a new GRH file of the rays which passed out of the source container, passed through the collimator or were absorbed in the detector. Each file is of identical format. It consists of a file header into which each program writes its geometric data in pre-allocated areas, and the bulk of the file contains all needed data on each gamma ray processed. These data are saved in the most compact form possible, to maximize the disk storage efficiency. Table 17.1 shows the data saved, their format and their precision.

All programs are written in VMS Fortran. The programs are run on one of several VAXstation† 4000/60 workstations in a VAXcluster configuration. The cluster has two 1 Gbyte and one 2 Gbyte SCSI disks and two Exabyte 8500 SCSI tape drives. A typical simple 20 cm water-filled cylinder 10 cm high phantom simulation, including positron range and noncollinearity effects, creates over 7000 positrons per minute and tracks both gamma rays until they are absorbed or emerge from a 50 cm diameter, 12 cm high cylinder.

†Digital Equipment Corporation.

17.3 SOURCE DISTRIBUTION SIMULATION

The program PHANTOM is used to generate positrons and annihilation photons and track them through the materials which contain the source. The source distribution consists of up to ten cylinders, spheres or boxes. The cylinders and spheres can be either hollow or solid, and placed anywhere in the scanner's field of view. Each source is assigned a relative activity concentration. The program calculates the volume of each region, and generates the correct fraction of positrons in each region. During the simulation blocks of 10 000 positrons are created and travel a distance based on their positron range, annihilate with an electron and produce two gamma rays offset from 180° by a random noncollinearity angle. The approximate attenuation and FDG distribution of a brain can be constructed by using these cylinders, and is illustrated in figures 17.1 and 17.2. This phantom is known as the Bart Simpson phantom because of the similarity to the TV show character.

Well established techniques are used [9] to produce uniformly distributed positron creation density throughout each region, and create uniform distribution in polar coordinates for the first gamma ray. The distance travelled by each positron is based on a bi-exponential distribution of positron ranges observed by Derenzo [14]. He measured the positron range for many isotopes in Styrofoam and calculated the range in water based on the relative electron densities. From the table in [14] we fitted the two exponential range constants to a function of energy of the form

$$d = A \, e^{-\alpha E} + (A - 1) \, e^{-\beta E} \qquad (17.2)$$

where A, α and β are the fitted values, E is the end-point positron energy and d is the distance travelled by the positron before annihilation. This distribution can be simulated by generating an exponentially biased random number using the standard property of random numbers and their inverse, i.e.

$$R[F(x)] = F^{-1}[R(X)] \qquad (17.3)$$

where $R(x)$ is a random number in the range $0 < R(x) \le 1.0$. In this case, we take the natural logarithm of the random number. We chose the exponent α if a second random number is greater than A, and the exponent β when the second random number is less than A. This gives the magnitude of the displacement before annihilation: the direction of this displacement is random. One of the rays' directions is changed by an angle derived from a Gaussian random number distribution with a full width at half maximum (FWHM) of 0.5° to account for the noncollinearity.

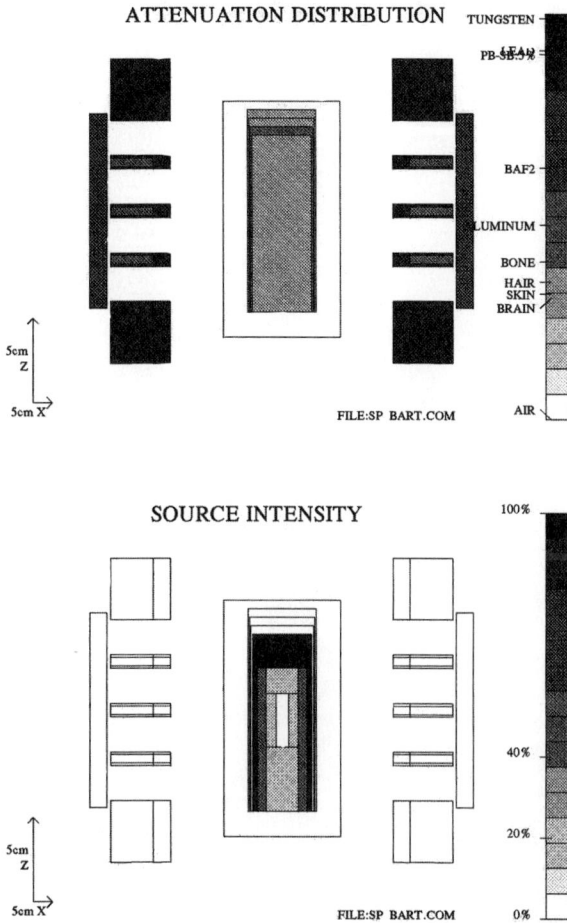

Figure 17.1 *Attenuation and source distribution for a simulated FDG brain study in the Super-PETT scanner. (Note the vertical scale is five times greater than the horizontal scale to show the septum detail.)*

17.4 SOURCE SCATTERING GEOMETRY SIMULATION

The source scattering geometry is defined independently of the source distribution. It is specified in terms of a set of nested (the first one must fit inside the next one) cylinders, spheres or boxes. The axis of cylinders is parallel to the scanner's axis. Each cylinder is described in terms of its outer radius, and the z coordinates of the top and bottom surfaces. Only the radius of spheres is used and they are assumed to be centred on the origin of the coordinate system. Boxes may be placed anywhere. The material used in the phantom must be specified. The material must be chosen from one in a

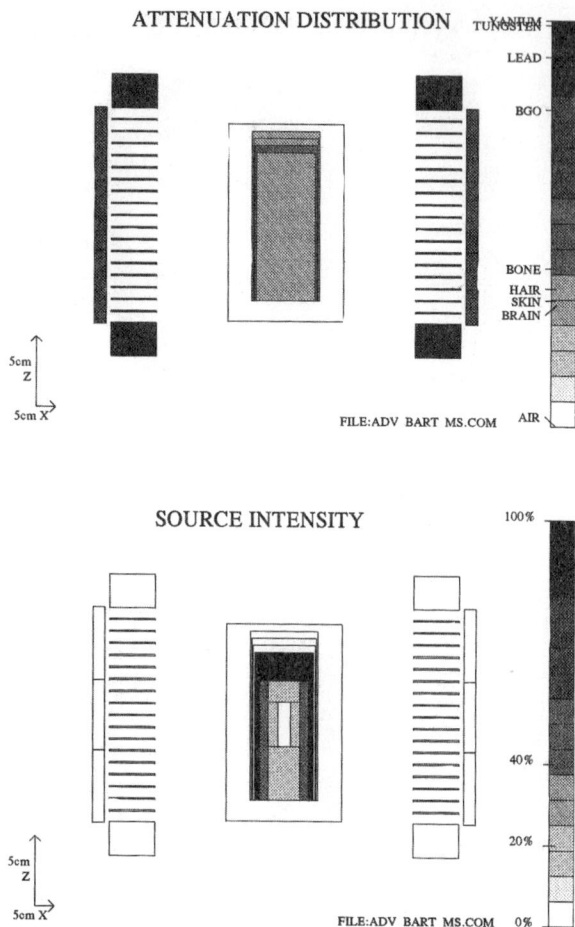

Figure 17.2 *Attenuation and source distribution for a simulated FDG brain study in the GE Advance scanner.*

table, as its name is used to find the coefficients. The electron density of the medium for calculation of the Compton scattering cross section, and parameters A_n, B_n and k which are used to fit the photo-electric cross section according to

$$C = A_1 E^{-B_1} \quad (E > k)$$
$$C = A_2 E^{-B_2} \quad (E \leq k)$$

(17.4)

are found for each material by fitting data produced by a program XGAM [15] available from the US National Institutes of Standards and Technology. k is the average energy of the K edge in the photo-electric cross section in

Table 17.2 *Inter-ring sensitivity matrix for double-tapered septa. Double tapered collimator 300 keV (one-off coinc); coincidence efficiency matrix for all eight slices. Numbers on the diagonal are for the direct slices; numbers above the diagonal are for valid cross slices; numbers below the diagonal would be rejected by the coincidence circuit. Efficiencies are in kcps $(\mu Ci\,cm^{-3})^{-1}$. Scatter percentages are 'scatter $\times$ 100/ (scatter + true)'.*

Slice		1	2	3	4	5	6	7	8
1	T_{coinc}	9.26	16.96	0.00	0.00	0.00	0.00	0.00	0.00
	Scat. (%)	12.1	13.7	0.0	0.0	0.0	0.0	0.0	0.0
2	T_{coinc}	0.00	12.17	19.0	0.00	0.00	0.00	0.00	0.00
	Scat. (%)	0.0	14.5	14.6	0.0	0.0	0.0	0.0	0.0
3	T_{coinc}	6.88	0.00	12.0	18.24	0.00	0.00	0.00	0.00
	Scat. (%)	21.2	0.0	12.4	16.6	0.0	0.0	0.0	0.0
4	T_{coinc}	2.01	7.62	0.00	12.28	18.99	0.00	0.00	0.00
	Scat. (%)	37.8	25.3	0.0	15.6	16.5	0.0	0.0	0.0
5	T_{coinc}	0.81	3.04	9.09	0.00	12.46	18.74	0.00	0.00
	Scat. (%)	58.3	37.4	22.7	0.0	14.0	16.1	0.0	0.0
6	T_{coinc}	0.24	0.79	2.76	8.64	0.00	12.87	19.02	0.00
	Scat. (%)	75.4	61.7	38.4	24.3	0.0	14.1	14.6	0.0
7	T_{coinc}	0.20	0.26	0.83	2.92	7.75	0.00	11.55	17.44
	Scat. (%)	72.4	77.9	59.8	38.8	22.3	0.0	13.3	13.0
8	T_{coinc}	0.24	0.22	0.27	0.73	1.90	6.84	0.00	9.54
	Scat. (%)	63.1	72.3	74.8	59.5	44.2	22.1	0.0	11.9
Singles efficiency		692.5	910	926	958	970	921	902	695

True efficiency of direct slice	11.52 kcps $(\mu Ci\,cm^{-3})^{-1}$
Scat. efficiency of direct slice	1.81 kcps $(\mu Ci\,cm^{-3})^{-1}$ or 13.6%
True efficiency of one-off cross slice	18.35 kcps $(\mu Ci\,cm^{-3})^{-1}$
Scat. efficiency of one-off cross slice	3.25 kcps $(\mu Ci\,cm^{-3})^{-1}$ or 15.1%
Total singles: true coincident ratio	75.7 for direct slice
Total singles: true coincident ratio	108.6 for one-off cross slice

kiloelectron volts. Tables 17.2 and 17.3 list the coefficients used for many biological and detector materials. The values A_1 and B_1 are used when the energy is above the K absorption edge. If the energy is lower, the values A_2 and B_2 are used [16].

Compton scattering is simulated by looking up a 100×100 table which contains a set of 100 equiprobable scattering angles for each gamma ray energy in the range 11–511 keV, in 5 keV energy increments. These data were tabulated from the Klein–Nishina cross section, and are stored as a common file. The same table is used by all programs requiring it. The current gamma ray energy and a random number between 1 and 100 are used to find the new ray's scattering angle. The new energy is then calculated. This angle defines the surface of the scattering cone; a random azimuthal angle

Table 17.3 *Photo-electric and Compton scattering cross sections for several biological materials.*

Material	Electrons (cm^{-3})	A_1	B_1	K edge	A_2	B_2
Air	3.622×10^{20}	7.680	3.169	0.0	7.680	3.169
Blood	3.513×10^{23}	7.110×10^3	3.178	0.0	7.110×10^3	3.178
Bone	5.267×10^{23}	3.989×10^4	3.075	0.0	3.989×10^4	3.075
Brain	3.438×10^{23}	7.031×10^3	3.171	0.0	7.031×10^3	3.171
Brain stem	3.500×10^{23}	7.727×10^3	3.169	0.0	7.727×10^3	3.169
Calcium	4.628×10^{23}	1.762×10^5	3.054	0.0	1.762×10^5	3.054
Cerebellum	3.428×10^{23}	7.044×10^3	3.172	0.0	7.044×10^3	3.172
Cerebrum	3.433×10^{23}	7.002×10^3	3.171	0.0	7.002×10^3	3.171
Eye lenses	3.635×10^{23}	6.464×10^3	3.195	0.0	6.464×10^3	3.195
Fat	3.061×10^{23}	3.725×10^3	3.202	0.0	3.725×10^3	3.202
Hair	4.202×10^{23}	7.412×10^3	3.074	0.0	7.412×10^3	3.074
Heart	3.415×10^{23}	6.658×10^3	3.178	0.0	6.658×10^3	3.178
Lung	8.607×10^{22}	1.779×10^3	3.173	0.0	1.779×10^3	3.173
Muscle	3.445×10^{23}	7.070×10^3	3.175	0.0	7.070×10^3	3.175
Skin	3.639×10^{23}	6.618×10^3	3.182	0.0	6.618×10^3	3.182
Water	3.343×10^{23}	6.869×10^3	3.192	0.0	6.869×10^3	3.192

is used to define a line on the scattering cone along which the new ray will travel.

The two rays are then tracked through the source distribution until they escape the source boundaries, or they are absorbed. If both rays escape the source boundaries, they are stored in the GRH file. If only one escapes, it will be stored if the next random number is greater than 0.9. This, in effect, saves only one single ray in ten. Since far more single rays emerge from a typical source distribution than coincident pairs, this saves storage space, without losing precision for dead time and random count estimation. The GRH files can be any length up to the space available on a single disk. At present the largest disk available is 2 Gbytes. After the source simulation is complete the data are usually copied to an Exabyte tape. Other programs can read either directly from disk or from tape. With the record length used to store data on tape just over 5 Gbytes can be saved on one standard-length video cassette tape.

17.5 COLLIMATOR SIMULATION

The collimator simulation reads the GRH file produced by the source simulation program, and tracks the photons through the scanner's collimator. It copies all the header information describing the source distribution and scattering geometry from the input file, and writes a description of the collimator structure into the new file's header. The seed of the random number generator is read in order to use a continuous list of random numbers. The parts of the collimator are considered as either nested cylinders or as fustra of cones (truncated cones). Each region is made out of air or a dense material, lead or tungsten alloy. In multislice (MS) PET systems the slices are usually separated by cylindrical or tapered septa. These are absent from positron volume imaging (PVI) systems. The axial extent of the field of view is normally shielded by thick lead cylinders. The collimator simulation may have two radially distinct regions. This allows systems such as the Superpett collimator structure to be simulated. The inter-plane septa in this system are made of a sandwich of lead on the outside and aluminium inside, except that there is an inner ring which is lead right through. This is shown in figure 17.1. (Note that the axial scale is much bigger than the radial scale to show the collimator details.) The GE Advance collimator geometry is illustrated in figure 17.2 showing the inter-plane septa in place.

In the collimator phase of the simulation both rays are first moved from their position on emerging from the source to the inner wall of the collimator along the path on which they escaped the source. The z coordinate at this point is used to find which collimator segment they enter, and tracking proceeds from there.

As each photon escapes the collimator region, a new entry is made in the output GRH file. The present coordinates, number of scatters, energy and a two-bit identifying code (coincident, single) are saved in the output file. The code identifies the photons as being either the first or second of a valid coincident pair, or a single photon whose mate was lost in the phantom or collimator or was not detected in the detector singled.

17.6 DETECTOR SIMULATION

The detector simulation reads the GRH file produced by the collimator simulation, and tracks the photons through the scanner's detectors. The detector geometry is normally defined by an annulus made up of segments of detector material, separated by gaps which may be filled with air, or dense material to absorb gamma rays scattered from one detector element to the next. Alternatively the detector can be described as a series of rectangular blocks separated by wedge-shaped septa. A special case, that of two rectangular blocks whose separation can be chosen independent of the detector size, is also available. This allows simulation of the HIDAC, Q-PET or PEM detector geometry to be simulated. The number of detectors corresponds to the number of encoded blocks rather than the number of discrete crystals, or regions within the block. If each crystal has a separate readout, then the number of crystals is equal to the number of blocks. For example the Donner 600 [17] would be simulated as having 600 detector elements; the Penn-PET

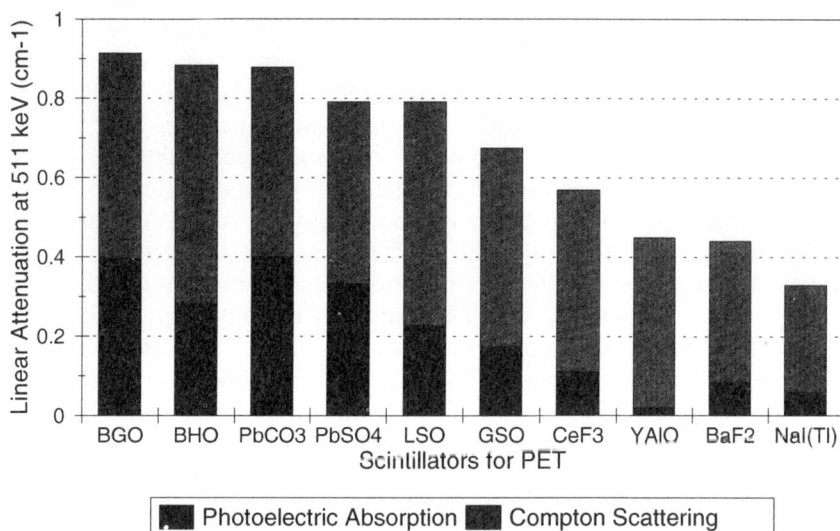

Figure 17.3 *Linear attenuation coefficients of scintillators used in PET.*

[18] has only six and can be simulated as six 'blocks' separated by wedge-shaped air gaps. This allows multiple interactions of a ray within the detector to be considered as one detected event. This is what happens in practice in a detector block. If a ray deposits more than an allowable energy in an adjacent detector, both interactions can be discarded for coincident detection purposes, but included for live-time purposes. This implements the 'adjacent-crystal veto' in the Donner 600 [19].

At present none of the scintillators used in PET have a photo-electric cross section higher than their Compton cross section. The linear attenuation of currently used scintillators at 511 keV is illustrated in figure 17.3. The total height of each bar is the total attenuation coefficient. This gives a visual indication of the relative stopping power of detectors, and the probability for additional blurring due to Compton scatter within the crystal.

The detector is composed of active and passive components [20]. The active components 'detect' gamma rays and the energy deposited in them is included in the output. The passive components 'absorb' gamma rays, and that energy is lost. The ray may scatter from an absorbing region to a detecting region, or vice versa. In all cases only the energy deposited in the active region is reported. This allows gamma rays to scatter from light guides or PMT entrance windows. The detector can be in several layers either in a radial direction, or in the vertical direction. This allows the resolution of systems with various depth of interaction encoding strategies to be simulated.

The detector program creates an output GRH file whose header contains all the source and collimator information as well as a description of the detector geometry. The body of the file contains the location of the centroid of all interactions of each ray within each detector block. This is the best possible localizing information which could eventually represent the real point of detection. The detector program updates the status bits used to identify the number of interactions each ray has had with the detector block. The two bits allocated for this are interpreted as corresponding to one, two, three or four and more interactions. This program also produces a file which is a histogram of the component parallel to the detector face in the X–Y plane of distance between the point of first interaction and the centroid of all interactions. This effect was studied in detail in [10]. Gamma rays which leave the back, top or bottom of the detector are counted, but discarded for further study. A knowledge of the fraction lost in this manner is useful in optimizing the detector depth, since a deeper crystal will be more efficient, but will cause more radial blurring, and be more expensive.

A typical source (approximating a typical FDG brain study) and scanner configuration is shown in figure 17.2. Diagrams such as this can be drawn by interpreting the data in the header of the GRH file, or from any command file which will execute the simulation programs in sequence. These can be displayed on the screen of a Vaxstation or terminal, and plotted on a laser or colour printer. In figure 17.2(a) the blackness is proportional to the linear attenuation coefficient of the medium at 511 keV. Figure 17.2(b) shows the source distribution, with the blackening proportional to the source intensity, and other elements outlined. On the workstation display or colour printer a 'rainbow' colour scale is used to represent the attenuation coefficient or source intensity.

17.7 ANALYSIS PROGRAM

This program reads the GRH file produced by the detector program, and creates in-plane and axial detection profiles, generates data from which NEC plots can be obtained, generates spectra of detected rays categorized by number of scatters and produces inter-ring sensitivity (IRS) matrices. The NEC and IRS data are the most useful in analysing the tradeoffs in collimator design.

The spectra of detected events allow one to set low- and high-energy discriminators to improve signal to noise in the final image. The ray's actual energy is blurred by a Gaussian whose FWHM, W_{511}, is given at 511 keV. At lower energies the FWHM, W, is given by

$$W = W_{511}\sqrt{E/511}. \tag{17.5}$$

The FWHM specified should be consistent with the known detector response. The spectra are separated by the number of scatters undergone by both coincident rays. These are useful in assessing the role of energy discrimination, or the use of multiple energy windows (holospectral) for scatter compensation [21]. It also allows the possibility of combining events from a lower energy window which is mostly rays which forward scatter from the crystal. This happens in the NaI crystals in the Penn-PET (17) [22] and provides a useful efficiency improvement.

The analysis program also generates profiles of the radial and axial system response to both true and scattered events. It also estimates the 'centroid of interaction' of each crystal, by building a map of the coordinates of the mean energy deposited in detecting each incoming coincident ray. We have found that this centroid moves slightly according to the position along each projection. It is also different for adjacent projections because of the effect of the inter-block septa used on most scanners. Using this information we have been able to improve both the spatial resolution and image distortion in the Scanditronix PC2048 [20].

The IRS matrix considers the detector array to be made up of several rings, and tabulates the inter-ring true coincidence efficiency in cps $(\mu Ci\,cm^{-3})^{-1}$ and scatter fractions. A square matrix is produced, with the valid coincident rings above the diagonal, and the inter-ring data which would be rejected by the system's coincidence circuit below the diagonal.

Many scanners can now operate in a high-efficiency mode where events from crystal rings which are offset by two rings are combined with the traditional direct slices. The three-off crystal rings are combined with the one-off rings to generate high-efficiency cross slices. These and other sensible combinations can be allowed, and analysed. This is especially useful in determining the optimal inter-ring septum length. A simple example is given in table 17.4, in which the collimator on the Scanditronix PC2048 was replaced by a double-tapered one in order to improve the efficiency of the cross slices.

17.8 RESOLUTION PROGRAM

It is in the resolution program that the number of individual crystals per block is used to assign any gamma ray to one crystal. The effects of individual blurring components can be plotted by this program. Because of the coding and tracking schemes described previously, it is possible to separate the effects of positron range, crystal penetration and detector quantization in this program. The effects on gamma ray noncollinearity cannot be removed, since one ray's path could have taken it to another detector, and it is too late to re-trace it.

This program can estimate the image blurring in the various detectors illustrated in figure 17.1. This illustrates the fate of a ray entering three crystal configurations and making two Compton and one photo-electric

Table 17.4 *Photo-electric and Compton scattering cross sections for several detector and collimator materials.*

Material	Electrons (cm^{-3})	A_1	B_1	K edge	A_2	B_2
Detector						
BGO	1.806×10^{24}	3.793×10^6	2.584	95.0	9.942×10^5	2.593
BHO	2.097×10^{24}	5.165×10^6	2.690	70.0	1.099×10^5	1.947
Cerium fluoride	2.597×10^{23}	3.117×10^6	2.758	45.0	5.752×10^5	2.780
Caesium fluoride	1.044×10^{24}	2.406×10^6	2.778	40.0	4.050×10^5	2.785
$FeSO_4.H_2SO_4$	3.412×10^{23}	7.706×10^3	3.177	0.0	7.706×10^3	3.177
Germanium	1.420×10^{24}	1.529×10^6	2.940	0.0	1.529×10^6	2.940
GSO	1.740×10^{24}	3.800×10^6	2.717	55.0	8.198×10^5	2.759
Lead carbonate	1.666×10^{24}	3.940×10^6	2.588	90.0	7.809×10^5	2.550
Lead comb	8.818×10^{23}	2.840×10^6	2.588	90.0	5.610×10^5	2.549
Lead glass	2.124×10^{23}	4.159×10^5	2.588	90.0	8.477×10^4	2.556
Lead sulphate	1.601×10^{24}	3.266×10^6	2.588	90.0	6.670×10^5	2.556
Lithium fluoride	7.341×10^{23}	2.082×10^4	3.179	0.0	2.082×10^4	3.179
LSO	1.968×10^{24}	3.680×10^6	2.670	65.0	9.294×10^5	2.727
Sodium iodide	9.429×10^{23}	1.999×10^6	2.791	35.0	3.217×10^5	2.791
Collimator						
Lead	2.704×10^{24}	8.720×10^6	2.588	90.0	1.722×10^6	2.549
Lead– 5% antimony	2.658×10^{24}	8.342×10^6	2.590	90.0	1.444×10^6	2.506
Tantalum	4.033×10^{24}	1.334×10^7	2.655	70.0	3.522×10^6	2.721
Tungsten	4.690×10^{24}	1.554×10^7	2.648	70.0	4.202×10^6	2.716
Yanium[a]	2.000×10	1.000×10^8	0.000	0.0	1.000×10^8	0.000

[a]Used for simulating 'opaque' collimators; has negligible Compton cross section and very high photo-electric cross section.

interactions. The best spatial resolution is obtained with the individual crystals if an adjacent-crystal veto is applied [10]. Dahlbom and Hoffman [23] have shown that the events recorded in block detectors tend to cluster in islands using suitable encoding of the PMT signals. In fact the incident gamma ray flux must be uniform on the block's face, except for interactions where Compton scattering occurs between adjacent blocks, and the effects of inter-plane septa. However the surface treatment of the crystals channels the light so that it appears to come from one crystal in the block. The best a coding scheme could do would be to identify the correct crystal in the block. This component is called the detector quantization in the resolution program.

17.9 EXAMPLES

17.9.1 Efficiency and scatter fraction dependence changes with ring offset

Table 17.2 shows the inter-ring sensitivity matrix for the double-tapered septum case studied. The 8×8 matrix shows the straight slices on the

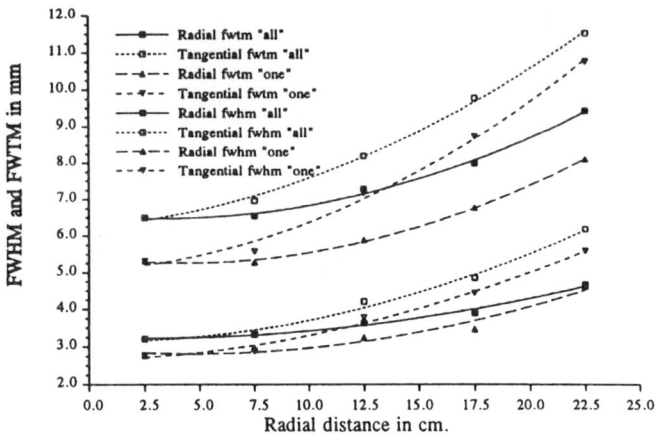

(a)

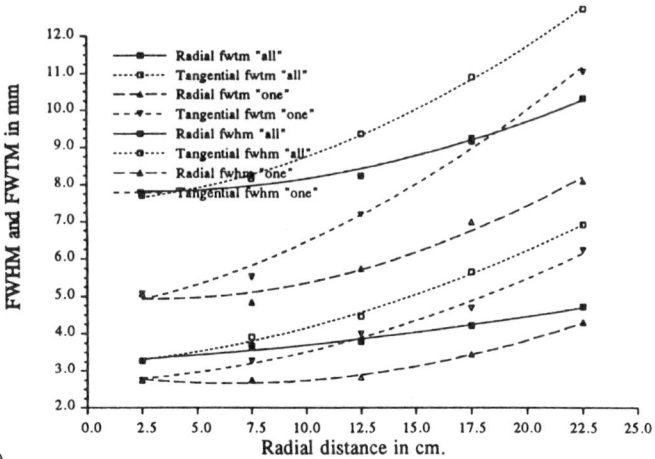

(b)

Figure 17.4 *The radial and tangential components for FWHM and FWTM for the GE Advance with (a) BGO and (b) GSO detectors.*

diagonal (ring *n* in coincidence with ring *n*). For this analysis only the one-off cross slices were selected by the coincidence circuit. The efficiencies for these cross slices appear above the diagonal, while those rejected by the coincidence circuit are below the diagonal. As one moves further from the diagonal the scatter fraction increases, and the true efficiency decreases.

17.9.2 Components of spatial resolution with crystal size

The resolution program allows the separation of image blurring into several causes. Coincident events in which each ray makes only one interaction with the detector can be distinguished from those which make any number of interactions. Figure 17.4 compares the spatial resolution expressed as both FWHM and full width at tenth maximum (FWTM) as a function of radial distance for the GE Advance with block detectors made from BGO or GSO in the geometry of the GE Advance with the septa extended. The simulation was made by specifying the source as an air-filled cylinder 50 cm in diameter. This eliminates scatter in the source giving good counting statistics, but the positron range for ^{18}F in water allows the range blurring to be shown. All curves include the effect of positron range and noncollinearity. The different line styles in figure 17.4 represent coincident events in which each ray makes any number of interactions ('all') and those in which each ray makes only one ('one'). The difference in FWTM under these conditions is greater with GSO than BGO, but the difference in FWHM is not as much. The blurring due to radial penetration is greater for GSO because of its lower attenuation coefficient.

ACKNOWLEDGMENTS

This work is supported specifically by the National Science and Engineering Research Council of Canada (NSERC) grant no OGP0036672. The support by the Medical Research Council of Canada for the MNI Research Computing Laboratory is gratefully acknowledged. General Electric Medical Systems (Pet Engineering) have contributed to the development of the PETSIM program.

REFERENCES

[1] Lammertsma A A and Jones T 1983 Correction for the presence of intra-vascular oxygen-15 in the steady state technique for measuring regional oxygen extraction ratio in the brain: description of the method *J. Cereb. Blood Flow Metab.* **3** 416-24

[2] Phelps M E, Huang S C, Hoffman E J, Selin C, Sokoloff L and Kuhl D E 1979 Tomographic measurement of local cerebral glucose metabolic rate in humans with (F-18)-2-fluoro-2-deoxy-D-glucose: validation of method *Ann. Neurol.* **6** 371-88

[3] Yamamoto M, Ficke D C and Ter-Pogossian M M 1982 Performance study of PETT VI, a positron computed tomograph with 288 cesium fluoride detectors *IEEE Trans. Nucl. Sci.* **NS-29** 529-33

[4] Cooke B E, Evans A C, Fantome E O, Alaire R and Sendyk A M 1983 Performance figures and images from the Therascan 3128 positron emission tomograph *IEEE Trans. Nucl. Sci.* **NS-31** 640-4

[5] Casey M E and Nutt R 1986 A multi-crystal two dimensional BGO detector system for positron emission tomography *IEEE Trans. Nucl. Sci.* **NS-33** 460-3

[6]   Thompson C J and Meyer E 1987 The effect of live time in components of a positron tomograph on image quantification *IEEE Trans. Nucl. Sci.* **NS-34** 337–43

[7]   Cho Z H, Juh S C, Friedenberg R M, Bunney W, Buchabaum M and Wong E 1990 A new approach to very high resolution mini-brain PET using a small number of large detectors—II Performance study *IEEE Trans. Nucl. Sci.* **NS-37** 842–51

[8]   Cho Z H and Juh S C 1991 High resolution PET with large detectors—II Performance study *IEEE Trans. Nucl. Sci.* **NS-38** 726–31

[9]   Lupton L R and Keller N A 1983 Performance study of single-slice positron emission tomography scanners by Monte Carlo method techniques *IEEE Trans. Med. Imaging* **TMI-2** 154–68

[10] Thompson C J 1990 The effects of detector material and structure on PET spatial resolution and efficiency *IEEE Trans. Nucl. Sci.* **NS-37** 718–24

[11] Thompson C J 1989 The effects of collimation on singles rates in multi-slice PET *IEEE Trans. Nucl. Sci.* **NS-36** 1072–7

[12] Thompson C J 1988 The effect of collimation on scatter fraction in multi-slice PET *IEEE Trans. Nucl. Sci.* **NS-35** 598–602

[13] Thompson C J 1991 The effects of collimation on PET image noise due to scatter, random counts and dead time *J. Cereb. Blood Flow Metab.* **11** 31–7

[14] Derenzo S 1979 Precision measurement of positron annihilation point spread functions for medically important positron emitters *Positron Annihilation* ed R R Hasiguti and Fujiwara (Sendai: Japan Institute of Metals) pp 819–23

[15] Berger M J 1990 X-ray and gamma-ray attenuation coefficients and cross sections *Database V 2.0* Standard Reference Data (abstract)

[16] Picard Y, Thompson C J and Marrett S 1992 Improving the precision and accuracy of Monte Carlo simulation in positron emission tomography *IEEE Trans. Nucl. Sci.* **NS-39** 1111–6

[17] Derenzo S, Huesman R H, Cahoon L J, Geyer A, Uber D, Vuletish V and Budinger T F 1987 Initial results from the DONNER 600 crystal positron tomography *IEEE Trans. Nucl. Sci.* **NS-34** 321–5

[18] Muehllehner G, Karp J S and Mankoff D A 1989 Design and performance of a new positron tomograph *IEEE Trans. Nucl. Sci.* **NS-35** 670–4

[19] Turko B T, Zizka G, Lo C C, Leskovar B, Cahoon L J, Huesman R H, Derenzo S, Geyer A and Budinger T F 1987 Scintillation photon detection and event selection in high resolution positron emission tomography *IEEE Trans. Nucl. Sci.* **NS-34** 326–9

[20] Thompson C J and Picard Y 1993 Two new strategies to increase the signal to noise ratio in positron volume imaging *IEEE Trans. Nucl. Sci.* **NS-40** 956–61

[21] Gagnon D, Todd-Pokropek A, Arsenault A and Dupras G 1989 Introduction to holospectral imaging in nuclear medicine for scatter subtraction *IEEE Trans. Med. Imaging* **MI-8** 245–50

[22] Thompson C J 1993 The problem of scatter correction in positron volume imaging *IEEE Trans. Med. Imaging* **MI-12** 124–32

[23] Dahlbom M and Hoffman E J 1987 An evaluation of a two-dimensional array detector for high resolution PET *IEEE Trans. Med. Imaging* **MI-7** 262–72

CHAPTER 18

MONTE CARLO IN QUANTITATIVE 3D PET: SCATTER

Robert S Miyaoka and
Robert L Harrison

18.1 INTRODUCTION

Three-dimensional positron emission tomography (3D PET) offers a factor of six to seven in true coincidence sensitivity over current 2D PET systems [1–3]. Higher coincidence sensitivity translates into better statistical quality of collected data or a reduction in scanning time or administered activity of studies while maintaining similar counting statistics. Since most PET studies are count limited, any increase in the true coincidence sensitivity is highly desirable. Ultra-high-resolution PET detectors for small-animal imaging systems are already being built with 2 mm axial slice spacing [4]. To take optimal advantage of the highest-resolution detector designs requires 3D data acquisition [1, 3, 5, 6]. The two factors that limit the practice of quantitative 3D PET in the body are inadequate scatter correction and transmission measurement time. Methods to accurately estimate scatter in 3D whole-body PET and to reduce the scan time for transmission imaging are active areas of research.

System characteristics are usually defined by global measurements (e.g. noise equivalent count (NEC) rate [7]) and are measured using homogeneous and symmetric phantoms [8]. Here, the Monte Carlo method provides accurate information about true, scatter and random events for the simulated imaging protocols and tomograph geometries. The method is also useful for performing detailed studies of the characteristics of true, scatter and random events for complex imaging environments. Because accurate methods exist to estimate random events [9, 10], the Monte Carlo method has not been required to study these effects. In contrast, when imaging complex, inhomogeneous objects it is difficult to distinguish scattered events from true events. The development of efficient Monte Carlo simulation packages allows the study of scatter for realistic activity distributions and attenuation maps for formulating and evaluating scatter correction techniques.

18.2 SCATTER CORRECTIONS FOR 3D PET

Scatter correction methods proposed for 3D PET, can be divided into six groups: (i) convolution subtraction; (ii) tail fitting; (iii) energy based; (iv) analytic; (v) a method requiring an auxiliary 2D scan and (vi) Monte Carlo. Monte Carlo simulation has been used to evaluate the accuracy of many of the proposed methods and also to understand the strengths and shortcomings of the various techniques. Incorporating scatter in an iterative reconstruction can also remove scatter from the activity estimate. The computational burden of 3D iterative reconstruction methods [11] excludes this method from practical consideration at this time.

The convolution subtraction or 'deconvolution' technique [12, 13] is widely used for 2D PET and has also been extended to 3D PET [14–18]. The method requires the empirical determination of the system's point source scatter response function for a standard attenuation object. The scatter distribution is assumed to depend only on the point source position on the selected projection plane and is usually modelled as a 2D Gaussian or a polynomial. This fast method explicitly accounts for the activity distribution. However, the depth-independent scatter response function is characterized for a specific object which may differ significantly from the object being imaged. Since the method does not account for activity outside the field of view (FOV), it is limited for 3D PET.

The second technique fits Gaussian [19, 20] or polynomial [5] functions to the lines of response (LORs) outside the object and works well for head-sized objects and with some success in an NaI(Tl)-based PET scanner for torso scans [5]. It has been less successful when applied to torso scans acquired using BGO PET systems [2] but the method is easy to implement; it can account for scatter from activity outside the field of view; and it does not require *a priori* information about the activity and attenuation distributions. Its main weakness is its estimation of the magnitude and shape of scatter from the scatter tails. Information in the tails may not be sufficient to account for complexity in the central part of the distribution.

Energy-based scatter corrections have been studied for SPECT imaging (see Chapter 13 and [21]). With the introduction of 3D PET, two dual-energy-window scatter correction techniques were proposed. In the first technique, data from a lower-energy window (200–380 keV) are used to estimate scatter in the photo-peak window (380–850 keV) [22]. In the second technique, data from an upper-energy window (550–850 keV), that are assumed to be scatter free, are used to estimate the contribution of trues in a wide photo-peak window (250–850 keV) [23]. Both techniques depend on an empirical factor relating the ratio between scatter and true events in each of the windows. A shortcoming of the first technique is that the spatial distribution of scatter in the lower window differs from the distribution of scatter in the photo-peak window [24, 25]. The shortcomings with the second method are limited statistics in the upper window and possible problems with pulse pileup and sensitivity to fluctuations in the detector/electronics. A triple-energy-window technique has also been proposed [26]. This technique works marginally better than the dual-energy-window method [22] when applied on data from an NaI(Tl)-based PET system. Holospectral (multiple-energy-window) techniques are currently under investigation [27, 28]. Energy-based techniques do not make any assumptions about the activity or attenuation distribution. Therefore, they can potentially correct for scatter from activity outside the FOV and for heterogeneous objects.

Analytic methods comprise the fourth class of scatter correction techniques [29–32]. Given the emission data and an accurate attenuation map, the single-scatter distribution can be calculated directly using the Klein–Nishina equation (1.7). The method does not require any empirical parameters and the correction does not add noise to the data. The limitation is that it can only account for scatter from activity within the tomograph's axial FOV. This technique is computationally demanding; however, methodologies have been proposed which make it feasible for commercial implementation [31, 32].

The fifth scatter correction technique requires a short auxiliary 2D scan [33]. The difference between normalized versions of the 2D and 3D scans is used to estimate the scatter in the 3D data. The requirement of an additional scan is a significant drawback for clinical use.

The final method is to use the Monte Carlo method to estimate scatter from the collected data [34]. This is the 'gold standard' for estimating scatter from activity within the FOV. Like the analytic methods this technique requires an accurate attenuation map and can only account for scatter from activity in the FOV. The present authors are unaware of any centres using full Monte Carlo simulations for scatter estimation. Tail fitting techniques provide acceptable performance for head-sized objects and the advantage the Monte Carlo method has over analytic methods for whole-body imaging is small compared to the large increase in computer processing required.

18.3 SCATTER IN 3D PET EMISSION IMAGING

Slice septa have traditionally been used to reduce scatter in PET. For axially distributed sources, scatter fraction increases approximately linearly with the axial angle of acceptance. Tungsten (or lead) septa as thin as 1.0 mm work, however, very well at rejecting scattered events. For a multislice tomograph with 8 mm axial detector spacing, the minimum path length through

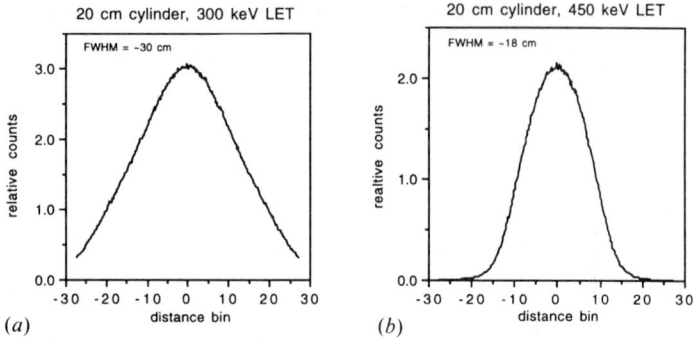

Figure 18.1 *Simulated scatter profiles of a point source centred in a water-filled 20 cm diameter by 40 cm long right circular cylinder, 300 keV LET (a) and 450 keV LET (b). Simulated tomograph geometry is similar to the Advance (GEMS) PET system. Events within the central third of the tomograph's axial FOV (5.2 cm) are included in the profile. Scatter profiles are summed over all angles.*

the septa (12 cm long, 1 mm thick) is 7.5 mm. This is approximately two attenuation path lengths in tungsten. Removing the septa increases the true coincidence sensitivity by a factor of six to seven. Unfortunately, removing the septa increases the sensitivity to scatter by factors of 20 or more. The scatter fraction† for a GE Advance PET system is approximately 10% for 2D imaging and approximately 34% for 3D imaging [1], following the PET NEMA protocol [8]. For 3D scans of the upper thorax or lower abdomen the scatter fraction can exceed 50%. For some LORs the scatter fraction can reach 75%. Badawi *et al* [35] report a scatter fraction of 50% for a 37 cm by 48 cm chest phantom using a 400 keV lower-energy-threshold (LET) setting on a Siemens ECAT 951R PET system.

By itself, scatter provides little information about the activity distribution and the attenuation object. This is different from SPECT, where scatter exhibits more structure. The scatter profile of a simulated point source centred in a water-filled 20 cm right circular cylinder is plotted in figure 18.1. The geometric parameters used for the simulation were similar to those of the Advance PET system. The full width at half maximum (FWHM) of the scatter profile is approximately 30 cm for a LET setting of 300 keV. Raising or lowering the LET changes the shape of the scatter profile; however, even with the LET raised to 450 keV the FWHM of the scatter response function is approximately 18 cm. In comparison, the intrinsic spatial resolution of current generation PET systems is 4–5 mm FWHM for reconstructed point or line sources [1, 5, 36]. Another characteristic of scatter in 3D PET is that scatter from off-centred sources does not align with the peak of the trues

†Scatter fraction is defined as scatter fraction = scatter/(trues + scatter).

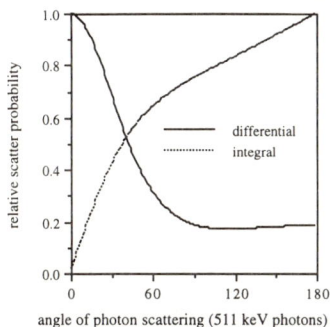

Figure 18.2   *The relative differential and integral scattering cross sections for 511 keV photons as a function of scattering angle.*

(illustrated in figure 18.10). This was demonstrated experimentally by Lercher *et al* [18] and using the Monte Carlo method by Barney *et al* [29] and Ollinger [31].

Scatter characteristics can be studied experimentally using simple phantoms; however, the Monte Carlo method is required to fully investigate scatter in realistic imaging environments. Investigators used the Monte Carlo method in the early 1980s to study the characteristics of scatter for single-slice tomographs [37–39]. Because of the higher scatter fractions, understanding scatter associated with 3D PET is even more critical for quantitative 3D imaging.

18.3.1 The effect of energy—thresholds and windows

Energy discrimination cannot completely eliminate scatter because of the finite energy resolution of all commercial PET detector systems. Energy settings of 380 and 450 keV place the energy threshold at the lower-energy tail of the photo-peak for detectors with 23% (BGO) and 10% (NaI(Tl)) energy resolution, respectively. A 511 keV photon can scatter 30° and still possess 450 keV of energy. The probability of scattering at a given angle can be determined using the Klein–Nishina equation (1.7). The relative differential and integral scattering cross sections are plotted as a function of scattering angle in figure 18.2. The differential cross-section plot illustrates that scatter is forward peaked for 511 keV photons. The integral cross-section plot shows that most 511 keV photons still possess over 400 keV of energy after a single Compton interaction.

The energy spectra for true and scattered coincidences from a uniformly filled 20 cm diameter cylindrical phantom have been simulated for both NaI(Tl)- and BGO-based 3D PET systems [40]. Plots of the energy spectra

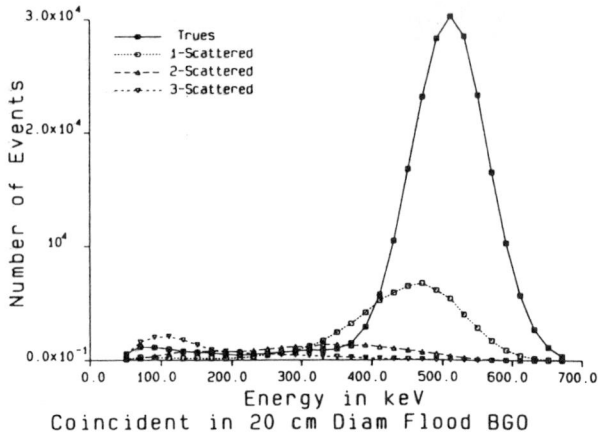

(a)

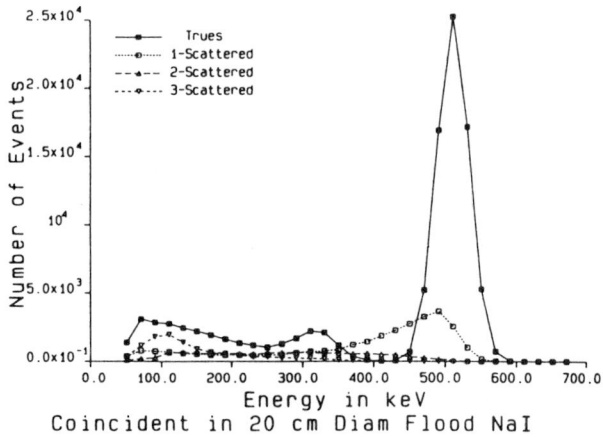

(b)

Figure 18.3   *The coincidence spectra from a 20 cm diameter flood source detected in a 50 cm diameter FOV by (a) BGO and (b) NaI(Tl) detector systems. Note that the ray from each coincident pair which scattered the highest number of times is used to make these spectra. (From Thompson C J 1993 The problem of scatter correction in PVI IEEE Trans. Med. Imaging MI-12 124–32. © 1993 IEEE.)*

are shown in figure 18.3. Most of the scattered coincidences have energies above the lower-energy tail of the true energy spectrum, 380 keV for BGO and 450 keV for NaI(Tl). One reason for this is that the tomograph endplates provide some collimation and discriminate against large-angle scatter. For a head-sized object, the scatter fraction is fairly insensitive to the LET setting once it is below approximately 380 keV. NEC measurements using the 20 cm PET performance phantom support this observation [1, 35]. This is not the

Table 18.1 *Scatter fraction versus LET for a 37 cm by 48 cm chest phantom (results taken from [35]).*

LET (keV)	Scatter fraction (%)
250	60
300	57
350	52
400	48
450	41

case for large torso-sized objects. Measured scatter fractions for a large torso phantom in multiple-energy windows are listed in table 18.1 [35]. Raising the LET from 250 to 400 keV reduces the scatter fraction from 60 to 48%. This translates into approximately 35% reduction in the absolute number of accepted scattered events. Work by others also supports the use of a high-energy threshold (400+ keV for BGO, 450 keV for NaI(Tl)) when imaging the upper thorax and abdomen [41, 42].

In a Monte Carlo simulation, Harrison *et al* [24, 25] have shown that scatter in a Compton window has a different spatial distribution than scatter in the photo-peak window. Scatter in a Compton window (200–400 keV) and a photo-peak (400–650 keV) window for a uniformly filled 20 cm diameter cylindrical phantom are shown in figure 18.4. Compton window events include both Compton–Compton and Compton–photo-peak coincidences.

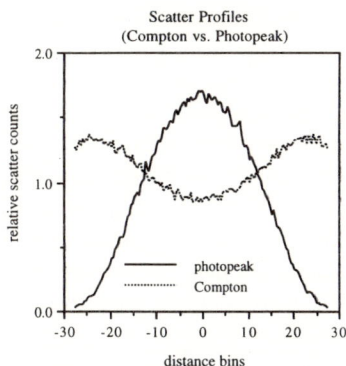

Figure 18.4 *Simulated scatter profiles for a 20 cm diameter uniformly filled right circular cylinder. The Compton window was 200–400 keV. The photo-peak window was 400–650 keV. Compton window events include both Compton–Compton and Compton–photo-peak coincidences. The majority of Compton window events are Compton–photo-peak coincidences.*

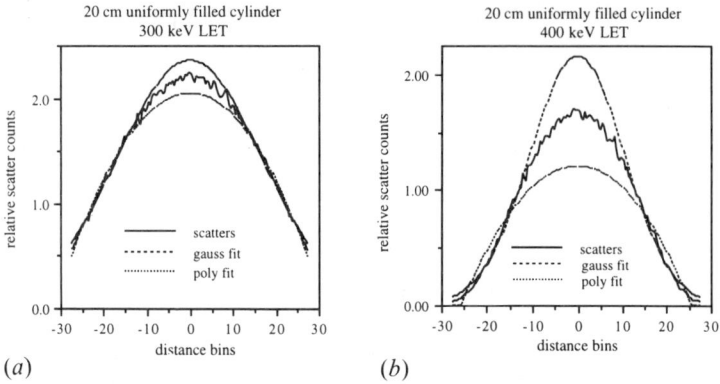

Figure 18.5 *Simulated scatter profiles, 300 keV LET (a) and 400 keV LET (b), for a 20 cm diameter uniformly filled right circular cylinder. The scatter profiles are estimated using Gaussian and polynomial function fits. Data from all LORs outside the object were used to fit the curves.*

Single scatter events in the Compton window have scattered through much larger angles than single scatter events in the photo-peak window. This leads to a broader spatial distribution of scattered events in the lower window. The dip in the window's scatter profile is a result of the Compton window's upper energy cut-off discriminating against small-angle scatters. This is in direct contrast to the LET for the photo-peak window, which discriminates against large-angle scatters but cannot discriminate against small-angle scatters. Because the spatial distribution of scatter varies with energy, scatter in one energy window cannot provide spatial information about scatter in a different window.

The LET setting affects the accuracy of a correction based on estimating the scatter profile from the scatter 'tails' (image background). Scatter profiles for a uniformly distributed source in 20 and 35 cm diameter cylindrical phantoms are illustrated in figures 18.5 and 18.6. Each of the profiles is fitted using a Gaussian and a second-order polynomial function. Data from all LORs that do not pass through the object are used to fit the curve. Using a Gaussian fit, scatter is overestimated in the central region of the phantom. A second-order polynomial fit underestimates scatter for the 20 cm phantom data, but overestimates scatter when applied to the 35 cm phantom data. Adjusting the region of the scatter 'tails' used to fit a function can improve the scatter estimate [5, 19, 20].

18.3.2 Single and multiple scatter

Experiments and simulations have been used to investigate the relationship between scatter and energy. Analytic scatter methods can only directly compute the distribution of single scatters. Understanding the magnitude and

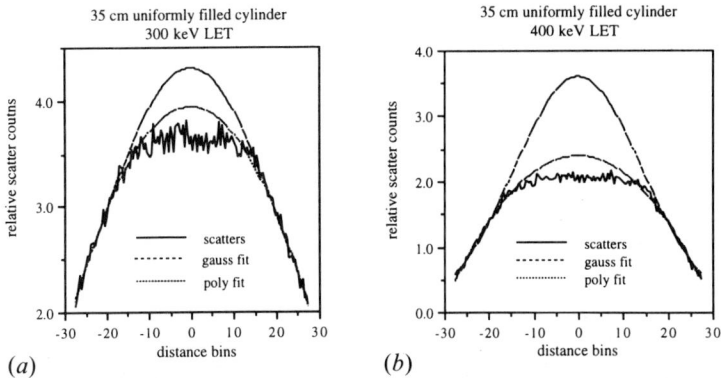

Figure 18.6 *Simulated scatter profiles, 300 keV LET (a) and 400 keV LET (b), for a 35 cm diameter uniformly filled right circular cylinder. The scatter profiles are estimated using Gaussian and polynomial function fits. Data from all LORs outside the object were used to fit the curves.*

spatial characteristics of multiple-scatter events is important in developing methods to account for these events when using analytic scatter corrections. Table 18.2 lists the percentage of single-interaction scatters for a point source in a 20 cm diameter and a 35 cm diameter water-filled cylinder. For the 20 cm diameter phantom the percentage of single scatters increases from 72 to 83% by raising the LET from 300 to 400 keV. The number of detected single scatter events in LORs that pass through the cylinder only decreases by 3%. Therefore, the increase in the percentage of single-scatter events is due almost entirely to the LET discriminating against multiple-scatter events.

Analytic scatter correction techniques require models to estimate the distribution of multiple-scatter events. Profiles of multiple-scatter coincidences versus single-scatter coincidences for uniformly filled 20 cm diameter and 35 cm diameter right circular cylinders are shown in figures 18.7 and 18.8. The results for the 35 cm phantom illustrate the importance of using a high

Table 18.2 *The percentage of single-scatter events (point source centred in water-filled right circular cylinder; only LORs which pass through object; BGO detector system).*

LET (keV)	20 cm diameter water-filled cylinder	35 cm diameter water-filled cylinder
300	72	51
350	77	60
400	83	70
450	89	79

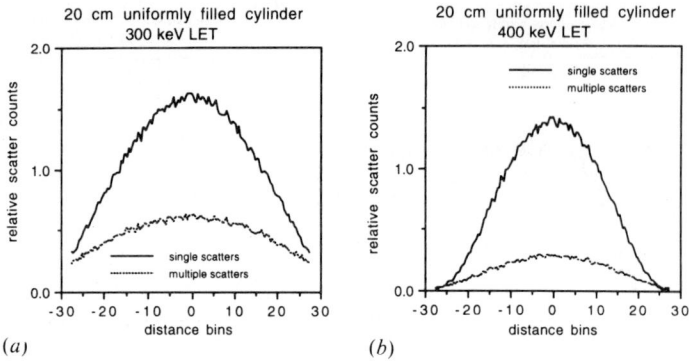

Figure 18.7 *Simulated scatter profiles, 300 keV LET (a) and 400 keV LET (b), for a 20 cm diameter uniformly filled right circular cylinder. Profiles for single-scatter and multiple-scatter events are plotted.*

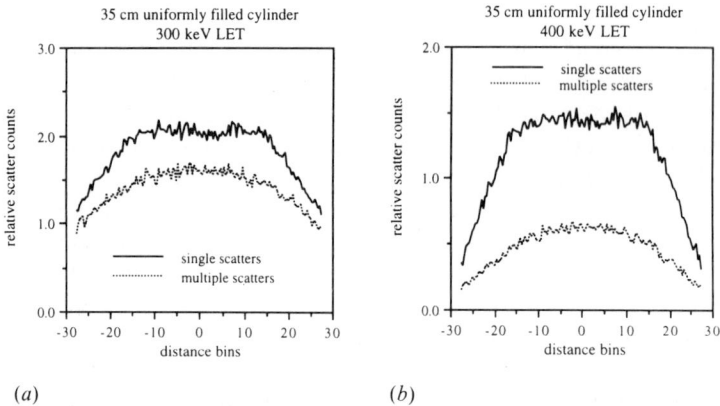

Figure 18.8 *Simulated scatter profiles, 300 keV LET (a) and 400 keV LET (b), for a 35 cm diameter uniformly filled right circular cylinder. Profiles for single-scatter and multiple-scatter events are plotted.*

LET to reduce the fraction of accepted multiple-scatter events when imaging large objects. The central region of the profiles of the single-scatter events is relatively flat for the 35 cm phantom but peaked for the 20 cm phantom. All of the profiles of the multiple-scatter events have a central peak and a broad distribution. The multiple-scatter distribution is well modelled as an integral transformation of the single-scatter distribution, where the magnitude of the kernel is object dependent [43].

18.3.3 Other considerations

The Monte Carlo method has been used to study other parameters affecting the characteristics of scatter. For example, the convolution subtraction

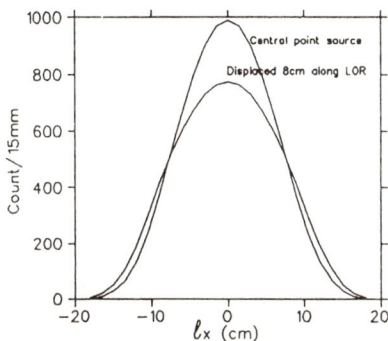

Figure 18.9 *Simulated profiles for a point source in the centre and displaced 8 cm along the central LOR of a 20 cm diameter water-filled cylinder (400 keV LET). (From Barney J S, Rogers J G, Harrop R and Hoverath H 1991 Object shape dependent scatter simulations for PET. IEEE Trans. Nucl. Sci. **NS-38** © 1991 IEEE.)*

techniques are based on the assumption that the scatter profile does not vary along the projection LOR or with object size. Barney *et al* used the Monte Carlo method to demonstrate the variation in scatter profiles of point sources at different depths along the same projection line in 3D PET [29] (see figure 18.9). The approximation that the scatter response is depth independent is acceptable for 2D PET, but will lead to more significant errors in 3D PET. Additionally, the scatter response varies significantly for

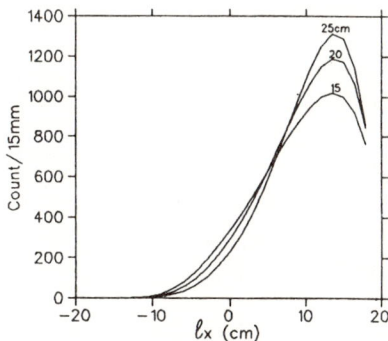

Figure 18.10 *Simulated profiles for a point source displaced 8 cm perpendicular to the LOR in water-filled right elliptical cylinders. The three cylinders have a 20 cm axis perpendicular to the LOR and 15, 20 or 25 cm axis lengths parallel to the LORs. Note that the peak of the scatter profile is shifted from the 8 cm axial displacement. (From Barney J S, Rogers J G, Harrop R and Hoverath H 1991 Object shape dependent scatter simulations for PET. IEEE Trans. Nucl. Sci. **NS-38** © 1991 IEEE.)*

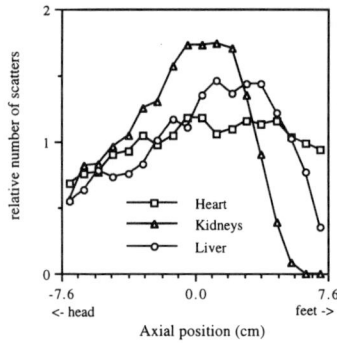

Figure 18.11 *Axial distribution of scatter by an organ in a simulated heart study. The values have been normalized to give an average value of one for each organ.*

right cylindrical ellipses of varying sizes, as shown in figure 18.10. Another characteristic of scatter in 3D PET is that for off-centred sources the peak of the scatter profile shifts with respect to the position of the source [18].

Analytic scatter corrections provide accurate estimates of the contribution of scatter from activity within the FOV. Using the Monte Carlo method, Harrison *et al* have shown that scatter from activity outside the field of view can affect quantitation in 3D PET [44] (figure 18.11). In some whole-body imaging procedures the liver, kidney or bladder can take up significant

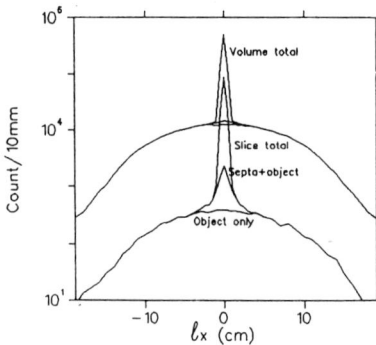

Figure 18.12 *Simulated profiles of an axial line source in a 20 cm diameter water-filled cylinder with septa in place (2D PET) and with septa retracted (3D PET). The lines represent all detected events; events scattered in the object plus septa and events scattered in the object only. (From Barney J S, Rogers J G, Harrop R and Hoverath H 1991 Object shape dependent scatter simulations for PET. IEEE Trans. Nucl. Sci.* **NS-38** *© 1991 IEEE.)*

amounts of activity relative to the objects of interest (e.g. primary or metastatic tumours). If these 'hot' organs reside just outside the tomograph's FOV, scatter from them can significantly impact the distribution of detected scatter. For these imaging situations endplate shielding, detector ring diameter and a good model for scatter from activity outside the FOV are important for accurate scatter correction.

The Monte Carlo method has also been used to discriminate between object scatter and scatter in the septa (for 2D PET), endplates, detectors and detector electronics (including the photomultiplier tubes) [29, 45]. Scatter in the slice septa makes a measurable contribution to the scatter under the true point response in 2D PET. In 3D PET the endplates contribute much less to the overall scatter fraction [29]. The contribution of septal and endplate scatter versus object scatter is shown in figure 18.12. Scatter in the detectors leads to a slight blurring of the intrinsic spatial resolution of a tomograph [46]; however, for current-generation commercial PET systems this contribution is minimal.

18.4 SCATTER IN TRANSMISSION IMAGING

Transmission imaging in PET has traditionally been performed with septa extended (2D) using rotating rod source(s) located just outside the FOV of the detector system. Scatter and random coincidences are discriminated against using a gating mask in the coincidence processing electronics [47–49]. An event is only accepted if the LOR passes through the transmission source. Before the use of a gating mask, scatter in coincidence transmission imaging affected the quantitative accuracy of the scan [50, 51]; however with the development of gated transmission imaging, the quantitative accuracy of transmission imaging is excellent.

The desire to build PET systems without septa (3D only PET systems), along with the desire to reduce transmission imaging times, has led to singles-based (noncoincidence) transmission imaging [52, 53]. Singles-based transmission imaging has a number of advantages over coincidence transmission imaging: (i) the counting rate is not limited by the near-detector dead time (which leads to a factor of five to seven in count rate capability); (ii) a ^{137}Cs source can be used instead of ^{68}Ge (^{137}Cs is less expensive than ^{68}Ge and has a longer half-life) and (iii) there are no random coincidences for a singles-based measurement. The main disadvantage is that a gating mask cannot be used to discriminate against scatter. Scatter in the transmission data set, as well as emission events from the object being imaged, leads to large errors in the measured attenuation path lengths [52]. Segmenting the attenuation image and substituting the appropriate linear attenuation coefficient for each tissue type has worked with some success [55, 56]. The main drawback to segmentation is that the segmentation procedure is not

Table 18.3   *Tomograph parameters.*

Effective ring diameter	94.2 cm
Number of crystals	672
Axial FOV	15.2 cm
Axial slice width	8.0 mm
Radial position of point source	31.25 cm
Energy resolution	20% (at 511 keV)
LET	variable
Lead endplates	1.2 cm × 12 cm

completely reliable and misclassification of single voxels can lead to image artifacts (hot or cold spots). In addition, motion artifacts caused by breathing, the heart motion and patient movement can lead to considerable partial-volume problems for attenuation imaging of the torso, especially the upper thorax. Finally it is difficult to assign an accurate attenuation coefficient value to lungs in a diseased state. Using an incorrect attenuation value will lead to bias and loss of quantitation in the reconstructed image. In general, segmented attenuation correction can lead to bias errors whereas measured attenuation correction factors amplify noise.

18.4.1 Experimental methods

The Monte Carlo method has been used to characterize scatter associated with singles-based transmission imaging [57, 58] from both 511 and 662 keV sources using the SimSET software (Chapter 7 and [59]). The tomograph parameters (table 18.3) used for the simulations were chosen to model the Advance PET system. The detectors were modelled as having perfect stopping power with 100% packing fraction. The effective ring diameter accounts for the average depth of interaction of a detected photon (the same value was used for both the 511 and 662 keV sources). Attenuation objects simulated include a 20 cm diameter right circular cylinder (70 cm long), a 32 cm by 24 cm right elliptical cylinder (70 cm long) and the upper thorax region of the voxel.man digital phantom [60]. The cylinders were centred in the tomograph. The voxel.man phantom was placed so that the heart and lungs were included in the middle slice of the tomograph.

The transmission source was simulated (511 or 662 keV single-photon point source) axially centred within the slice of interest and positioned at a radius of 31.25 cm. Simulations of the source at a single position were run for the right circular cylinder. Simulations for two source positions, 90° from each other (figure 18.13), were run for the elliptical cylinder and the voxel.man digital phantom. At each source position data were stored for events detected in the opposing 336 detectors (half the detector ring) directly opposite the point source. The LORs between the point source and detectors formed a fan. For each LOR events were binned according to the number

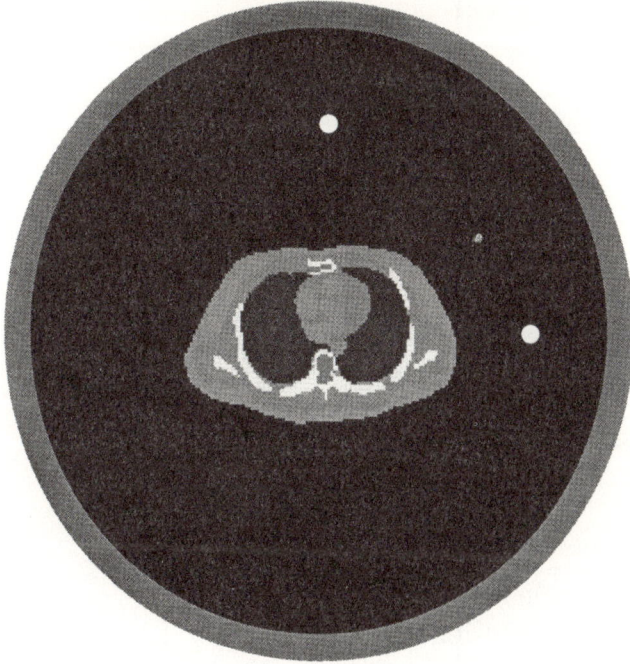

Figure 18.13 *The location of the singles transmission source for simulations.*

of scatters and energy. The weights of the detected events were normalized using a calculated blank scan. The scatter fraction was calculated only using the LORs which passed through the attenuation object.

18.4.2 The effect of energy—thresholds and windows

Table 18.4 lists the simulated scatter fractions as a function of LET for each of the attenuation objects. The scatter fraction at 511 keV is similar to the scatter fraction associated with 3D PET emission scans for similarly sized objects [2, 35]. The average scatter fraction in all of the simulations is less than 50%; however, for a 511 keV source and a 300 keV LET some LORs can have 2.5 times as many scatter events as true events. In all cases, for similar relative LETs (number of FWHMs of the detector's energy resolution from the photo-peak), the scatter fraction is lower for the 662 keV source than for the 511 keV source.

The scatter distribution (662 keV source, 550 keV LET) for the top and side views of the voxel.man digital phantom are shown in figure 18.14. Two observations can be made about scatter from the top view of the voxel.man phantom. First, while the distribution of trues is very structured, the scatter

Table 18.4 *Scatter fraction: $S/(T+S)$ (%).*

LET (keV)	20 cm cylinder average	20 cm cylinder max. LOR	32 cm × 24 cm ellipse short axis average	32 cm × 24 cm ellipse short axis max. LOR	32 cm × 24 cm ellipse long axis average	32 cm × 24 cm ellipse long axis max. LOR	voxel.man short axis average	voxel.man short axis max. LOR	voxel.man long axis average	voxel.man long axis max. LOR
300[a]	30	47	41	61	37	66	45	72	43	76
375[a]	25	42	34	54	30	61	36	65	36	69
425[a]	19	35	26	45	23	53	28	55	28	61
450[b]	16	32	21	40	19	47	24	48	23	55
500[b]	21	36	27	46	27	53	31	55	31	62
550[b]	17	30	21	38	21	45	24	46	25	53
600[b]	12	24	15	28	15	37	17	35	18	43

[a] 511 keV source.
[b] 662 keV source.

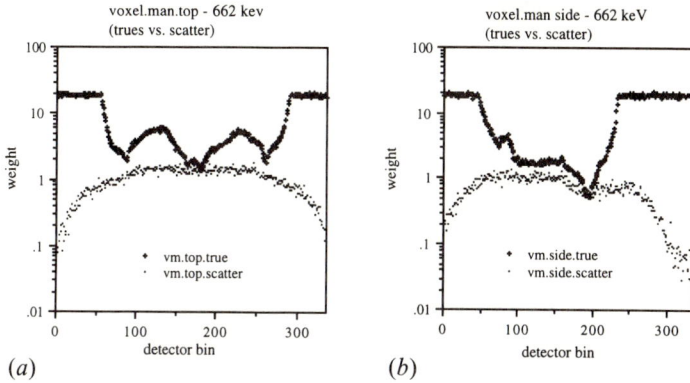

Figure 18.14 *Simulated profiles of true and scatter profiles for the top view (a) and side view (b) of the voxel.man phantom (662 keV transmission source and 550 keV LET).*

distribution is very smooth. Second, unlike emission imaging, where there are no true events in LORs outside the object, almost all of the detected transmission events in LORs outside the object are unscattered (i.e. true events). The side view of the voxel.man phantom shows a slight dip in the scatter distribution, corresponding to the longest attenuation paths through the object. Similar structure is seen along the longest path lengths through the 32 cm by 24 cm right elliptical phantom (not shown). These results elucidate the relationship between scatter and the attenuation object. Our simulations seem to indicate that the scatter response begins to dip along a LOR when the attenuation correction factor for a LOR is greater than approximately 20.

The scatter distributions (662 keV source) for a 300–550 keV Compton window and a 550–800 keV photo-peak window are shown in figure 18.15 for the top and side views of the voxel.man digital phantom. The vertical lines on the plots represent the LORs at the object boundaries. The shape of the scatter in the Compton window differs significantly from that in the photo-peak window. In addition, the characteristic differences between the scatter windows are radically different for the two views. Similar results can be attained for a 200–375 keV Compton window and a 375–650 keV photopeak window using a 511 keV source (not shown).

18.4.3 Single and multiple scatter

The number of scatter interactions versus LET results for each of the attenuation objects are listed in table 18.5. The ratios of single-scatter events versus multiscatter events are determined only using LORs that pass through the attenuation object. If the LET is kept above 375 keV at least 74% of the detected scatter events Compton scatter only once. In all cases, the fraction

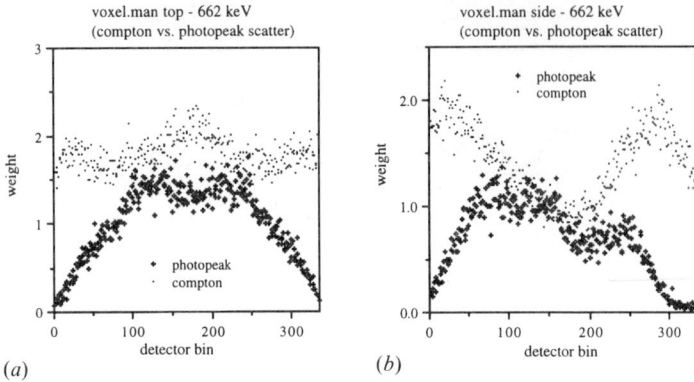

Figure 18.15 *Simulated scatter profiles for top (a) and side (b) views of the voxel.man phantom (662 keV photons). The Compton window was 300–550 keV. The photo-peak window was 550–800 keV.*

of single Compton events increases as the LET is raised. For similar relative LETs (percentage of source photon energy) the fraction of single interaction events using the 662 keV source always exceeds that of the 511 keV source.

The spatial distribution for single- and multiscatter events (662 keV source, 550 keV LET) is shown for the two views of the voxel man phantom in figure 18.16. The plots illustrate that the single-scatter events contain some spatial information. The apparent dip in both plots corresponds to the LORs with the largest attenuation correction factors. While there appears to be little if any spatial information provided by the multiscatter events, the multiscatter events might be modelled as a smoothed and scaled version of the single-scatter events.

18.5 CONCLUSION

This chapter has focused on how Monte Carlo simulation has been used to study scatter associated with 3D PET imaging. Because of the finite energy resolution of PET detector systems, it is difficult to discriminate between true and scatter events. Most experimental investigations of scatter have been limited to homogeneous and symmetric phantoms. The development of efficient Monte Carlo simulation packages has allowed the study of scatter for realistic activity distributions and attenuation maps and according to a variety of physical parameters. Monte Carlo investigations have been used to study the spatial characteristics of scatter versus energy. In addition, the Monte Carlo method is uniquely able to provide information about multiple- versus single-scatter events. Knowledge gained from Monte Carlo investigations has contributed to the formulation and evaluation of new scatter correction techniques.

Table 18.5 *Percentage of single scatters:* S_1/S_{tot} (%).

LET (keV)	20 cm cylinder average	20 cm cylinder max. LOR	32 cm × 24 cm ellipse short axis average	32 cm × 24 cm ellipse short axis min. LOR	32 cm × 24 cm ellipse long axis average	32 cm × 24 cm ellipse long axis min. LOR	voxel.man short axis average	voxel.man short axis min. LOR	voxel.man long axis average	voxel.man long axis min. LOR
300[a]	74	62	69	60	65	53	69	64	65	54
375[a]	82	71	78	70	74	62	78	72	74	64
425[a]	87	72	85	74	81	69	84	79	81	70
450[b]	89	73	87	75	85	71	87	80	84	72
500[b]	86	74	83	75	80	68	83	77	80	68
550[b]	90	75	88	78	85	73	88	80	85	71
600[b]	93	75	91	79	89	76	91	82	89	75

[a] 511 keV source.
[b] 662 keV source.

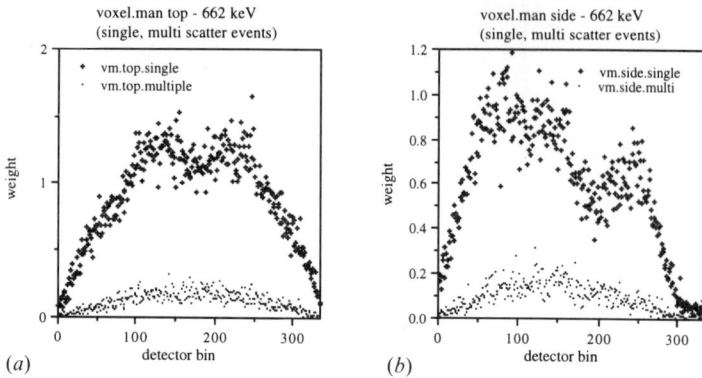

Figure 18.16 *Simulated scatter profiles for top (a) and side (b) views of voxel.man phantom (662 keV photons, 550 keV LET). Profiles for single-scatter and multiple-scatter events are plotted.*

ACKNOWLEDGMENTS

Work by the authors presented in this chapter was supported by PHS grant CA42593. The authors would also like to thank Ms Wendy L Swan, MS, for assistance in editing the text.

REFERENCES

[1] DeGrado T R, Turkington T G, Williams J J, Stearns C W, Hoffman J M and Coleman R E 1994 Performance characteristics of a whole-body PET scanner *J. Nucl. Med.* **35** 1398–406
[2] Lewellen T K, Kohlmyer S G, Miyaoka R S, Kaplan M S, Stearns C W and Schubert S F 1996 Investigation of the performance of the General Electric ADVANCE positron emission tomograph in 3D mode *IEEE Trans. Nucl. Sci.* **NS-43** 2199–206
[3] Spinks T J, Jones T, Bailey D L, Townsend D W, Grootoonk S, Bloomfield P M, Gilardi M-C, Casey M E, Sipe B and Reed J 1992 Physical performance of a positron tomograph for brain imaging with retractable septa *Phys. Med. Biol.* **37** 1637–55
[4] Cherry S, Shao Y, Siegel S, Silverman R W, Mumcuoglu E, Meadors K and Phelps M E 1996 Optical fiber readout of scintillator arrays using a multichannel PMT: a high resolution PET detector for animal imaging *IEEE Trans. Nucl. Sci.* **NS-43** 1932–7
[5] Karp J S, Muehllehner G, Mankoff D A, Ordonez C E, Ollinger J M, Daube-Witherspoon M E, Haigh A and Beerbohm D 1990 Continuous-slice PENN-PET: a positron tomograph with volume imaging capability *J. Nucl. Med.* **31** 617–27
[6] Townsend D W, Wensveen M, Byars L G, Geissbuhler A, Tochon-Danguy H J, Christin A, Defrise M, Bailey D L, Grootoonk S, Donath A and Nutt R 1993 A rotating PET scanner using BGO block detectors: design, performance and applications *J. Nucl. Med.* **34** 1367–76

[7] Strothers S C, Casey M E and Hoffman E J 1990 Measuring PET scanner sensitivity: relating count rates to image signal-to-noise ratios using noise equivalent counts *IEEE Trans. Nucl. Sci.* **NS-37** 783–8

[8] Karp J S *et al* 1991 Performance standards in positron emission tomography *J. Nucl. Med.* **32** 2342–50

[9] Casey M E and Hoffman E J 1986 Quantitation in positron emission computed tomography: a technique to reduce noise in accidental coincidence measurements and coincidence efficiency calibration *J. Comput. Assist. Tomogr.* **10** 845–50

[10] Knoll G F 1989 *Radiation Detection and Measurement* (New York: Wiley) pp 636–8

[11] Kinahan P E, Matej S, Karp J S, Herman G T and Lewitt R M 1995 A comparison of transform and iterative reconstruction techniques for a volume-imaging PET scanner with a large axial acceptance angle *IEEE Trans. Nucl. Sci.* **NS-42** 2281–7

[12] King P H, Hubner K, Gibbs W and Holloway E 1981 Noise identification and removal in positron imaging systems *IEEE Trans. Nucl. Sci.* **28** 148–50

[13] Bergstrom M, Eriksson L, Bohm C, Blomqvist G and Litton J 1983 Correction for scattered radiation in a ring detector positron camera by integral transformation of the projections *J. Comput. Assist. Tomogr.* **7** 42–50

[14] Shao L and Karp J S 1991 Cross-plane scatter correction-point source deconvolution in PET *IEEE Trans. Med. Imaging* **MI-10** 234–9

[15] Townsend D W, Geissbuhler A, Defrise M, Hoffman E J, Spinks T J, Bailey D L and Gilardi M C 1991 Fully three-dimensional reconstruction for a PET camera with retractable septa *IEEE Trans. Med. Imaging* **MI-10** 499–504

[16] McKee B T, Gurvey A T, Harvey P J and Howse D C 1992 A deconvolution scatter correction for a 3D PET system *IEEE Trans. Med. Imaging* **MI-11** 560–9

[17] Bailey D L and Meikle S R 1993 A convolution-subtraction scatter correction method for 3D PET *Phys. Med. Biol.* **39** 411–24

[18] Lercher M H and Wienhard K 1994 Scatter correction in 3D PET *IEEE Trans. Med. Imaging* **MI-13** 649–57

[19] Cherry S R and Huang S-C 1995 Effects of scatter on model parameter estimates in 3D PET studies of the human brain *IEEE Trans. Nucl. Sci.* **NS-42** 1174–9

[20] Stearns C W 1995 Scatter correction method for 3D PET using 2D fitted Gaussian functions *J. Nucl. Med.* **36** 105P

[21] Luo J-Q, Koral K F, Ljungberg M, Floyd C E and Jaszczak R J 1995 A Monte Carlo investigation of dual-energy-window scatter correction for volume-of-interest quantification in ^{99m}Tc SPECT *Phys. Med. Biol.* **40** 181–99

[22] Grootoonk S, Spinks T J, Jones T, Michel C and Bol A 1991 Correction for scatter using a dual-energy-window technique with a tomograph operating without septa *Proc. IEEE Nuclear Science Symp. Medical Imaging Conf. (Sante Fe, NM, 1991)* (New York: IEEE) pp 1569–73

[23] Bendriem B, Trebossen R, Frouin V and Syrota A 1993 A PET scatter correction using simultaneous acquisitions with low and high lower energy thresholds *Proc. IEEE Nuclear Science Symp. Medical Imaging Conf. (San Francisco, CA, 1993)* (New York: IEEE) pp 1779–83

[24] Harrison R L, Haynor D R and Lewellen T K 1991 Dual-energy-window scatter corrections for positron emission tomography *Proc. IEEE Nuclear*

Science Symp. Medical Imaging Conf. (Santa Fe, NM, 1991) (New York: IEEE) pp 1700–4

[25] Harrison R L, Haynor D R and Lewellen T K 1992 Limitations of energy-based scatter corrections for quantitative PET *Proc. IEEE Nuclear Science Symp. Medical Imaging Conf. (Orlando, FL, 1992)* (New York: IEEE) pp 862–4

[26] Shao L, Freifelder R and Karp J S 1994 Triple-energy-window scatter correction technique in PET *IEEE Trans. Med. Imaging* **MI-13** 641–8

[27] Bentourkia M, Msaki P, Cadorette J and Lecomte R 1995 Energy dependence of scatter components in multispectral PET imaging *IEEE Trans. Med. Imaging* **MI-14** 138–45

[28] Haynor D R, Kaplan M S, Miyaoka R S and Lewellen T K 1995 Multiwindow scatter correction techniques in single-photon imaging *Med. Phys.* **22** 5–24

[29] Barney J S, Rogers J G, Harrop R and Hoverath H 1991 Object shape dependent scatter simulations for PET *IEEE Trans. Nucl. Sci.* **NS-38** 719–25

[30] Hiltz L G and McKee B T A 1994 Scatter correction for three-dimensional PET based on an analytic model dependent on source and attenuating object *Phys. Med. Biol.* **39** 2059–71

[31] Ollinger J M 1996 Model-based scatter correction for fully 3D PET *Phys. Med. Biol.* **41** 153–76

[32] Watson C C, Newport D and Casey M E 1996 A single scatter simulation technique for scatter correction in 3D PET *Three-dimensional Image Reconstruction in Radiology and Nuclear Medicine* ed P Grangeat and J L Amans (Dordrecht: Kluwer) pp 255–68

[33] Cherry S R, Meikle S R and Hoffman E J 1993 Correction and characterization of scattered events in three dimensional PET using scanner with retractable septa *J. Nucl. Med.* **34** 671–8

[34] Levin C S, Dahlbom M and Hoffman E J 1995 A Monte Carlo correction for the effect of Compton scattering in 3D PET brain imaging *IEEE Trans. Nucl. Sci.* **NS-42** 1181–5

[35] Badawi R D, Marsden P K, Cronin B F, Sutcliffe J L and Maisey M N 1996 Optimization of noise-equivalent count rates in 3D PET *Phys. Med. Biol.* **41** 1755–76

[36] Weinhard K, Dahlbom M, Eriksson L, Michel C, Bruckbauer T, Pietrzyk U and Heiss W D 1994 The ECAT EXACT HR: performance of a new high resolution positron scanner *J. Comput. Assist. Tomogr.* **18** 110–8

[37] Bruno M F 1983 The evaluation of errors due to Compton scattering in gamma-ray emission imaging *Master's Thesis* University of California at Berkeley

[38] Logan J and Bernstein H J 1983 A Monte Carlo simulation of Compton scattering in positron emission tomography *J. Comput. Assist. Tomogr.* **7** 316–20

[39] Lupton L R and Keller N A 1983 Performance study of single-slice positron emission tomography scanners by Monte Carlo techniques *IEEE Trans. Med. Imaging* **MI-2** 154–68

[40] Thompson C J 1993 The problem of scatter correction in positron volume imaging *IEEE Trans. Med. Imaging* **MI-12** 124–32

[41] Smith R J, Karp J S and Muehllehner G 1994 The count rate performance of the volume imaging PENN-PET scanner *IEEE Trans. Med. Imaging* **MI-13** 610–8

[42] Ollinger J M 1996 The effect of energy threshold on image variance in fully 3D PET *Three-dimensional Image Reconstruction in Radiology and Nuclear*

Medicine ed P Grangeat and J L Amans (Dordrecht: Kluwer) pp 269–76

[43] Goggin A S and Ollinger J M 1994 A model for multiple scatters in fully 3D PET *Proc. IEEE Nuclear Science Symp. Medical Imaging Conf. (Norfolk, VA, 1994)* (New York: IEEE) pp 1609–13

[44] Harrison R L, Vannoy S D, Kohlmyer S, Sossi V and Lewellen T K 1994 The effect of scatter on quantitation in positron volume imaging of the thorax *Proc. IEEE Nuclear Science Symp. Medical Imaging Conf. (Norfolk, VA, 1994)* (New York: IEEE) pp 1335–8

[45] Adam L-E, Bellemann M E, Brix G and Lorenz W J 1996 Monte Carlo-based analysis of PET scatter components *J. Nucl. Med.* **37** 2024–9

[46] Bentourkia M, Msaki P, Cadorette J and Lecomte R 1996 Nonstationary scatter subtraction-restoration in high-resolution PET *J. Nucl. Med.* **37** 2040–6

[47] Thompson C J, Dagher A, Lunney D N, Strother S C and Evans A C 1986 A technique to reject scattered radiation in PET transmission scans *Proc. Int. Workshop on Physics and Engineering of Computerized Multidimensional Imaging and Processing (Newport Beach, 1986)* (Bellingham, WA: SPIE) vol 671, pp 244–53

[48] Huesman R, Derenzo S E, Cahoon J L, Geyer A B, Moses W W, Uber D C, Vuletich T and Budinger T F 1988 Orbiting transmission source for positron tomography *IEEE Trans. Nucl. Sci.* **NS-35** 735–9

[49] Kubler W, Ostertag H, Hoverath H, Doll J, Ziegler S and Lorenz W J 1988 Scatter suppression by using a rotating pin source in PET transmission measurements *IEEE Trans. Nucl. Sci.* **NS-35** 749–52

[50] Daube-Witherspoon M, Carson R E and Green M V 1988 Post-injection transmission attenuation measurements for PET *IEEE Trans. Nucl. Sci.* **NS-35** 757–61

[51] Digby W M and Hoffman E J 1989 An investigation of scatter in attenuation correction *IEEE Trans. Nucl. Sci.* **NS-36** 1038–42

[52] deKemp R A and Nahmias C 1994 Attenuation correction in PET using single photon transmission measurement *Med. Phys.* **21** 771–8

[53] Jones W, Vaigneur K, Young J, Reed J, Moyers C and Nahmias C 1995 The architectural impact of single photon transmission measurements on full ring 3D positron tomography *Proc. IEEE Nucl. Science Symp. Medical Imaging Conf. (San Francisco, CA, 1995)* (Piscataway, NJ: IEEE) pp 1026–30

[54] Karp J S, Muehllehner G, He Q and Xiao H-Y 1995 Singles transmission in volume-imaging PET with a ^{137}Cs source *Phys. Med. Biol.* **40** 929–44

[55] Tai Y C, Lin K P, Dahlbom M and Hoffman E J 1996 A hybrid attenuation correction technique to compensate for lung density in 3D total body PET *IEEE Trans. Nucl. Sci.* **NS-43** 323–30

[56] Xu M, Luk W K, Cutler P D and Digby W M 1994 Local threshold for segmented attenuation correction of PET imaging of the thorax *IEEE Trans. Nucl. Sci.* **NS-41** 1523–7

[57] Miyaoka R S, Costa W L S, Harrison R L and Lewellen T K 1995 Investigation of scatter in singles-based PVI transmission imaging *Proc. IEEE Nucl. Science Symp. Medical Imaging Conf. (San Francisco, CA, 1995)* (Piscataway, NJ: IEEE) pp 1771–5

[58] Miyaoka R S, Costa W L S, Harrison R L and Lewellen T K 1996 Scatter in singles-based PVI transmission imaging *J. Nucl. Med.* **37** 172P

[59] Harrison R L, Vannoy S V, Haynor D R, Gillispie S B, Kaplan M S and Lewellen T K 1993 Preliminary experience with the photon history generator module of a public-domain simulation system for emission tomography

Proc. IEEE Nuclear Science Symp. Medical Imaging Conf. (San Francisco, CA, 1993) (New York: IEEE) pp 1154–8

[60] Zubal G, Gindi G, Lee M, Harrell C and Smith E 1990 High resolution anthropomorphic phantom for Monte Carlo analysis of internal radiation sources *Proc. 3rd Ann. IEEE Symp. on Comp.-Based Med. Sys. (Chapel Hill, NC, 1990)* (Los Alamitos, CA: IEEE) pp 540–7

CHAPTER 19

APPLICATION OF MONTE CARLO METHODS IN 3D PET DESIGN

Magnus Dahlbom and Lars Eriksson

19.1 INTRODUCTION

To fully utilize the potential of positron emission tomography (PET), an instrument with both high resolution and sensitivity is required. With current detector technology based on bismuth germanate (BGO) block detectors [1–3], a spatial resolution better than 3.5 mm in all spatial dimensions is achievable. However, to attain this high resolution, very small detector elements are required (≈ 3 mm). At this resolution the sensitivity of the instrument is very poor when conventional two-dimensional (2D) acquisition and reconstruction techniques are used which results in images with very high levels of noise. The main reason for this is the low count rate in the individual detector channel. Since only a fraction of all available coincidence planes are used in 2D, one can improve the overall sensitivity by collecting all possible combinations. This requires that the scatter reducing inter-plane septa, located between the detector rings, are removed [4, 5]. This mode of operation is generally referred to as 3D PET. The necessity of improving detection efficiency on these high-resolution systems has been recognized by the PET manufacturers, who have implemented retractable septa on their latest-generation systems to allow 3D acquisitions in addition to the conventional mode of operation. For instance, the Siemens/CTI ECAT EXACT and ECAT EXACT HR [6, 7] are both 24-detector-ring systems with retractable septa and a 15–16 cm axial field of view (FOV). Another example is the GE Advance [8] with 18 detector rings and 15 cm axial FOV. These systems all allow the entire brain or the heart to be studied simultaneously with the option to use a full 3D acquisition mode of operation.

When the septa are retracted and all lines of response are acquired, the sensitivity is improved by a factor of six to seven compared to 2D [6, 7, 9, 10]. In addition to the increased sensitivity, the scatter also increases by a factor of three to four; thus, some sort of scatter correction of the data is required. Furthermore, the increase in sensitivity is not supported by an equal increase in the speed of the data acquisition and processing system. Dead-time problems start to become severe at administered doses three to five times lower than previously met with 2D data acquisition techniques [9, 11]. Data losses in current PET systems mainly originate from the signal processing in the block detectors and from the transfer and storage of coincidence data. The contribution to the system dead time by the block detector may be reduced by using smaller detector blocks with a smaller solid angle to reduce the singles count rate [12]. Dead-time contributions due to data processing are elevated in 3D mode, caused by an order of magnitude increase in data flow. The ECAT EXACT HR system, for instance, requires 6.2 Mbyte of data storage of the 47 sinograms produced when operating in the conventional 2D mode. The full 3D mode of operation requires more than ten times more storage space, 76 Mbyte of data [7].

In this chapter, the problem of scatter and count rate performance is studied in a high-resolution PET configuration with a large axial field of view (FOV), designed mainly, but not exclusively, for brain activation studies. The aim of the study is twofold: first, to find a optimal system geometry which provides the best signal-to-noise ratio in the final images; second, to identify and possibly also solve potential problems in the design of the next-generation high-resolution PET system. The starting point for this study is the use of Monte Carlo simulations to estimate true (unscattered), scattered, random and singles events for different system geometries. These numbers are used both to study the spatial distribution of the events for a particular system geometry and also to provide the input for calculating the relative count rate contribution at different activity concentrations.

19.2 THE BLOCK DETECTOR DESIGN

The proposed block detector design used in this simulation study is shown in figure 19.1. It consists of a $25 \times 25 \times 30$ mm^3 BGO block into which an 8×8 detector matrix is cut, resulting in 64 $3 \times 3 \times 30$ mm^3 crystal elements. The BGO block is coupled to four photomultiplier tubes (PMTs), such as the Hamamatsu tube R2102, a square-shaped PMT with a 12.5×12.5 mm^2 window area (Hamamatsu Photonics KK, Japan). It is assumed that the identification capability is precise enough to provide an intrinsic resolution of at least 3.3 mm full width at half maximum (FWHM) at a detector separation of 80 cm. The separations between the individual crystals are assumed to be small enough to provide a crystal–crystal distance of 3.15 mm. This would provide a linear sampling distance of 1.58, which fairly well matches the intrinsic resolution of 3.2 mm and should yield images free of linear sampling artifacts.

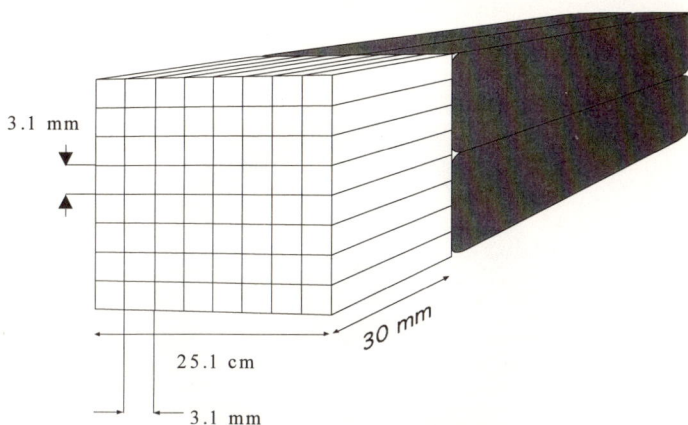

Figure 19.1 *A schematic diagram of the block detector design in the proposed high-resolution PET system. Each crystal element is $3 \times 3 \times 30$ mm³ and they are separated by a 0.1 mm gap.*

19.3 GEOMETRICAL DESIGN OF THE POSITRON CAMERA SYSTEMS

Only a cylindrical system geometry is considered, with six different diameters between 50 and 100 cm and five different axial FOVs between 10 and 50 cm. A cylindrical FOV of 25 cm in diameter is chosen, thus mainly focusing on brain applications. To find the number of required detector combinations to cover a 25 cm diameter transaxial FOV, the view angle α is first determined, with which the 25 cm cylindrical FOV will project itself onto the periphery of the detector ring seen from the side of one of the detectors of the ring (figure 19.2). The ratio

$$f_\alpha = \frac{2\alpha}{360°} \qquad (19.1)$$

gives the fraction of detectors, f_α, with which a given detector must be able to record coincidences to yield a 25 cm FOV. If the number of detectors per detector ring is denoted by N, the number of actual detectors to cover the FOV is

$$N_\alpha = f_\alpha N. \qquad (19.2)$$

The number of coincidence combinations per detector ring is then

$$N_{\text{comp}} = \frac{f_\alpha N^2}{2}. \qquad (19.3)$$

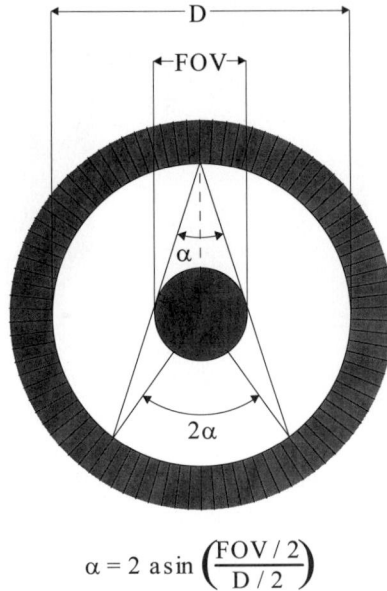

$$\alpha = 2 \, a\sin \left(\frac{FOV/2}{D/2} \right)$$

Figure 19.2 *An illustration of the relationship between the system diameter D, the FOV and the view angle α.*

Table 19.1 gives the N_α, N and N_{comb} for the six different ring diameters considered in this study. The main reason for the dramatic increase in data flow in 3D PET is the quadratic relationship between the number of detector plane combinations and the axial FOV. In table 19.2 the number of possible sinogram combinations for the axial FOVs of the considered PET system geometries is shown.

Table 19.1 *The relationship between ring diameters, number of blocks per block ring, number of detectors per detector ring, number of detectors within the selected viewing angle α relevant for a 25 cm FOV and total number of detector combinations per ring (sinogram). If one byte is assigned to each detector combination, the sinogram size is then N_{comb} bytes.*

Ring diameter (cm)	Actual diameter (cm)	No of blocks per block ring	N (detectors/ ring)	N_α (detectors within the peripheral angle α)	N_{comb} (detector combinations/ ring)
50	51.0	64	512	170	43 520
60	60.5	76	608	164	49 856
70	70.0	88	704	164	57 728
80	79.6	100	800	162	64 800
90	90.7	114	912	162	73 872
100	100.3	126	1008	162	81 648

Table 19.2 *The axial extension is increased by adding more block rings. The table gives the relationship between the axial dimensions, the number of block rings and the resulting number of sinograms.*

Axial dimension (cm)	Number of block rings	Number of detector rings	$N_{\text{sinograms}}$
10	4	32	1 024
20	8	64	4 096
30	12	96	9 216
40	16	128	16 384
50	20	160	25 600

19.4 MONTE CARLO SIMULATIONS

The comparison of the performance of the different system geometries is based on Monte Carlo simulation of the true and scatter distribution [3, 4, 13]. In the simulations of the system geometries described above, a 20 cm diameter sphere with uniform activity served as the only source of the annihilation events and as the only scattering medium ($\mu = 0.095$ cm^{-1}). The simulations were performed by generating a large number of annihilation events throughout the phantom, all emitted isotropically. Each photon was followed until fully absorbed in the phantom or hitting the detector assembly. If the photon reached the detector assembly the event was recorded if the energy was above 250 keV, thus assuming a 100% detection efficiency. Information was recorded whether one or both of the two annihilation photons had scattered prior to being detected. Sinograms for each detector ring combination were created into which all recorded events were sorted and histogrammed, where the true (i.e. unscattered) and scattered events were sorted separately. The storage of the events in sinograms allows analysis of the spatial distribution of the scatter. In addition to the coincidence events, the number of photons hitting each detector element was also recorded (i.e. single events). This allows the randoms count rate to be estimated from the product of the singles rates and the coincidence time window. All simulations were performed without the presence of inter-plane septa (i.e. 3D mode).

The Monte Carlo program was validated by comparing data acquired on the ECAT 831 [9, 14] and ECAT EXACT/HR [7] PET systems, both operated in 3D mode, with simulations of these two system geometries. In the measurements a standard 20 cm diameter and 20 cm tall uniform cylinder phantom filled with ^{68}Ge was used. Since the actual PET systems only provide measurements of trues + scatter, the validation of the simulations was performed by comparing the measured and simulated trues + scatter profiles. Figure 19.3 shows the comparison of the measured and simulated sinogram projections for the 20 cm diameter cylinder phantom. In order to improve statistics in the measured profiles, all angles in the sinogram were

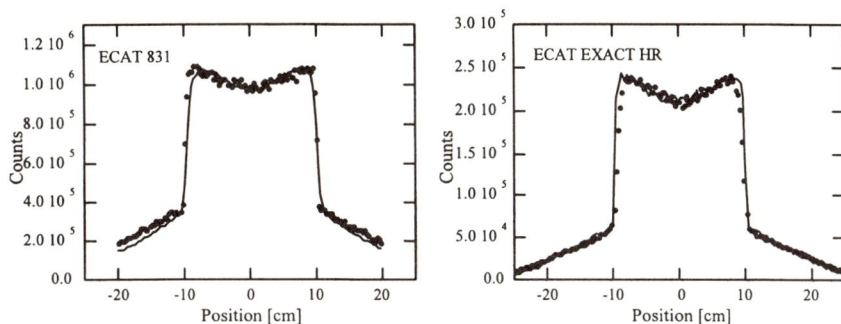

Figure 19.3   *A comparison of the Monte Carlo simulations (●) and the measurements (——) on the ECAT 831 (left) and the ECAT EXACT HR (right).*

added together. To adjust for any positioning offsets from the centre of the FOV, all sinogram profiles were realigned to a common central point prior to the angular summation. No normalization was applied to the measured data. The curves in figure 19.3 show a very good agreement between the measured and simulated data; however, there is a slight deviation in the scatter tails. This is most likely due to the fact that in the simulation the energy threshold is set at a well defined energy, whereas in the measurements the threshold for each detector channel is somewhat diffuse [7]. Another factor that is not included in the simulation is the geometrical detector efficiency profile, which is not identical in the measurement and the simulation.

19.5 SCATTER DISTRIBUTION

The 2D scatter distribution for the 20 cm spherical phantom for a given plane difference was generated from the sinogram data. Since all simulations were performed for a sphere placed at the centre of the system, all projection views for a given coincidence plane combination are identical. This rotational symmetry was used to improve the counting statistics, by adding all projection views in the sinogram. The 2D projection view of the object can then be formed by combining the 1D projection views from the axial coincidence planes. In the left-hand column of figure 19.4, the trues + scatter distributions are shown for the 50, 80 and 100 cm diameter systems. For this source distribution, it was found that the scatter component could fairly accurately be described by a 2D Gaussian distribution

$$s(x, y, z, \theta) = p_1(\theta)\, e^{-(p_2(\theta)x^2 + p_3(\theta)z^2)} + p_4(\theta) \qquad (19.4)$$

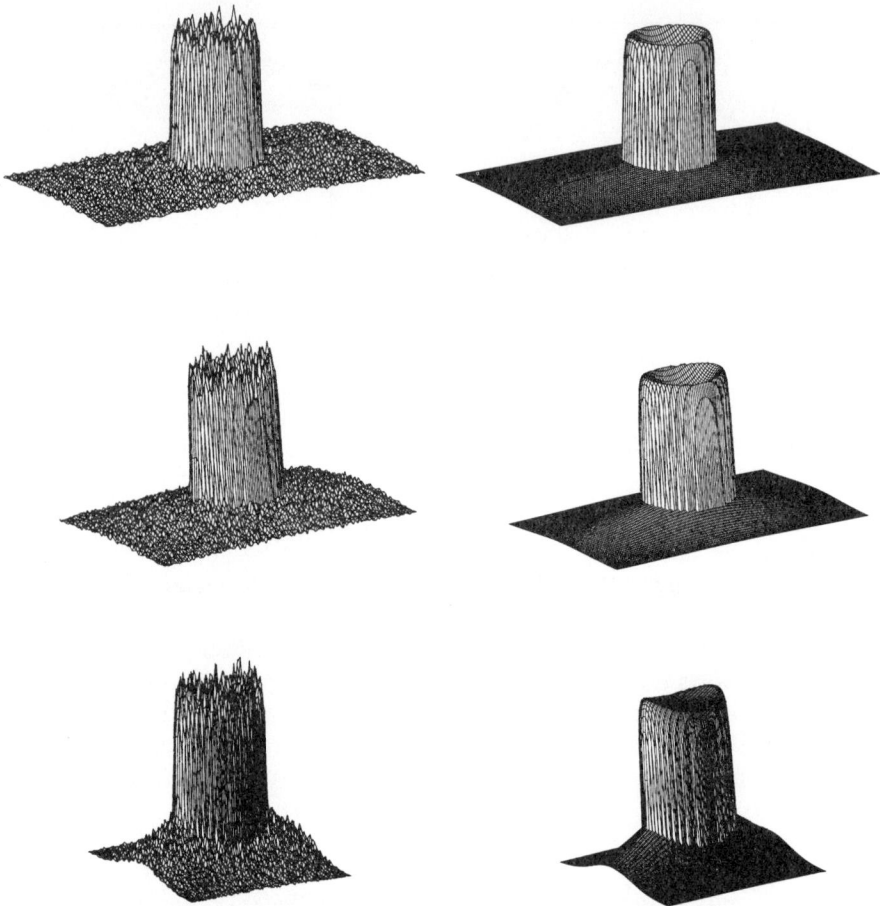

Figure 19.4 *2D trues + scatter distributions for the 50 cm (top), 80 cm (middle) and 100 cm (bottom) diameter PET systems. The figures to the left are the simulated distributions and the figures to the right are the corresponding fits according to equations (19.4) and (19.5).*

where x and z are the coordinates in the FOV and q is the axial polar angle [9]. Since the activity distribution is in the shape of a geometrical object (sphere), the unscattered activity distribution can be described by an analytical function

$$t(x, y, z, \theta) = \begin{cases} p_s 2\sqrt{R^2 - \rho^2}\, e^{-2\mu\sqrt{R^2 - \rho^2}} & \text{for } x^2 + y^2 \leqslant R^2 \\ 0 & \text{otherwise} \end{cases} \quad (19.5)$$

where R is the radius of the sphere and $\rho = \sqrt{x^2 + y^2}$. Fits of the simulated trues + scatter distributions using equations (19.4) and (19.5) are shown in the right-hand column in figure 19.4.

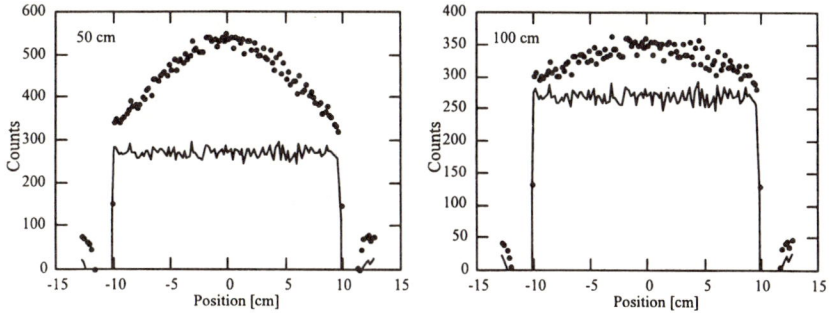

Figure 19.5. *The effect of scatter on the reconstructed image. The curves are profiles through a 3D reconstruction of the Monte Carlo data for the 50 cm (left) and 100 cm (right) diameter systems. The dots represent the reconstruction of the trues + scatter and the solid line the trues only.*

The projected scatter distributions shown in figure 19.4 can be used to study the effects of the scatter in the reconstructed image. An example of this is shown in figure 19.5. These curves are profiles through the simulated 20 cm diameter sphere, with and without scatter correction. The images were reconstructed using a 3D filtered backprojection algorithm [15]. The curve to the left shows the profile for the 50 cm diameter system and the curve to the right that for the 100 cm system. This example illustrates the severity of scatter in 3D PET and the need for an accurate scatter correction method. For the small-radius system, there is an overestimation in the activity level of about 100% in the centre of the object and the distribution is not uniform. For the 100 cm diameter system, the scatter contamination is less severe, but still produces a 30% overestimation in activity in the centre of the object. The uniformity is slightly improved due to the broader scatter distribution at larger system radii (cf figure 19.4), but there is still a significant deviation from a uniform activity distribution.

These results indicate that as a first-order correction for scatter in situations where both the activity and attenuation distributions are relatively uniform (e.g. ^{18}FDG or ^{15}O–H$_2$O brain scans) one could fit a simple Gaussian function to the scatter tails of the sinogram projections [16, 17].

19.6 MODELLING OF THE ELECTRONIC PERFORMANCE

In the modelling of the electronic performance of the PET system, the basic assumption is that there are two main sources of dead time in the system: first, the dead time in the processing of the single events in the block detector and, second, the dead time in the coincidence event histogramming system. For simplicity a paralysing dead-time description is assumed, which has

previously been found to fairly well describe the experimental live-time fractions in the ECAT EXACT and ECAT EXACT HR systems [18]. The block detector live-time fraction (BlockLive) can be approximated by

$$\text{BlockLive} = e^{-64 S_i \tau_{\text{block}}} \tag{19.6}$$

assuming 64 detectors per detector block. τ_{block} is the dead-time constant of the block detector electronics, which is about $2\,\mu$s for the ECAT EXACT system, and S_i is the average single count rate per detector element. The live-time fraction of the data processing part (DataLive) can be described in a similar way by an exponential function, i.e.

$$\text{DataLive} = e^{-\text{Coinc.Load}\,\tau_{\text{sys}}} \tag{19.7}$$

where τ_{sys} is the time constant and Coinc.Load is the data load experienced by the data handling system. For instance, in the ECAT EXACT HR this time constant is approximately 200 ns. The actual count rate, Coinc.Load, experienced by the data processing part is

$$\text{Coinc.Load} = \text{prompts BlockLive}^2 \tag{19.8}$$

where 'prompts' is the summed count rate of the trues, scatter and randoms. In the case of a system that simultaneously records delayed coincidences (randoms), these must be added into the prompts. The term BlockLive^2 describes the coincidence operation of the block detectors. The overall live-time fraction, Sys.Live, is then the product of the two live-time fractions

$$\text{Sys.Live} = \text{BlockLive}^2\,\text{DataLive}. \tag{19.9}$$

By multiplying the system live-time fraction at different tracer concentrations with the values of trues, scatter and randoms obtained from the simulations, the expected observed count rates of these quantities can be estimated. In the simulations only the trues, scatter and singles are produced as the output. The overall randoms rate must therefore be estimated from the average singles rate S_i and the total number of detectors in the system. The total number of detector combinations is

$$\text{Tot.Comb} = N_{\text{comb}}\,N_{\text{sinograms}} \tag{19.10}$$

where N_{comb} and $N_{\text{sinograms}}$ can be found in tables 19.1 and 19.2, for systems with 50, 80 and 100 cm diameter. The overall random rate is then

$$\text{randoms} = \text{Tot.Comb}\,S_i^2\,2\tau \tag{19.11}$$

where 2τ is the coincidence time window. A fit of the system live-time model used to actual experimental data was performed for the ECAT EXACT HR

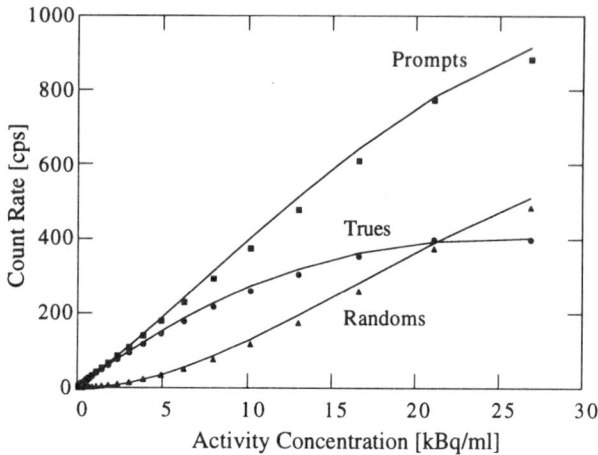

Figure 19.6   *A fit (—) of the dead-time model to measured data (●, ▲, ■) from the ECAT EXACT HR PET system and the measurements.*

[18] as shown in figure 19.6. The model fits are found to be in good agreement with the experiments for activity concentrations below 37 kBq ml^{-1}, which is the maximum activity concentration in typical brain flow studies.

19.7 NOISE EQUIVALENT COUNTS

As a figure of merit for the evaluation of the overall system electronic performance, the concept of noise equivalent counts (NECs) is used [19]. NEC is defined as

$$NEC = \frac{T^2}{(T+S+2kR)}. \tag{19.12}$$

The constant k is defined as the ratio between the diameter d of the phantom and the diameter of the image field (25 cm). A conservative choice of 0.8 was chosen for the calculations. NEC can be interpreted in two ways. First it can be seen as simply the square of the signal-to-noise ratio, and second as the equivalent true count rate in the absence of randoms and scatter.

To determine the best choice of system diameter and axial extension of the systems, NEC curves were generated using values for τ_{block} of 1.5 μs and τ_{sys} of 50, 100 and 150 ns. In order to relate the simulated specific activities in the 20 cm diameter sphere to actual subject studies, we limit ourselves to allowing for a maximum of a 740 MBq injection of [^{15}O]-butanol in a blood flow study. Of the injected activity, around 15% or 111 MBq, is extracted into the brain, which corresponds to an activity concentration of approximately 26 kBq ml^{-1} in the 20 cm diameter sphere. This is set to be the upper concentration limit for the simulated systems. However, for other tracers, for example a receptor ligand such as [^{11}C]-raclopride, a 740 MBq injection would correspond to an activity concentration of only 11 kBq ml^{-1} in the phantom and for [^{18}F]-FDG it would be even less.

19.8 COUNT RATE SIMULATIONS

The results from the count simulations are summarized in table 19.3. The table shows the overall true coincidence, trues + scatter $(T+S)$ coincidence count rates, scatter fraction (sf) (defined as $S/T+S$) and the average single count rate (S_i) per detector at an activity concentration of 37 kBq ml^{-1} in a 20 cm diameter spherical phantom positioned at the centre of the systems. $(T+S)$ and sf are given for both the full FOV and a 20 cm FOV which is the FOV covered by the phantom. The absolute calibration into counts kBq^{-1} ml^{-1} was extrapolated from the simulations of the ECAT EXACT HR system. The packing fraction squared was taken into account, ranging from 0.76 for the 50 cm diameter system to 0.81 for the 100 cm diameter system. With the average single count rate per detector available, the number of randoms was calculated. The random coincidence count rate (R) is based on the average single rate, S_i, of the detectors, the coincidence time window, chosen to be 12 ns, and the total number of detector combinations for a 25 cm diameter FOV. In the calculation of the NEC curves a linear relationship was assumed between T, S, S_i and activity concentration.

The resulting NEC results are summarized in figures 19.7–19.9. In figure 19.7, the ideal NEC curves are shown. These curves were calculated for the 50, 80 and 100 cm system diameters with a fixed axial FOV of 30 cm, by setting the dead-time constants, τ_{block} and τ_{sys}, to zero. Figure 19.8 shows the resulting NEC curves where the axial FOV was varied between 10 and 50 cm for the 50, 80 and 100 cm diameter systems and τ_{sys} was fixed to 1.5 μs and to 150 ns. We also found that the NEC curves are actually fairly insensitive to the choice of τ_{block}. The curves in figure 19.9 show the NEC rate for a fixed axial FOV of 30 cm in the 50, 80 and 100 cm diameter systems with τ_{sys} varied from 50 to 150 ns.

19.9 DISCUSSION

The smaller-diameter systems have, as expected, a higher sensitivity, but are also more sensitive to randoms and scatter. The global scatter fraction is not dramatically reduced with increased system radius. However, the scatter distribution is more localized under the area of the imaged object for the

Table 19.3 *Counts in kcps for the simulated configurations including scatter fractions and the average single count rates. These data are used for the simulations of the NEC curves. Scatter fractions (%) defined as scatter/(trues + scatter) measured over the indicated FOV.*

System diameter (axial FOV)	Trues (kcps kBQ^{-1} ml^{-1})	Trues + scatter (kcps kBQ^{-1} ml^{-1}, full FOV)	Scatter fraction (full FOV)	Trues + scatter (kcps kBQ^{-1} ml^{-1}, 20 cm FOV)	Scatter fraction (20 cm FOV)	Average singles per detector (kcps kBQ^{-1} ml^{-1})
50 cm						
10 cm	32.3	58.8	45.1	50.1	35.6	41.5
20 cm	116.5	210.1	44.6	176.3	34.0	40.4
30 cm	203.4	378.6	46.2	304.3	33.1	38.7
40 cm	273.0	523.9	47.9	404.4	32.5	36.4
50 cm	326.4	639.3	48.9	479.3	31.9	33.9
60 cm						
10 cm	27.2	46.9	42.1	39.0	30.3	30.8
20 cm	99.2	170.8	42.0	139.8	29.1	30.4
30 cm	176.7	316.4	44.1	247.1	28.5	29.7
40 cm	242.3	450.8	46.2	336.4	28.0	28.6
50 cm	295.8	546.4	47.7	407.8	27.5	27.4
70 cm						
10 cm	23.5	38.8	39.3	31.8	25.9	24.0
20 cm	86.6	143.0	39.5	115.4	25.0	23.8
30 cm	156.3	269.4	42.0	207.2	24.6	23.6
40 cm	217.5	391.9	44.5	286.9	24.2	23.1
50 cm	269.5	502.1	46.3	353.6	23.8	22.5
80 cm						
10 cm	20.8	32.9	36.8	26.8	22.4	19.3
20 cm	77.0	122.5	37.2	98.2	21.6	19.2
30 cm	140.2	233.4	39.9	178.1	21.3	19.2
40 cm	197.2	344.4	42.7	249.7	21.0	19.1
50 cm	247.2	448.2	44.8	311.8	20.7	18.8
90 cm						
10 cm	18.5	28.2	34.5	22.9	19.4	15.9
20 cm	68.6	105.6	35.1	84.5	18.8	15.9
30 cm	125.6	202.7	38.0	154.3	18.6	16.0
40 cm	178.2	302.0	41.0	218.2	18.4	16.0
50 cm	225.4	397.8	43.4	275.1	18.1	15.9
100 cm						
10 cm	17.0	25.2	32.5	20.5	16.8	13.4
20 cm	63.5	94.9	33.1	75.8	16.3	13.4
30 cm	116.7	182.9	36.2	138.7	16.2	13.5
40 cm	166.4	274.4	34.9	198.2	16.0	13.6
50 cm	211.9	364.7	41.9	251.8	15.8	13.6

smaller ring diameter, as evident from figure 19.4. This is also reflected in the scatter fraction, where only the object area is taken into account in the scatter fraction calculation. In this case, the scatter fraction is reduced by a factor of two, when increasing the system radius from 50 to 100 cm (table 19.3).

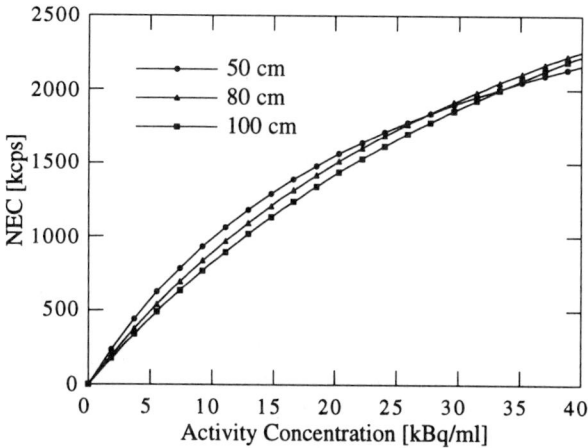

Figure 19.7 *Ideal NEC curves generated with* $\tau_{block}=0$ *and* $\tau_{sys}=0$ *for systems with a 30 cm axial FOV and diameters 50, 80 and 100 cm.*

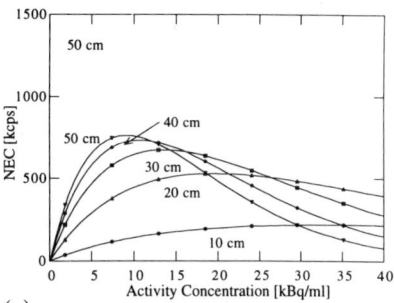

(*a*)

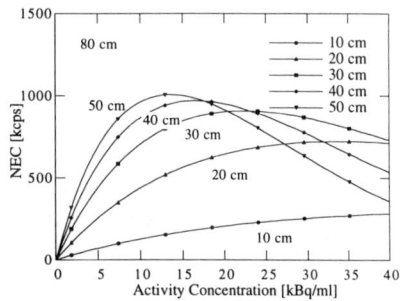

(*b*)

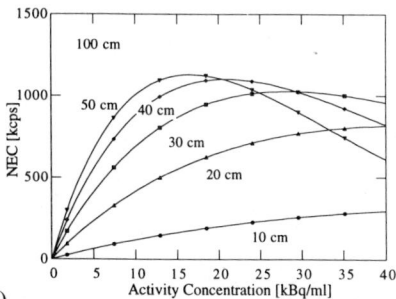

(*c*)

Figure 19.8 *NEC rates for different axial FOVs for 50 cm (a), 80 cm (b) and 100 cm (c) system diameters.* τ_{block} *and* μ_{sys} *were fixed to 1.5 μs and 100 ns, respectively.*

In the absence of a count rate limiting data acquisition system, the NEC curves only reflect the noise increase when correcting for randoms and scatter. This is illustrated in figure 19.7, where the ideal NEC curves are shown for the 50, 80 and 100 cm system diameters with a fixed axial FOV of 30 cm. Due to a smaller fraction of scatter and random coincidences, the larger-diameter systems are favoured at higher activity concentrations giving almost the same NEC rate at 31 kBq ml^{-1}. The tendency of the smaller systems to be less efficient will become much more accentuated when finite live times of the detector blocks and the data acquisition system are considered, which is also demonstrated in figures 19.8 and 19.9.

Figure 19.8 shows the NEC rates for 50, 80 and 100 cm diameter systems with axial FOVs between 10 and 50 cm, obtained with a τ_{sys} of 100 ns. Due to the limited capacity of the data acquisition system, the NEC curves for all systems saturate at high concentration values. This saturation occurs, however, earlier for larger axial FOVs (figure 19.8). The gain in NEC on adding more block detector rings will therefore be smaller and smaller. A compromise may be a choice of a 30 cm axial FOV as a practical upper limit. For the system with a 50 cm diameter, the 30 cm axial FOV NEC curve exceeds the 40 and 50 cm FOV NEC curves even at a concentration of 13 kBq ml^{-1}. For the 80 cm diameter system the same situation occurs above 18.5 kBq ml^{-1}. Furthermore, the requirement of an upper limit of 26 kBq ml^{-1} rules out axial FOVs higher than 30 cm.

The best choice of diameter is also based on the generated NEC curves. Figure 19.9 shows a comparison between NEC data for three different system diameters (50, 80 and 100 cm), all with an axial FOV of 30 cm, up to an activity concentration of 26 kBq ml^{-1}, generated with τ_{sys} of 150, 100 and 50 ns, respectively. For a τ_{sys} of 150 ns, the NEC of the 50 cm system crosses the 80 and the 100 cm data below 7 kBq ml^{-1}. At the selected end point of 26 kBq ml^{-1} the NEC curve for the 100 cm diameter system is the highest, although close to the 80 cm data. This trend is maintained for the faster data acquisition systems, shown in the two other graphs, with a reduced separation between the 100 and the 80 cm diameter systems at 26 kBq ml^{-1}. The choice of τ_{sys}, the speed of the data acquisition system, is critical for the choice of ring diameter, while the influence of the block dead time is fairly insensitive to the NEC performance due to the small block size. As an example we may take an 80 cm diameter system with a 30 cm axial FOV. With a τ_{sys} of 150 ns, the block dead time τ_{block} may vary by ± 1 μs from a nominal value of 1.5 μs causing the maximum NEC performance to vary by only 2%, going up to a 4% variation if the system dead time τ_{sys} is lowered to 50 ns.

With a τ_{sys} of 50 ns the 80 cm system is, on average, the most efficient system within the whole range from 0 to 27 kBq ml^{-1}, while for τ_{sys} of 150 ns the 100 cm diameter system may be a better choice, but only for the upper range from 15 to 26 kBq ml^{-1}. The smaller-ring-diameter systems begin to compete below 11 kBq ml^{-1}.

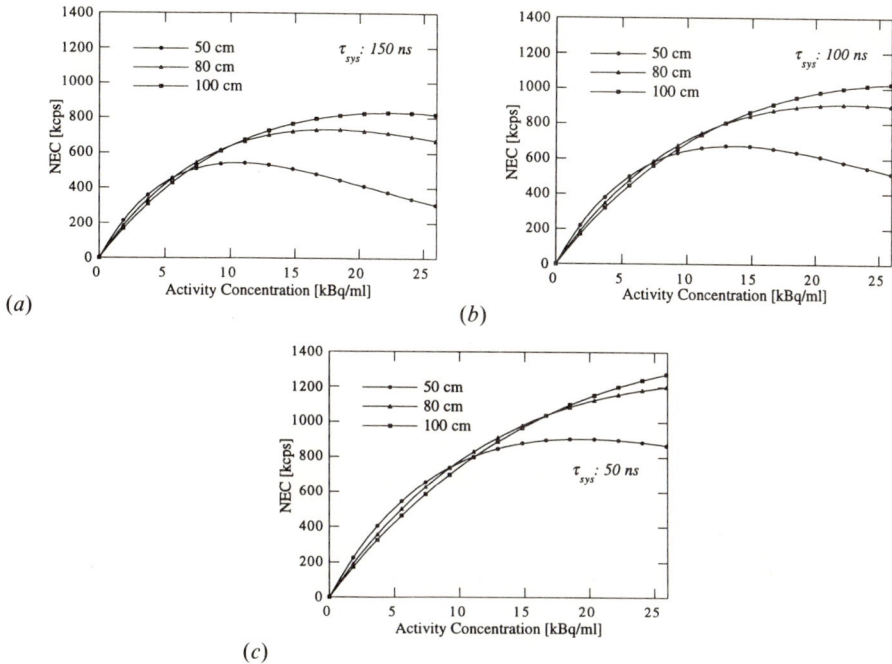

Figure 19.9   *NEC rates for the 50, 80 and 100 cm diameter systems with τ_{sys} set to 150 ns (a), 100 ns (b) and 50 ns (c). τ_{block} was set to 1.5 μs.*

19.10 SUMMARY

In this chapter we have shown how Monte Carlo simulations can be used in conjunction with more conventional mathematical modelling techniques to study the performance of a high-resolution PET system. This may provide valuable information in the design of future PET systems and also for the understanding of the behaviour of a large and complex data acquisition system. Furthermore, the use of Monte Carlo techniques in studying the behaviour of scatter in 3D PET may provide guidance in the design of new accurate scatter correction schemes.

REFERENCES

[1]   Casey M E and Nutt R 1986 A multicrystal two dimensional BGO detector system for positron emission tomography *IEEE Trans. Nucl. Sci.* **NS-33** 460–3

[2]   Dahlbom M and Hoffman E J 1988 An evaluation of a two-dimensional array detector for high-resolution PET *IEEE Trans. Med. Imaging* **MI-7** 264–72

[3]   Digby W M, Dahlbom M and Hoffman E J 1990 Detector shielding and geometric design factors for a high resolution PET system *IEEE Trans. Nucl. Sci.* **NS-37** 664–70

[4]   Dahlbom M, Eriksson L, Rosenqvist G and Bohm C 1989 A study of the possibility of using multislice PET systems for 3D imaging *IEEE Trans. Nucl. Sci.* **NS-36** 1066–71

[5]   Townsend D W, Spinks T, Jones T, Geissbüler A, DeFrise M, Gilardi M C and Heather J 1989 Three dimensional reconstruction of PET data from a multi-ring camera *IEEE Trans. Nucl. Sci.* **NS-36** 1056–65

[6]   Wienhard K, Eriksson L, Grootoonk S, Casey M, Pietrzyk U and Heiss W D 1992 Performance evaluation of the positron scanner ECAT EXACT *J. Comput. Assist. Tomogr.* **16** 804–13

[7]   Wienhard K, Dahlbom M, Eriksson L, Michel C, Bruckbauer T, Pietrzyk U and Heiss W D 1994 The ECAT EXACT HR: performance of a new high resolution positron scanner *J. Comput. Assist. Tomogr.* **18** 110–8

[8]   DeGrado T R, Turkington T G, Williams J J, Stearns C W, Hoffman J M and Coleman R E 1994 Performance characteristics of a whole-body PET scanner *J. Nucl. Med.* **35** 1398–406

[9]   Cherry S R, Dahlbom M and Hoffman E J 1991 3D PET using a conventional multislice tomograph without septa *J. Comput. Assist. Tomogr.* **15** 655–68

[10]   Townsend D W, Geissbühler A, DeFrise M and Fully 1991 Three-dimensional reconstruction for a PET camera with retractable septa *IEEE Trans. Med. Imaging* **MI-10** 505–12

[11]   Dahlbom M, Cherry S R, Eriksson L, Hoffman E J and Wienhard K 1993 Optimization of PET instrumentation for brain activation studies *IEEE Trans. Nucl. Sci.* **NS-40** 1048–53

[12]   Germano G and Hoffman E J 1990 A study of data loss and mispositioning due to pileup in 2D detectors in PET *IEEE Trans. Nucl. Sci.* **37** 671–5

[13]   Dahlbom M 1987 An investigation of the physical factors affecting the implementation of high resolution PET *Dissertation* University of California–Los Angeles

[14]   Hoffman E J, Digby W M, Germano G, Mazziotta J, Huang S C and Phelps M E 1986 Performance of a neuroPET system employing 2D modular detectors *J. Nucl. Med.* **29** 983–4

[15]   Colsher J G 1980 Fully three-dimensional positron emission tomography *Phys. Med. Biol.* **25** 103–15

[16]   Lercher M and Wienhard K 1994 Scatter correction in 3D PET *IEEE Trans. Med. Imaging* **MI-13** 649–57

[17]   Cherry S R and Huang S-C 1995 Effects of scatter on model parameter estimates in 3D PET studies of the human brain *IEEE Trans. Nucl. Sci.* **NS-42** 1174–9

[18]   Eriksson L, Wienhard K and Dahlbom M 1994 A simple data loss model for positron camera systems *IEEE Trans. Nucl. Sci.* **NS-41** 1566–70

[19]   Strother S C, Casey M E and Hoffman E J 1990 Measuring PET scanner sensitivity: relating countrates to image signal-to-noise ratios using noise equivalent counts *IEEE Trans. Nucl. Sci.* **NS-37** 783–8

CHAPTER 20

SUMMARY

Michael Ljungberg, Sven-Erik Strand and Michael A King

Monte Carlo (MC) methods have become an important tool for the investigation of different imaging parameters in nuclear medicine. This book gives the background and a summary of the techniques used today. The physics and technology behind imaging systems and several MC simulation programs are described. Examples of simulated imaging parameters are given in the MC-related chapters. It is shown that despite the complexity the proposed computer models can predict measured data with high accuracy.

It is clear from the topics presented that the MC method plays a significant role in evaluation and development of different imaging systems. Programs that simulate single-photon emission computed tomography (SPECT), positron emission tomography (PET) or both have been presented. All of these programs have advantages and disadvantages and the scope of this book is to show the broad spectrum of programs that can be available to researchers. Some of the key features of the MC codes described are summarized in table 20.1

Another important field in nuclear medicine for MC calculations is dosimetry in both diagnostic and therapeutic application. Here, quantification of the activity distribution in the patient often relies on correction methods developed as described in several chapters in this book. The concept of S values, used to estimate the absorbed dose, is derived from MC calculation of photon transport in an anthropomorphic phantom. Current research on patient-specific dosimetry for internal radionuclide therapy is also moving towards a full three-dimensional (3D) MC dose calculation using photon and electron transport codes. The reason for this is the fact that individual humans can be modelled using, for example, CT-based information about the attenuated tissue. Here, the general photon/electron packages, such as

Table 20.1   *The key features of the MC codes described are summarized in the table below.*

Feature	EGS	ITS	MCNP	SIMSET
SPECT	no[a]	no	no	yes
PET	no[a]	no	no	yes
Transmission SPECT/PET	no[a]	no	no	yes
Voxel-based phantoms	no[a]	no	no	no
Source code	yes	yes	yes	yes
Collimator	no[a]	no	no[b]	yes
Electron modelling	yes	yes	yes	no
Language	Mortran[c]	Fortran	Fortran	C
Available	yes	yes	yes	yes

Feature	MCMATV	SIMIND	SIMSPECT	PETSIM
SPECT	yes	yes	yes	no
PET	no	no	no	yes
Transmission SPECT/PET	no	yes	yes	yes
Voxel-based phantoms	yes	yes	yes	no
Source code	yes	yes	yes	yes
Collimator	yes	yes[d]	yes	yes
Electron modelling	no	no	yes	no
Language	Fortran	F90	MCNP/C	Fortran
Available	yes	yes	yes	yes

[a] Since EGS mostly handles particle interaction, the feature can be simulated but requires an extensive amount of user programming.
[b] Not explicit but can be modelled by a combinatorial geometry package.
[c] Mortran is a pre-processor that generates Fortran code.
[d] Geometrical calculations or routine including scatter and penetration (Chapter 10).

EGS4 and MCNP, will play an important role, because these include very sophisticated models for the more difficult electron modelling.

Because of the fast development in computing processes, the vision of fully 3D MC-based correction methods may not be far away. Therefore, good knowledge about the basics of MC and its application will be essential for the future. In this book, we hope that we have laid the ground for such knowledge.

BIOSKETCHES

Ljungberg, Michael (http://www.radfys.lu.se)
In 1979, Michael Ljungberg began studying physics, mathematics and radiation physics at the Lund University towards a BSc degree gained in 1983. He started his research in the Monte Carlo field through a project aimed at developing an MC code that could calibrate whole-body counters. He later changed track to general nuclear medicine imaging and SPECT and in 1985 in parallel with his continuing development of the MC code, he started working with quantitative SPECT and the problem of attenuation and scatter. He graduated with his PhD degree in 1990 and then gained a Research Assistant position at the Department of Radiation Physics, Lund University, where he continued working on his MC code and quantitative SPECT. In 1994, he became an Associate Professor at the Lund University. In 1997, Dr Ljungberg started working part time as a hospital physicist in Helsingborg at the Department of Clinical Physiology. His current research includes an extensive ongoing project in oncological nuclear medicine, where he is developing methods—based on quantitative SPECT, Monte Carlo absorbed dose calculations and the development of coregistration methods—for an accurate 3D dose planning scheme for internal radionuclide therapy. He is also involved in undergraduate education and guiding PhD students. Dr Ljungberg is a member of the American Association of Physicists in Medicine (AAPM) and the Society of Nuclear Medicine (SNM).

Strand, Sven-Erik (http://www.radfys.lu.se)
Sven-Erik Strand was born 1946, received his Masters of Science 1972, his PhD in 1979 and his Medical Bachelor in 1981. He became Associate Professor in 1981 and Professor in 1995. Since 1981 he has been a lecturer at the University of Lund and teaching director at the Department of Radiation Physics. Since 1997 he has been director of Jubileum Institute and Vice Dean for the Mathematical Physics Faculty. From September 1991 until December 1992 he spent his sabbatical at the Nuclear Medical Department, Brookhaven National Laboratory. He has published about 150 regular articles, four book chapters and holds two patents. He has served as member of various international committees, predominantly the AAPM Nuclear Medicine Task Groups on dosimetry. He has served as reviewer for peer-written

manuscripts in several journals and organized international symposia and conferences.

Sven-Erik Strand's research activities are in the field of Systemic Radiation Therapy Physics including: registration of time-sequence 3D pre-therapy radionuclide images from SPECT/PET with CT/MRI anatomical information for improved image quantification and to provide the basis for a 3D radiotherapy planning method; the application of radioactivity quantification methods to determine tumour and non-target tissue radionuclide uptake; the use of regional radionuclide pharmacokinetics from the time-course quantitative SPECT/PET images, Monte Carlo methods to produce 3D radiation dose distributions. Production of dose volume histograms from registered radiation dose contour maps to provide input into tumour-control probability models for individual patient dose planning; Enhanced tumour-to-normal-tissue residence time ratio, based on a novel method with extra-corporeal immunoadsorption (ECIA).

King, Michael A

Michael A King received his BA from SUNY at Oswego in 1969 where he majored in physics. He received his MS in physics from SUNY at Albany in 1972, and in 1978 he received his PhD in radiation biology and biophysics from the University of Rochester. The title of his thesis was 'Radiation Bone Damage and Its Imaging.' Dr King was a postdoctoral fellow in Medical Physics under the direction of Dr Gary Barnes of Radiology at the University of Alabama in Birmingham from 1977 to 1979. In 1979, he joined the department of Nuclear Medicine at the University of Massachusetts Medical School in Worcester Massachusetts as an instructor. Dr King is currently a tenured professor of Nuclear Medicine and direction of the medical physics imaging group within this department. Dr King is the author or co-author of more than one hundred articles on medical imaging. His current research interests include: tomographic reconstruction; single photon emission computed tomographic (SPECT) imaging; *in vivo* activity quantitation; restoration filtering; use of observer detection studies to compare imaging strategies; and image segmentation.

Andreo, Pedro (http://www.fysik.lu.se/~radiofys/pedro.htm)

Pedro Andreo graduated in Theoretical Physics in 1974 and became Doctor in Physical Sciences in 1982 at the University of Zaragoza (Spain). He moved to the Department of Medical Radiation Physics, Karolinska Institute, University of Stockholm, Sweden, as Research Fellow in 1987, becoming Associated Professor at the University of Stockholm in 1989. He was appointed Full Professor in radiation physics (radiotherapy) at the University of Lund (Sweden) in 1993. In 1995 he was appointed Head of the Dosimetry and Medical Radiation Physics Section, Division of Human Health, at the International Atomic Energy Agency (IAEA). By virtue of his position, he is the

Secretary of the IAEA/WHO Network of Secondary Standards Dosimetry Laboratories. Since his PhD on applications of the Monte Carlo method to dosimetry, his scientific activities have emphasized the use of the MC method in radiotherapy physics, mainly absolute dosimetry and treatment planning with electron and photon beams. His cv includes approximately 200 publications, scientific papers, and congress abstracts and proceedings. Pedro Andreo has been co-author of the IAEA Codes of Practice for the Dosimetry of Therapeutic Beams (IAEA TRS-277, 1987) and the Use of Parallel-Plate Ionization Chambers (IAEA TRS-381, 1997). He is a member of the ICRU Report Committee on Absorbed Dose to Water Standards for Photon Dosimetry and Consultant for the ICRU Report on Proton Dosimetry (ICRU 59). Recently, his research activities have been addressed to the dosimetry of therapeutic proton beams and the use of 'direct' Monte Carlo calculations for radiotherapy treatment planning.

Belanger, Marie-Jose
Marie-Jose Belanger received a BSc degree in Engineering Physics from Queen's University, Canada, in 1990. She is currently a PhD candidate in the Medical Engineering/Medical Physics program of the Harvard-MIT Division of Health Sciences and Technology pursuing research in the detection of arterial lesions with Nuclear Medicine. Other research interests include: development of radiation transport simulation code and methods of statistical image analysis. Ms Belanger was awarded a fellowship from the Medical Research Council of Canada and a Poitras pre-doctoral fellowship from the Massachusetts Institute of Technology.

Dahlbom, Magnus (http://www.nuc.ucla.edu/)
Magnus Dahlbom received his BSc degree in physics from the University of Stockholm in 1982 and his PhD in medical physics from the University of California, Los Angeles, in 1987. During 1987–1989 he was a research scholar at the Karolinska Institute in the Department of Radiation Physics. In 1989 he joined the faculty of the Department of Molecular and Medical Pharmacology at the University of California where he is now an Associate Professor. His interests are in nuclear medicine instrumentation, tomographic image reconstruction and image processing. He has authored and co-authored more than 60 papers that have been published in scientific journals. Dr Dahlbom is a member of the IEEE, the Society of Nuclear Medicine and the AAPM.

de Vries, Daniel J
Daniel de Vries received his bachelor's degree in mathematics computer science from Barrington College, Rhode Island, USA, in 1984 and his PhD in Biomedical Engineering from Worcester Polytechnic Institute (WPI), Massachusetts, USA, in 1997. The research for his Master of Science degree

(also from WPI, 1989) was on the subject of simulating gamma camera collimator interactions, for which he received the 1989 Master's Research Award from the WPI Chapter of Sigma Xi. The subject of his PhD research was the development of scatter subtraction for SPECT imaging and the evaluation of its effect on the accuracy of lesion detection and activity quantitation. He has authored or co-authored 18 papers that have been published in conference proceedings or journals. Dr de Vries is a research associate in the Department of Nuclear Medicine at the University of Massachusetts Medical School, USA.

Dobrzeneicki, Andy B
Andy B Dobrzeneicki was born in Chicago, Illinois, USA, in 1958 and received the SB in Physics and the SM in Nuclear Engineering from MIT in 1987. During his education, from 1980 to 1985, he worked as a staff engineer at Stone and Webster in the area of power-plant design and construction. Dr Dobrzeneicki received a PhD in Artificial Intelligence from MIT in 1992, with a dissertation on 'Model Based Reasoning for Complex Physical Systems'. He joined the Harvard/MIT Division of Health Sciences and Technology in 1992 as a Research Associate and a faculty member at the Harvard Medical School, with a focus on computer-based modelling and visualization methods applied to medicine. Since 1996 Dr Dobrzeneicki has been an Assistant Research Professor in the Department of Mathematical Modelling at the Danish Technical University in Copenhagen. His research, conducted at the 3D Laboratory for Image Processing and Reconstruction at the University Hospital of Denmark, involves building computer systems for simulating medical procedures, and modelling disease processes and treatments. His major projects involve simulation of otological surgery, virtual reality-based rehabilitation and assessment of stroke, methods for diabetes screening, and Monte Carlo modelling of medical imaging systems. Dr Dobrzeneicki also does extensive work within industry, and is currently the Manager of Technology Development at Torsana A/S, a medical start-up firm in Denmark. He is a member of Phi Beta Kappa (MIT), Sigma Xi, IEEE, and AAPM (pending).

Eriksson, Lars
Lars A Eriksson was born in 1938 in Stockholm, Sweden, completed his BSc in 1964 and his PhD in nuclear physics in 1967. He belonged to the research staff in nuclear physics at the Institute of Physics, University of Stockholm until 1974, when he was appointed Associate Professor (Docent) in Nuclear Physics. Between the years of 1974 and 1976 he was a visiting scientist at UCLA and participated in building one of the first ring-detector systems for positron emission tomography. During 1976–1978 he joined the Karolinska Institute in Stockholm and participated in a multi-centre collaboration for the first Swedish positron camera system. In 1985 he

became a professor at the Karolinska Institute. During 1978–1988 he developed, together with the Swedish company Scanditronix, commercial positron camera systems. In 1997 he joined the CTI in Knoxville, Tennessee, and is now responsible for the detector R&D.

Erlandsson, Kjell
Kjell Erlandsson received his MSc in radiation physics in 1989 and his PhD in radiation physics in 1996, both at Lund University, Sweden. The title of his thesis was 'PET with 3D reconstruction'. KE worked part-time as an assistant university teacher and as an assistant hospital physicist in Lund from 1987 to 1996, and is currently a postdoctoral research fellow at the Joint Department of Physics at The Institute of Cancer Research and The Royal Marsden NHS Trust in Sutton, UK. His major research interest is 3D tomographic image reconstruction and quantitation for positron emission tomography (PET) and single photon emission tomography (SPECT). He has collaborated with researchers at Brookhaven National Laboratory (New York, USA), Columbia Presbyterian Hospital (New York, USA) and Gunma University (Japan), and is co-author of a dozen articles in refereed scientific journals.

Esser, Peter D (http://cpmcnet.columbia.edu/dept/radiology/)
In 1972 Dr Esser received his PhD based on studies of the optical absorption and thermoluminescence of potassium chloride containing substitutional thallium in the Solid State Physics Department, Brookhaven National Laboratory, Long Island. His Postgraduate training consisted of a fellowship in the Department of Nuclear Medicine at Brookhaven and was followed by an appointment at the Radiology Department at Columbia University where he is now a Professor of Clinical Radiology and Chief Physicist, Nuclear Medicine and PET, Columbia–Presbyterian Medical Center. His special interests include computer applications in nuclear medicine and radiology. He has edited four books and written numerous papers. In addition he has certifications from the American Board of Science in Nuclear Medicine and the American Board of Medical Physics. Dr Esser still lives on Long Island.

Harrison, Robert L
Robert Harrison received a BA in drama and mathematics from the University of California, San Diego, in 1978 and an MA in applied mathematics from the University of Washington in 1984. In 1982 he became a research assistant in the Division of Nuclear Medicine at the University of Washington, and he has continued to work there was a research scientist since receiving his MA. He has authored or co-authored over 50 papers published in conference proceedings and technical journals, including many on scatter

in emission tomography and simulation of emission tomography. Robert is a member of the IEEE.

Haynor, David

David Haynor is a neurologist at the University of Washington. He has three children. His research interests include image processing and reconstruction, scatter correction, object-oriented video, and computational molecular biology.

Kijewski, Marie Foley

Marie Foley Kijewski received her BS in Physics in 1971 from the University of Detroit, and her MS and ScD in Medical Radiological Physics in 1979 and 1984 from the Harvard School of Public Health. Since 1984 she has been Associate Physicist at Brigham and Women's Hospital, and she is currently Assistant Professor of Radiology at Harvard Medical School. Her primary research interests are the physics and mathematics of nuclear medicine imaging systems, and assessment of medical imaging systems. Dr Kijewski is a member of the American Association of Physicists in Medicine (AAPM), the Society of Nuclear Medicine (SNM), and the IEEE.

Koral, Kenneth F

Ken Koral received his Bachelor of Science degree in physics from Case Institute of Technology and his PhD in nuclear physics from its successor, Case Western Reserve University, Cleveland, Ohio, United States. His dissertation advisor was Philip R Bevington, author of *Data Reduction and Error Analysis for the Physical Sciences*. Dr Koral started his research career with the US National Aeronautics and Space Administration and moved into medical physics during a year's postdoctoral in the United Kingdom. He was recruited to the University of Michigan Medical Center by William H Beierwaltes. His current title there is Senior Research Scientist. He is the principal holder of a patent for a method of correction for Compton scattering of gamma rays within patients. He was joined in that invention by long-time collaborators W Leslie Rogers and Neal H Clinthorne. His main research goal has been activity quantification in nuclear-medicine single-photon emission computed tomography. He is the principal investigator of a US National Cancer Institute grant whose purpose is to pursue accurate tumour dosimetry for tumours in ^{131}I monoclonal-antibody therapy. Dr Koral belongs to a family consisting of himself, his wife Mary and their three adopted children, one each from Viet Nam, India and South Korea. They enjoy movies, swimming and hiking. He is a member of the Society of Nuclear Medicine and of the American Association of Physicists in Medicine.

Lewellen, Tom (http://totally.rad.washington.edu/nm/default.html)
Tom Lewellen is a Professor of Radiology, Adjutant Professor of Electrical
Engineering, and Adjutant Professor of Bioengineering at the University of
Washington. He earned his BA from Occidental College in 1967 and his
PhD in experimental nuclear physics from the University of Washington in
1972. He then switched fields and spent one year working on particle optics
for a neutron therapy project. In 1974 he became a member of the Division
of Nuclear Medicine at the University of Washington. He is currently the
director of the Nuclear Medicine Physics Group and associate director of
Nuclear Medicine. His major research interests are in PET detector design
and quantitative PET and SPECT imaging.

Miyaoka, Robert S
Robert Miyaoka received his general engineering degree in 1983 from Harvey Mudd College (Claremont, California). He received his Masters Degree
in 1987 and his PhD in 1992 (on the subject of PET detector designs for
small animal imaging), both from the University of Washington (Seattle,
Washington). He is currently a Research Scientist with the Imaging Research
Laboratory, University of Washington Medical Center (Seattle, Washington). His interests are and/or have been PET detector development, dual-
head coincidence imaging, and singles-based attenuation imaging. He has
authored or co-authored more than 30 papers that have been published in
conference proceedings and technical journals. Dr Miyaoka is a member of
the Society of Nuclear Medicine (SNM).

Moore, Stephen C (http://www.med.harvard.edu/JPNM/JPNM.html)
Stephen Moore received his BS in Physics in 1972 from Bucknell University,
and his PhD in physics in 1978 from Brandeis University, where he remained
for a year as a postdoctoral Research Associate in high-energy particle physics. After two years in industry as Research Manager for Union Carbide
Imaging Systems, he returned to academia, working in the Department of
Radiology at Harvard Medical School and Brigham and Women's Hospital
as Research Associate, Instructor, and then Assistant Professor of Radiology. From 1986 to 1992, he was Associate Professor of Biomedical Engineering at the Worcester Polytechnic Institute, in Worcester, Massachusetts.
In 1992, he returned to Harvard Medical School as Associate Professor of
Radiology, with clinical appointments through Harvard's Joint Program in
Nuclear Medicine, first, at the Veterans Affairs Medical Center in West
Roxbury, Massachusetts and, currently, at the Brigham and Women's Hospital. Although his primary research interests are related to the physics of
nuclear medicine imaging and performance assessment of imaging systems,
he has also published scientific articles in the fields of high-energy physics,
computed tomography, and nuclear magnetic resonance imaging and spec-
troscopy. Dr Moore is a member of the American Association of Physicists

in Medicine (AAPM), the Society of Nuclear Medicine (SNM), and the IEEE.

Mueller, Stefan P

Stefan P Mueller attended Hannover Medical School and was licensed as a physician in Germany in 1979. From 1979 to 1983 he was resident in the Department of Nuclear Medicine in Special Biophysics in Hannover Medical School. In 1984 he received his certification in nuclear medicine. From 1983 to 1986 he was Research Fellow in Radiology at Brigham and Women's Hospital and Harvard Medical School. From 1987 up to the present he has been Senior Attending Physician in Nuclear Medicine at the Klinik für Nuklearmedizin, Universitaetsklinikum Essen, Germany, as well as Research Associate in Radiology at Brigham and Women's Hospital and Harvard Medical School. Aside from his medical practice, his research interests are in the physics and mathematics of nuclear medicine imaging and the assessment of imaging systems.

Ohlsson, Tomas (http://www.radfys.lu.se)

Tomas Ohlsson received his radiation physics degree in 1985 and his PhD degree in 1996 (on the subject of positron emission tomography), both from the University of Lund, Sweden. In 1996 he became research assistant at the department of Radiation Physics, Lund. Since 1985 he has also been working part-time as a medical physicist in nuclear medicine. His interests are positron emission tomography, modelling in nuclear medicine and dose planning for radionuclide therapy.

Penney, Bill

Bill Penney (also known as Bill O'Brien-Penney) received his BS in Electrical Engineering in 1973, MS in BioMedical Engineering (BME) in 1976, and PhD in BME in 1979, all from Worcester Polytechnic Institute (WPI). He remained at WPI as an Assistant Professor of BME for four years. He then joined the University of Massachusetts Medical Centre and conducted image processing research in nuclear medicine until 1994, as Assistant and then Associate Professor of Nuclear Medicine and Surgery. Since then he has been with the University of Chicago as a Clinical Physicist and Associate Professor of Clinical Radiology. His research interests have included a number of SPECT topics: (i) restoration filters; (ii) iterative reconstruction; (iii) attenuation correction; and (iv) scatter correction. Dr Penney is certified by the American Board of Radiology in Medical Nuclear Physics and is a member of the Society of Nuclear Medicine (SNM).

Picard, Yani

Yani Picard obtained his MSc and PhD degrees at the Montreal Neuro-logical Institute under the supervision of Dr Chris Thompson. As well as his numerous contributions to the development and applications of the PETSIM Monte Carlo program, he has developed a technique for monitoring subject movement during PET scans using two video cameras. The technique allows a sequence of PET images to be re-aligned should the subject move during a long dynamic scan. This is particularly important in the measurement of biological rate constants in patients with movement disorders like Parkin-son's disease.

Smith, Mark F

Mark F Smith was born in Baltimore, Maryland, USA, in 1957 and received an AB from Washington University (St Louis) in 1979 with majors in phys-ics, mathematics and Germanic languages and literature. From 1980 to 1985 he worked as a geophysicist for Gulf Oil Corporation in Houston, Texas, in the area of exploration seismology. Dr Smith received a PhD in earth sciences in 1989 from the University of California, San Diego. His Doctoral Dissertation was 'Imaging the Earth's Aspherical Structure with Free Oscil-lation Frequency and Attenuation Measurements'. He joined the National Science Foundation/Engineering Research Center in Emerging Cardiovas-cular Technologies at Duke University, Durham, North Carolina, as a Post-doctoral Fellow in 1989 and worked in the SPECT Research Laboratory at Duke University Medical Center. Dr Smith was Assistant Research Pro-fessor in the Department of Biomedical Engineering (1992–1997) and in the Department of Radiology (1995–1997) at Duke University. He is presently a physicist in the Nuclear Medicine Department of the Clinical Center at the National Institutes of Health, Bethesda, Maryland, USA. His research interests are in the area of quantitative SPECT and include photon transport modelling, pinhole tomography, generalized matix inverse reconstruction and collimator design. The major clinical focuses of his research are cardiac and tumour imaging. Dr Smith is a member of the Institute of Electrical and Electronics Engineers (IEEE), the American Association of Physicists in Medicine (AAPM) and the Society of Nuclear Medicine (SNM).

Thompson, Chris J

Chris Thompson has been working at the Montreal Neurological Institute in the field of PET instrumentation since 1975. He was the first to design a PET scanner using bismuth germanate crystals, and to demonstrate the feasibility of simultaneous emission and transmission scans in PET. He is the author of many papers on the use of Monte Carlo simulation to optimize the performance of PET scanners, and evaluate new geometries for PET systems. He has recently developed an instrument for detecting breast cancer using two planar PET detectors which fit in a conventional mammography system and provide metabolic images of the breast co-registered with the X-ray image.

Tin-Su Pan

Tin-Su Pan received BS and MS degrees in electrical engineering from the National Tsing-Hua University, Shin-Chu, Taiwan, in 1983 and 1985, respectively, and a PhD degree in electrical engineering and computer science from the University of Michigan, Ann Arbor, MI, in 1991. From 1988 to 1991, he was a research assistant in the SPECT research laboratory at the University of Michigan Medical Center, Ann Arbor, MI. From 1991 to 1996, he carried out his postdoctoral work in SPECT at the University of Massachusetts Medical Center, Worcester, MA. Currently, he is a senior physicist in the Applied Science Laboratory of General Electric Medical Systems, Milwaukee, WI. His research interests include tomographic image reconstruction and corrections in both SPECT and CT. Dr Pan is a member of ABSNM, IEEE, AAPM, and Phi Tau Phi.

Vannoy, Steven

Steven Vannoy began developing data acquisition software for PET systems in 1987. He completed his Bachelors degree at the University of Washington in 1990. From 1991 through 1992 he was involved in the development of commercial system utilities software. Since 1993 he has been implementing the SimSET simulation software. He is responsible for coordinating the design, implementation, and testing of the code as well as the development of documentation. He provides user support at all levels, from helping with the installation to creating custom modules for very specific but generally useful features.

Yanch, Jacquelyn C

Jacquelyn Yanch is Associate Professor of Nuclear Engineering at the Whitaker College of Health Sciences and Technology at MIT in Cambridge, Massachusetts. She carried out her doctoral research at the Royal Marsden Hospital in the UK in the area of scatter correction methodology in SPECT and earned her PhD degree in 1988. Since coming to MIT she has continued working in various aspects of nuclear medicine, concentrating on Monte Carlo simulation methods. In addition, she works on a number of topics related to the computer simulation of radiation as applied to radiation therapy of cancer and non-malignant disease.

Zubal, I George

Dr Zubal received a Bachelor's degree in Physics at the Ohio State University (1972). He began his graduate work at OSU and focused on medical applications for his Master's thesis (1974) in which he developed position-sensitive semiconductor detectors for nuclear medicine applications. He went on to study abroad and earned his PhD degree (1981) from the Universitaet des Saarlandes, Homburg, while developing Monte Carlo based dosimetry maps overlaid onto CT images for therapy planning with fast neutrons at the German Cancer Research Center in Heidelberg, Germany. He returned to the USA to work as a post-doc in the Medical Department of the Brookhaven National Laboratories, Upton, New York. In 1984 he was recruited into industry (Picker International, Northford, CT) and worked as the Senior Scientist for nuclear medicine camera and computer systems (PCS-512). He re-entered the academic world in 1986 and became Technical Director of the Section of Nuclear Medicine at Yale/New Haven Hospital and currently holds the position of Associate Professor in the Department of Diagnostic Imaging, Yale University School of Medicine. He has continued his work in Monte Carlo simulations of nuclear medicine patient geometries and has extended his image processing techniques to evaluate functional disorders—notably for localizing seizures in the brains of epilepsy patients.

INDEX

Numbers in **bold** refer to **figures** and those in *italics* refer to *tables*.